Methods in Molecular Biology™

For further volumes:
http://www.springer.com/series/7651

Single-Stranded DNA Binding Proteins

Methods and Protocols

Edited by

James L. Keck

Department of Biomolecular Chemistry, School of Medicine and Public Health, University of Wisconsin, Madison, WI, USA

Editor
James L. Keck
Department of Biomolecular Chemistry
School of Medicine and Public Health
University of Wisconsin
Madison, WI, USA

ISSN 1064-3745 ISSN 1940-6029 (electronic)
ISBN 978-1-62703-031-1 ISBN 978-1-62703-032-8 (eBook)
DOI 10.1007/978-1-62703-032-8
Springer New York Heidelberg Dordrecht London

Library of Congress Control Number: 2012945406

Printed on acid-free paper

Humana Press is a brand of Springer
Springer is part of Springer Science+Business Media (www.springer.com)

Preface

The genomes of cellular organisms are organized as double-stranded DNAs with the information content (nucleotide bases) packed into the interior of the protective double helix. This structure must be unwound to provide DNA replication, recombination, and repair machinery access to genomic information. DNA unwinding comes with inherent risks, however—single-stranded (ss) DNA can self-associate to create impediments to genome maintenance and is prone to chemical and nucleolytic attack. To help mediate these risks, bacterial, archael, and eukaryotic cells have evolved protective ssDNA-binding proteins (SSBs) that bind ssDNA with high affinity and specificity. As such, SSB-coated ssDNA comprises the *bonafide* cellular nucleoprotein substrates upon which genome maintenance processes ultimately must act. Accordingly, SSBs from all kingdoms of life directly interact with protein components that are central in DNA replication, recombination, repair, and replication restart.

This volume assembles protocols and methods developed over the past 20 years for examining the fundamental properties of SSBs and for exploiting the biochemical functions of SSBs for their use as in vitro and in vivo reagents. The chapters are organized into several themes: (1) an introduction to the structures and functions of SSBs, (2) protocols for studying SSB/DNA complexes, (3) methods for studying SSB/heterologous protein complexes, (4) protocols for interrogating post-translational modifications of SSBs, and (5) uses of fluorescently labeled SSBs for in vitro and in vivo studies of genome maintenance processes. Together, these chapters assemble a rich introduction for investigators who are interested in this fascinating family of DNA-binding proteins and for exploiting their unique and highly-adapted biochemical functions to new uses.

WI, USA *James L. Keck*

Contents

Preface . . . . *v*
Contributors . . . . *ix*

1 Functions of Single-Strand DNA-Binding Proteins in DNA Replication, Recombination, and Repair 1
Aimee H. Marceau

2 Structural Diversity Based on Variability in Quaternary Association. A Case Study Involving Eubacterial and Related SSBs 23
S.M. Arif and M. Vijayan

3 SSB Binding to ssDNA Using Isothermal Titration Calorimetry 37
Alexander G. Kozlov and Timothy M. Lohman

4 SSB–DNA Binding Monitored by Fluorescence Intensity and Anisotropy 55
Alexander G. Kozlov, Roberto Galletto, and Timothy M. Lohman

5 Single-Molecule Analysis of SSB Dynamics on Single-Stranded DNA 85
Ruobo Zhou and Taekjip Ha

6 Sample Preparation Methods to Analyze DNA-Induced Structural Changes in Replication Protein A 101
Chris A. Brosey, Susan E. Tsutakawa, and Walter J. Chazin

7 Structural Studies of SSB Interaction with RecO 123
Mikhail Ryzhikov and Sergey Korolev

8 Investigation of Protein–Protein Interactions of Single-Stranded DNA-Binding Proteins by Analytical Ultracentrifugation 133
Natalie Naue and Ute Curth

9 Ammonium Sulfate Co-precipitation of SSB and Interacting Proteins 151
Aimee H. Marceau

10 Analyzing Interactions Between SSB and Proteins by the Use of Fluorescence Anisotropy 155
Duo Lu

11 Far Western Blotting as a Rapid and Efficient Method for Detecting Interactions Between DNA Replication and DNA Repair Proteins 161
Brian W. Walsh, Justin S. Lenhart, Jeremy W. Schroeder, and Lyle A. Simmons

12 Methods for Analysis of SSB–Protein Interactions by SPR 169
Asher N. Page and Nicholas P. George

13 Use of Native Gels to Measure Protein Binding to SSB 175
Jin Inoue and Tsutomu Mikawa

14 Identification of Small Molecules That Disrupt SSB–Protein Interactions Using a High-Throughput Screen 183
Douglas A. Bernstein

15 Detection of Posttranslational Modifications of Replication Protein A 193
Cathy S. Hass, Ran Chen, and Marc S. Wold

16 Detecting Posttranslational Modifications of Bacterial SSB Proteins 205
Dusica Vujaklija and Boris Macek

17 Fluorescent SSB as a Reagentless Biosensor for Single-Stranded DNA 219
Katy Hedgethorne and Martin R. Webb

18 Fluorescent Single-Stranded DNA-Binding Proteins Enable In Vitro and In Vivo Studies 235
Piero R. Bianco, Adam J. Stanenas, Juan Liu, and Christopher S. Cohan

19 Use of Fluorescently Tagged SSB Proteins in In Vivo Localization Experiments 245
Rodrigo Reyes-Lamothe

Index *255*

Contributors

S.M. Arif • *Molecular Biophysics Unit, Indian Institute of Science, Bangalore, India*
Douglas A. Bernstein • *Whitehead Institute for Biomedical Research, Cambridge, MA, USA*
Piero R. Bianco • *Center for Single Molecule Biophysics, Department of Microbiology and Immunology, University at Buffalo, Buffalo, NY, USA*
Chris A. Brosey • *Departments of Biochemistry and Chemistry, Center for Structural Biology, Vanderbilt University, Nashville, TN, USA*
Walter J. Chazin • *Departments of Biochemistry and Chemistry, Center for Structural Biology, Vanderbilt University, Nashville, TN, USA*
Ran Chen • *Department of Biochemistry, Carver College of Medicine, University of Iowa, Iowa City, IA, USA*
Christopher S. Cohan • *Department of Pathology and Anatomical Sciences, University at Buffalo, Buffalo, NY, USA*
Ute Curth • *Hannover Medical School, Institute for Biophysical Chemistry, Hannover, Germany*
Roberto Galletto • *Department of Biochemistry and Molecular Biophysics, Washington University School of Medicine, St. Louis, MO, USA*
Nicholas P. George • *Department of Biomolecular Chemistry, University of Wisconsin School of Medicine and Public Health, Madison, WI, USA*
Taekjip Ha • *Howard Hughes Medical Institute and Department of Physics and Center for the Physics of Living Cells, University of Illinois, Urbana-Champaign, IL, USA*
Cathy S. Hass • *Department of Biochemistry, Carver College of Medicine, University of Iowa, Iowa City, IA, USA*
Katy Hedgethorne • *MRC National Institute for Medical Research, London, UK*
Jin Inoue • *Cellular & Molecular Biology Unit, RIKEN Advanced Science Institute, Yokohama, Japan*
Sergey Korolev • *Department of Biochemistry and Molecular Biology, St. Louis University School of Medicine, St. Louis, MO, USA*
Alexander G. Kozlov • *Department of Biochemistry and Molecular Biophysics, Washington University School of Medicine, St. Louis, MO, USA*
Justin S. Lenhart • *Department of Molecular, Cellular, and Developmental Biology, University of Michigan, Ann Arbor, MI, USA*
Juan Liu • *Center for Single Molecule Biophysics, Department of Microbiology and Immunology, University at Buffalo, Buffalo, NY, USA*
Timothy M. Lohman • *Department of Biochemistry and Molecular Biophysics, Washington University School of Medicine, St. Louis, MO, USA*
Duo Lu • *Institute for Neurodegenerative Diseases, University of California, San Francisco, CA, USA*
Boris Macek • *Proteome Center, Interfaculty Institute for Cell Biology, University of Tuebingen, Tuebingen, Germany*
Aimee H. Marceau • *Department of Biomolecular Chemistry, University of Wisconsin School of Medicine and Public Health, Madison, WI, USA*

TSUTOMU MIKAWA • *Cellular & Molecular Biology Unit, RIKEN Advanced Science Institute, Yokohama, Japan*
NATALIE NAUE • *Hannover Medical School, Institute for Biophysical Chemistry, Hannover, Germany*
ASHER N. PAGE • *Department of Biochemistry, University of Wisconsin, Madison, WI, USA*
RODRIGO REYES-LAMOTHE • *Department of Biochemistry, University of Oxford, Oxford, UK*
MIKHAIL RYZHIKOV • *Department of Biochemistry and Molecular Biology, St. Louis University School of Medicine, St. Louis, MO, USA*
JEREMY W. SCHROEDER • *Department of Molecular, Cellular, and Developmental Biology, University of Michigan, Ann Arbor, MI, USA*
LYLE A. SIMMONS • *Department of Molecular, Cellular, and Developmental Biology, University of Michigan, Ann Arbor, MI, USA*
ADAM J. STANENAS • *Center for Single Molecule Biophysics, Department of Microbiology and Immunology, University at Buffalo, Buffalo, NY, USA*
SUSAN E. TSUTAKAWA • *Lawrence Berkeley National Laboratory, Berkeley, CA, USA*
M. VIJAYAN • *Molecular Biophysics Unit, Indian Institute of Science, Bangalore, India*
DUSICA VUJAKLIJA • *Head of the Laboratory for Molecular Genetics, Division of Molecular Biology, Rudjer Boskovic Institute, Zagreb, Croatia*
BRIAN W. WALSH • *Department of Molecular, Cellular, and Developmental Biology, University of Michigan, Ann Arbor, MI, USA*
MARTIN R. WEBB • *MRC National Institute for Medical Research, London, UK*
MARC S. WOLD • *Department of Biochemistry, Carver College of Medicine, University of Iowa, Iowa City, IA, USA*
RUOBO ZHOU • *Department of Physics and Center for the Physics of Living Cells, University of Illinois, Urbana-Champaign, IL, USA*

Chapter 1

Functions of Single-Strand DNA-Binding Proteins in DNA Replication, Recombination, and Repair

Aimee H. Marceau

Abstract

Double-stranded (ds) DNA contains all of the necessary genetic information, although practical use of this information requires unwinding of the duplex DNA. DNA unwinding creates single-stranded (ss) DNA intermediates that serve as templates for myriad cellular functions. Exposure of ssDNA presents several problems to the cell. First, ssDNA is thermodynamically less stable than dsDNA, which leads to spontaneous formation of duplex secondary structures that impede genome maintenance processes. Second, relative to dsDNA, ssDNA is hypersensitive to chemical and nucleolytic attacks that can cause damage to the genome. Cells deal with these potential problems by encoding specialized ssDNA-binding proteins (SSBs) that bind to and stabilize ssDNA structures required for essential genomic processes.

SSBs are essential proteins found in all domains of life. SSBs bind ssDNA with high affinity and in a sequence-independent manner and, in doing so, SSBs help to form the central nucleoprotein complex substrate for DNA replication, recombination, and repair processes. While SSBs are found in every organism, the proteins themselves share surprisingly little sequence similarity, subunit composition, and oligomerization states. All SSB proteins contain at least one DNA-binding oligonucleotide/oligosaccharide binding (OB) fold, which consists minimally of a five stranded beta-sheet arranged as a beta barrel capped by a single alpha helix. The OB fold is responsible for both ssDNA binding and oligomerization (for SSBs that operate as oligomers). The overall organization of OB folds varies between bacteria, eukaryotes, and archaea.

As part of SSB/ssDNA cellular structures, SSBs play direct roles in the DNA replication, recombination, and repair. In many cases, SSBs have been found to form specific complexes with diverse genome maintenance proteins, often helping to recruit SSB/ssDNA-processing enzymes to the proper cellular sites of action. This clustering of genome maintenance factors can help to stimulate and coordinate the activities of individual enzymes and is also important for dislodging SSB from ssDNA. These features support a model in which DNA metabolic processes have evolved to work on ssDNA/SSB nucleoprotein filaments rather than on naked ssDNA.

In this volume, methods are described to interrogate SSB-DNA and SSB-protein binding functions along with approaches that aim to understand the cellular functions of SSB. This introductory chapter offers a general overview of SSBs that focuses on their structures, DNA-binding mechanisms, and protein-binding partners.

Key words: Single-strand DNA-binding protein, Replication protein A, DNA replication, DNA repair, DNA recombination, DNA replication restart, Protein interactions, DNA binding, OB domain

James L. Keck (ed.), *Single-Stranded DNA Binding Proteins: Methods and Protocols*, Methods in Molecular Biology, vol. 922, DOI 10.1007/978-1-62703-032-8_1,

1. Structural Diversity Among SSB Proteins

The structural organization of ssDNA-binding proteins (SSBs) enables simultaneous ssDNA and heterologous protein binding. Outside of the observations that most SSBs contain at least one conserved DNA-binding OB domain, SSBs' from different kingdoms of life are strikingly structurally divergent (1, 2). The OB fold is responsible for SSB ssDNA binding and, in cases where the SSB has a functional quaternary structure, for oligomerization. OB folds can vary in length from 70 to 150 amino acids, each forming, minimally, a beta barrel consisting of five antiparallel beta strands capped by an alpha helix. The primary sequence of the SSB OB fold is not well conserved, and the loops connecting the β strands can vary in length and organization (3).

The majority of SSBs can be characterized based on their phylogenetic distribution as being "bacterial" or "eukaryotic." The major eukaryotic SSB proteins are also called Replication Protein A, or RPA. Archaeal SSBs have representatives from both divisions. For example *Sulfolobus solfataricus* (4) has a bacterial-like SSB, while *Methanococcus jannaschii* (5) has an SSB that is more similar to RPA. Cellular SSBs appear to function as oligomers. For example, most bacterial SSBs form homotetramers in which each protomer contains a single OB fold per polypeptide and function as homotetramers (6). Interestingly, the presence of a small number of SSBs that encode two OB folds per polypeptide and function as homodimers indicates that there is some diversity within the bacterial protein class (7). In contrast to bacterial SSBs, eukaryotic RPAs generally function as heterotrimers with multiple OB folds distributed throughout the protomers (8, 9).

While bacterial SSBs function as either homotetramers (e.g., *Escherichia coli* SSB, ecSSB) (Fig. 1a) or more rarely as homodimers (e.g., *Deinococcus radiodurans* SSB, drSSB), both subtypes contain a total of four OB folds. The homodimers contain two OB folds per monomer and have been found thus far only in the *Deinococcus-thermus* genera (7, 10, 11). Each bacterial SSB OB fold is capable of binding to ssDNA in an arrangement in which the DNA wraps around the outside of the oligomeric protein (Fig. 1a) (6). EcSSB is the best studied bacterial SSB and its DNA binding has been shown to be surprisingly complex. Depending on the in vitro conditions and protein:ssDNA ratio, ecSSB tetramers can utilize either two or four OB folds to bind ssDNA. The interaction of SSB with ssDNA involves both hydrophobic stacking interactions as well as ionic phosphate backbone contacts (12–14). In addition to their OB fold, bacterial SSBs contain a highly conserved amphipathic unstructured C-terminus (SSB-Ct) that interacts with heterologous proteins (described in more detail below) (10, 15).

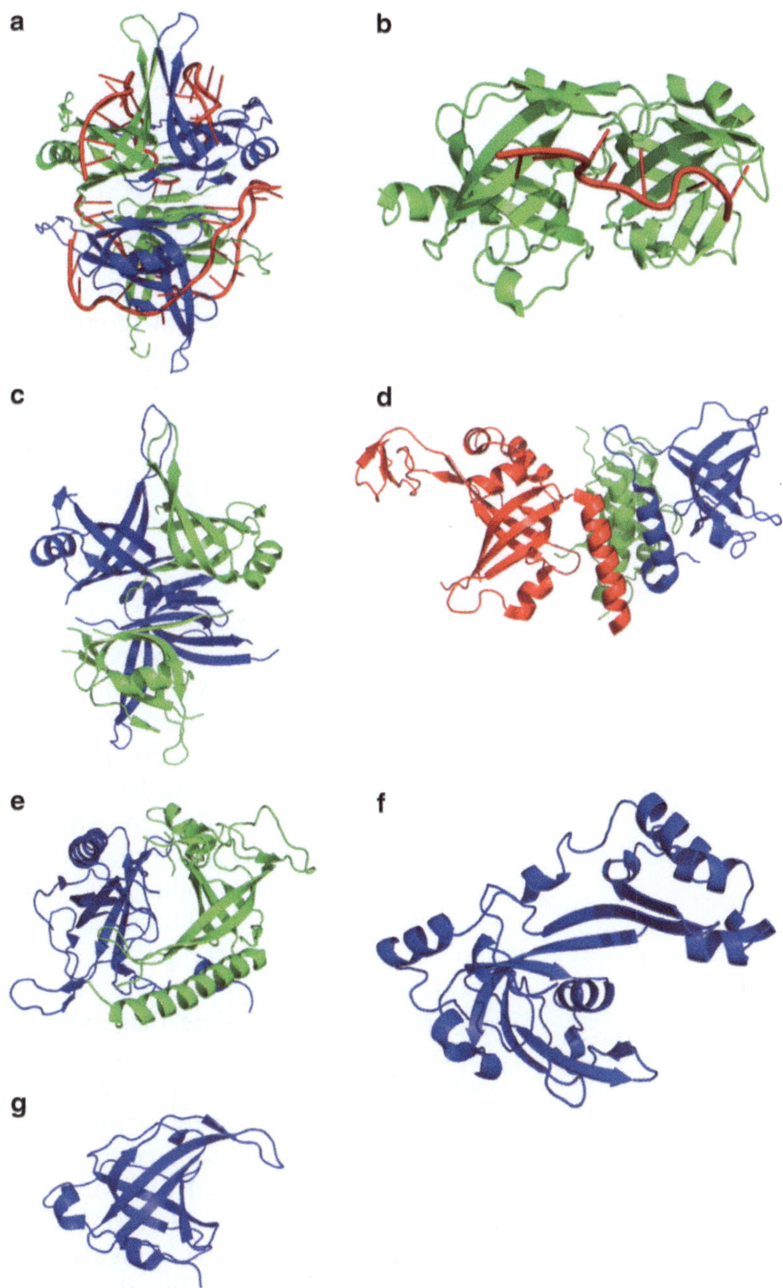

Fig. 1. Structures of SSB and RPA from all domains of life. (**a**) The structure of *E. coli* SSB bound to single-strand DNA (ssDNA) (*red*) (pdb 1EYG) (6). The structure is shown as each monomer colored *green* or *blue*. (**b**) RPA70 DNA binding domain A and DNA binding domain B are shown in *green* bound to ssDNA (*red*) (pdb 1JMC) (8). (**c**) The structure of human mitochondria SSB (pdb 2DUD). The structure is shown as each monomer colored *green* or *blue*. (**d**) The trimerization core RPA (pdb 1L10). RPA70 is shown in *red*, RPA32 in *green*, and RPA14 in *blue*. The three proteins interact primarily through a three-helical bundle (19). (**e**) The structure of the gp2.5 dimer, each monomer is shown in either *green* or *blue* (pdb 1JE5) (26). (**f**) The structure of gp32 core protein (pdb 1GPC) (25). (**g**) The structure of *Sulfolobus solfataricus* SSB (pdb 1O7I). The structurally dynamic C-terminus was not ordered in the crystal structure (23).

Eukaryotic RPAs generally function as heterotrimers and are typified by the human protein, which comprises 70, 32, and 14 kDa subunits, RPA70, RPA34, and RPA14, respectively (Fig. 1d). RPA contains a total of six OB folds among its three subunits, each of which is referred to as a "DNA-binding domain" (DBD-A through DBD-F). This nomenclature is somewhat misleading however, as only four of the OB folds appear to be responsible for binding to ssDNA. Three of these four (DBD-A through C) are located within RPA70, while the fourth (DBD-D) is located within RPA32 (8,16–19). In total, RPA70 contains four OB folds, three of which are involved in binding DNA, whereas RPA32 and RPA14 each contain a single OB fold. The three proteins interact with one another via OB domains (DBD-C, DBD-D, and DBD-E) and an additional three-helix bundle (Fig. 1d) (19). As with bacterial SSBs, RPA binds to ssDNA via stacking interactions of aromatic residues with individual DNA bases and interactions between the side chains of RPA and both the phosphate backbone and individual bases of the ssDNA (Fig. 1b) (1, 3, 8).

RPA lacks the SSB-Ct element found in bacterial SSBs but retains the ability to interact with genome maintenance proteins through multiple sites on RPA70 and RPA32. Additionally RPA32 contains a large N-terminal region that is phosphorylated in response to DNA damage and appears to play a role in regulating protein–protein interactions. RPA14 contains DBD-E and plays an essential role in stabilizing the structure of the complex (18). Human RPA (hsRPA) also contains a zinc finger motif on RPA70 that regulates ssDNA binding in a reduction/oxidation-sensitive manner (20, 21). ssDNA binding by RPA is enhanced over tenfold in reducing conditions compared to oxidizing conditions (21). The cytoplasm is generally more of a reducing environment than the nucleus. It is possible that increased DNA binding by RPA under reducing conditions serves as a defense mechanism against foreign DNA, specifically viral DNA.

The archaeal SSBs are particularly interesting because they share qualities with both bacterial SSB and eukaryotic RPA. For example, the crenarchaeal *S. solfataricus* SSB (ssSSB) contains an N-terminal OB fold and a C-terminal tail similar to bacterial SSB, but unlike bacterial SSB, it binds ssDNA as a monomer. The ssSSB OB fold structurally resembles the eukaryotic RPA70 DBD-B more closely than the bacterial OB folds; however the C-terminal tail is thought to function in protein/protein interactions like ecSSB (Fig. 1g) (4, 22, 23). In contrast, the SSB of the euryarcheal *M. jannaschii* (mjSSB) is more similar to that of full-length hsRPA70 than its crenarchaeal relatives. The sequence contains four tandem predicted OB folds and a putative zinc finger motif. MjSSB functions as a monomer in solution (5). Unlike ssSSB and mjSSB, the RPA from *Pyrococcus furiosus* (pfRPA) is composed of three subunits, RPA41, RPA32, and RPA14. RPA41 has some similarity to the eukaryotic RPA70, whereas the other two subunits lack homology to known proteins (24). Overall, SSBs found in archaea have diverse subunit composition and potentially divergent structures.

Bacteriophage SSBs can function either as monomers (e.g., T4 gp32) or as dimers (e.g., T7 gp2.5) (25, 26). T4pg32 consists of three domains: the amino-terminal 17 residues are essential for cooperativity in its ssDNA binding functions, the core domain directly binds ssDNA, and the C-terminal 46 residues contain the protein interaction site (Fig. 1f). The core domain contains a zinc-stabilized OB fold (10, 25). Each T7 gp2.5 monomer is composed of two domains: an N-terminal OB fold and a C-terminal tail. Both the OB fold and C-terminus are required for dimer formation and ssDNA binding. The C-terminal tail also mediates protein interactions (Fig. 1e) (10, 26, 27).

In summary, while sharing a common structural fold, SSBs from all domains of life exhibit diverse subunit arrangements. Despite this diversity, all SSBs possess common ssDNA- and protein-binding functions in the cell. It is possible that the differences among eukaryotic, prokaryotic, and archaeal SSBs are a result of the divergent evolutionary steps taken to satisfy the common ssDNA- and heterologous protein-binding needs during genome maintenance.

2. Strategies for Binding ssDNA in SSBs

SSBs bind to ssDNA with such high affinity that, barring SSB depletion, any exposed ssDNA in the cell is almost certain to be coated by SSB. Despite this common high-affinity binding, SSBs from different domains of life interact with ssDNA using surprisingly distinct mechanisms.

The most extensively studied bacterial SSB is the protein from *E. coli*. The four OB folds present in the ecSSB structure allow the protein to bind ssDNA in two distinct modes (Fig. 1a). These modes are named by the number of nucleotides (nt) occluded by the ecSSB tetramer: $(SSB)_{65}$ and $(SSB)_{35}$. In the limited-cooperativity $(SSB)_{65}$ mode, all four OB domains in each tetramer interact with ssDNA (Fig. 1a) (6, 10, 12). In the unlimited-cooperativity $(SSB)_{35}$ mode, two OB folds per SSB tetramer interact with the ssDNA. The DNA binding in the $(SSB)_{35}$ mode is highly cooperative, stimulating SSB to form long nucleoprotein filaments that coat ssDNA (12, 28). The stability of each binding mode is influenced in vitro by the concentration of monovalent salts, Mg^{2+}, spermidine, and spermine. Under high ionic conditions (>200 mM NaCl) and low SSB:ssDNA ratios, the $(SSB)_{65}$ mode is favored, whereas the $(SSB)_{35}$ mode is more stable under lower ionic strengths and high SSB:ssDNA conditions (12, 13). The $(SSB)_{35}$ binding mode has been proposed to function in DNA replication, whereas the $(SSB)_{65}$ mode may function under conditions favorable for homologous recombination (29).

The published structure of ecSSB bound to ssDNA provided the first structural models explaining how DNA wraps around the protein (Fig. 1a). The DNA makes several direct contacts with Trp40, Trp54, and Phe60 in the crystal structure (6). These results are consistent with Trp quenching studies and mutational analysis (14, 30, 31). In addition, electrostatic interactions utilize several Lys residues (43, 62, 73, and 87) and the N-terminus of SSB to bind the ssDNA backbone (6). The hydrophobic and ionic interactions together allow SSB to recognize and bind ssDNA independent of the sequence.

RPA binds to ssDNA using the same ssDNA-binding pockets in the OB domain but align along the ssDNA in a very different manner from the preformed tetrameric bacterial SSB DNA-binding core. RPA's four DNA-binding OB folds bind ssDNA in a sequential manner with 5′ to 3′ polarity, resulting in RPA70 being localized 5′ to RPA32 on the ssDNA. First, DBD-A of RPA70 initiates binding to 8–10 ssDNA nucleotides. This allows for DBD-B ssDNA binding, occluding a total of ~10 nucleotides (Fig. 1b). The ssDNA binding of the RPA70 A and B domains, which are separated by a short flexible linker, is coordinated with a conformational change in RPA70. Two loops within domains A and B enclose the ssDNA, which next allows DBD-C to bind, occluding an overall site of 12–23 nucleotides. Lastly, DBD-D from RPA32 binds to ssDNA leading to full binding that occludes 28–30 nucleotides (8, 9, 16, 32–34). Each of these different binding intermediate states is detectible in vitro, and have been termed globular, elongated contracted, and elongated extended (16). In yeast RPA, the transition from lower to higher binding site size is salt dependent (18, 35, 36). Unlike bacterial SSBs, which wrap ssDNA around the four relatively static OB folds, the conformation of RPA changes upon binding resulting in an elongated ssDNA substrate (Fig. 1b). Additionally, RPA can bind to multiple ssDNA substrates concurrently, allowing it to function as an ssDNA bridge in the process of DNA repair (19, 37). When RPA binds to ssDNA the conformational change allows the N-terminus of RPA32 to be more readily phosphorylated (16) (discussed in a later section).

Compared to bacterial and eukaryotic SSBs, far less is known about archaeal SSB DNA binding. These SSBs are similar in structure to the ssDNA binding domains from eukaryotic RPA, with the structure of the archaeal ssSSB being most similar to RPA70 DBD-B. The current model for ssSSB–ssDNA binding is based on this similarity. The model suggests that when ssSSB binds to ssDNA it undergoes a conformational change from an open to a closed state. Because the binding site consists of only 4–5 nucleotides per monomer (4, 22), it is highly unlikely that ssDNA wraps around the protein. The binding site size and the similarity to RPA suggest that ssSSB binds to ssDNA with SSB monomers multimerized along the substrate (23). MjSSB is similar to RPA70 in that it contains four predicted OB folds in tandem. DNA binding by mjSSB occludes between 15 and 20 nucleotides, which is slightly smaller than RPA (5). These are the only two archaeal SSBs with published ssDNA binding information. Without ssDNA bound protein structural information, it is difficult to compare these proteins to bacterial SSB and eukaryotic RPA.

Bacteriophage T7 gp2.5 binds to ssDNA in an auto-regulated manner, with the C-terminus of the protein competing with ssDNA for the binding site. The exact manner of its ssDNA binding is unknown but the C-terminus is proposed to act as a switch that prevents the DNA binding cleft from interacting with improper negatively charged surfaces. The OB fold contains aromatic and positively charged residues important for ssDNA interaction. T4 gp32 binds cooperatively to ssDNA via its N-terminal domain (gp32–gp32 interaction) and its central core (ssDNA binding). Both gp2.5 and gp32 have two clear binding modes (26, 27). The proteins first weakly interact nonelectrostatically with dsDNA backbone. Then, the proteins slide along dsDNA, which increases the association rate of the second binding mode in which the proteins bind to ssDNA (25).

In summary, all SSBs bind ssDNA with high affinity, thereby protecting any exposed ssDNA in the cell. In general, SSBs associate with ssDNA via both hydrophobic stacking and ionic backbone interactions. Bacterial SSBs and RPA have multiple binding modes that are detectible in vitro that may have in vivo significance. Bacterial SSBs bind to ssDNA by wrapping the DNA around the outside of the homotetramer/homodimer utilizing two to four OB folds per SSB (Fig. 1a). RPA binds to ssDNA in a sequential 5′ to 3′ manner, resulting in an extended arrangement of the protein along the ssDNA (Fig. 1b). The differences in ssDNA binding by SSBs from various domains of life may reflect the distinct genomic needs for ssDNA protection and utilization as a substrate.

3. Protein Interactions with SSBs

In addition to their core DNA-binding activities, SSBs have a second essential role in the organization, localization, and stimulation of genome maintenance proteins through direct physical interactions. Bacterial SSBs interact with heterologous proteins through their evolutionarily conserved SSB-Ct elements, which include both acidic and hydrophobic residues (Asp-Asp-Asp-Ile-Pro-Phe in *E. coli* SSB). Select bacterial SSBs from *E. coli* and *Bacillus subtilis* have well-characterized interaction networks, which highlight how integrated SSBs are in genome biological processes. Eukaryotic RPAs similarly interact with a large number of proteins but, unlike their bacterial counterparts, RPAs appear to have multiple protein-interaction sites on RPA70 and RPA32. Very little is known about the protein interactions of archaeal SSBs. The few known protein partners reveal potentially interesting connections between transcription and genome maintenance.

Bacterial SSBs have been shown to interact with over a dozen different proteins in *E. coli* and *B. subtilis.* These binding partners are involved in a wide range of functions in DNA metabolism. In all cases characterized to date, SSB's binding partners bind to the SSB-Ct. Unlike the well-ordered OB domains of bacterial SSBs, the C-terminus is structurally dynamic and can be readily proteolyzed. Interestingly, removal of the SSB-Ct in ecSSB alters its ssDNA binding properties (10, 38, 39), which could indicate that protein occupancy at the SSB-Ct could similarly affect DNA binding. Mutations in the SSB-Ct are detrimental to cell survival and, at least in *E. coli*, removal of the SSB-Ct is lethal. One well-characterized SSB-Ct mutation (*ssb113*) encodes for a variant in which the penultimate Pro is changed to a Ser. This change leads to temperature-sensitive lethality in *ssb113*-mutant *E. coli* (40). The SSB113 protein itself retains apparent wild-type ssDNA-binding functions, but at nonpermissive temperatures DNA replication is no longer supported. In addition, this strain exhibits DNA damage hypersensitivity at all temperatures (40–42). DNA damage hypersensitivity and temperature-dependent replication are the direct results of loss of SSB interaction with partner proteins involved in DNA repair and replication (43–45).

3.1. Bacterial SSB/DNA Replication Interfaces

During *E. coli* DNA replication, SSB interacts directly with at least two key proteins: the χ subunit of the replicative DNA polymerase III holoenzyme and primase (43, 44, 46–51). Variants of χ that fail to interact with SSB have been incorporated into the holoenzyme and assayed both in vivo and in vitro (52). In vitro the loss of the interaction results in a salt-dependent decoupling of leading- and lagging-strand DNA synthesis. In vivo the cells exhibit slower replication rates and temperature-sensitive growth. The interaction between χ and SSB appears to be primarily important for stabilization of the replisome but is not absolutely required for replication (44, 51, 52). During replication, the replisome cannot initiate DNA synthesis de novo—instead it extends preformed RNA primers generated by primase (DnaG). Primase interacts with both DnaB (replicative helicase) and SSB (43, 53). The precise role of the SSB/primase interaction is not well understood. It is possible that the interaction stabilizes primase on the RNA–DNA duplex, to aid in primase protection of the hybrid from degradation or dissociation prior to Okazaki fragment synthesis initiation (10, 43, 54). In *B. subtilis*, one of the essential DNA polymerases (DnaE) also directly interacts with the SSB C-terminus, although the role of this interaction has not been investigated (15).

3.2. Bacterial SSB/DNA Replication Restart Interfaces

The process of DNA replication does not always proceed smoothly, and the replisome can stall and/or dissociate from the fork when it encounters DNA damage or bound proteins that block progression on the DNA. In instances where the replisome dissociates, it must be reloaded by proteins referred to collectively as the DNA replication restart primosome. SSB is intimately involved in this process through its interaction with PriA, a 3′-to-5′ DNA helicase that initiates the major bacterial restart processes (15, 55–57). PriA appears to be constitutively bound to the replication fork in both *E. coli* and *B. subtilis* via its interaction with the SSB-Ct of the respective SSBs. Upon recognition of a stalled DNA replication fork structure, PriA nucleates assembly of the primosomal complex. Interaction with SSB stimulates PriA helicase activity, and PriA can displace SSB from ssDNA allowing primosome assembly, and ultimately DNA replication proteins, onto the collapsed fork (55, 56, 58).

3.3. Bacterial SSB/DNA Recombination Interfaces

SSBs from both *E. coli* and *B. subtilis* interact with several proteins involved in DNA recombinational repair via the SSB-Ct, localizing the proteins to the site of repair and in many cases stimulating their activity. Two key recombination initiation proteins from the RecF recombination pathway, RecQ DNA helicase and RecJ nuclease, both bind to and are stimulated by SSB. SSB-stimulated RecQ helicase activity enhances unwinding efficiency (59), whereas SSB-stimulated RecJ 5′ to 3′ exonuclease activity creates an ssDNA template for downstream RecA-mediated recombination (60–63).

One common challenge in RecA-mediated recombination is that RecA nucleation onto ssDNA/SSB substrates is impeded by SSB. Specialized proteins called RecA mediators catalyze RecA nucleation by displacing SSB. One such mediator, RecO, plays an important role in this process by directly binding to SSB (64–66). In addition to its RecA supporting activity, RecO binds ssDNA and dsDNA and has a DNA annealing activity that is also stimulated by SSB (67, 68). In addition to RecO, MgsA has recently been shown to interact with SSB and to facilitate RecA loading at stalled replication forks (15, 69). RecG is a monomeric DNA helicase important in remodeling stalled replication forks and unusual DNA structures (70, 71). RecG's interaction with SSB stabilizes its binding to various DNA structures and stimulates its ATPase activity (57, 72). RecS, a putative helicase found in *B. subtilis*, interacts with SSB in complex with YpbB. Its exact function in the cell is unknown. In *B. subtilis*, these proteins have been shown to localize to the replication forks in a manner dependent on the SSB-Ct (15).

3.4. Bacterial SSB/DNA Repair Interfaces

DNA repair is essential for maintaining the fidelity of genomes. Not surprisingly, SSB interacts with several enzymes involved in repair and stimulates their activities as well as localize them to the site of repair. Exonuclease I (ExoI) processively degrades ssDNA in a 3′-to-5′ direction. This activity is stimulated by binding to SSB, specifically the SSB-Ct, and is important for methyl-directed DNA mismatch repair (10, 73, 74). SSB also interacts directly with uracil DNA glycosylase (UDG), which catalyzes the first step in base excision repair. This interaction between UDG and SSB is widely conserved from eukaryotes to viruses (10, 15, 75, 76). This interaction localizes UDG to the replication fork potentially to aid in removal of dUTP misincorporation. DNA polymerase II (Pol II) is a DNA repair polymerase that is induced early in the SOS response (77). This polymerase is involved in synthesis over lesions and repair of damaged DNA (10, 78). SSB supports binding of Pol II to ssDNA and stimulates its polymerase activity (10, 79, 80). Pol II has poor processivity, but in complex with SSB and processivity factors (clamp loader complex and the β subunit), the processivity is greatly increased, which is required for bypass of abasic sites (81, 82). DNA polymerase V (Pol V) is another mutagenic polymerase that interacts with SSB via its SSB-Ct (83). Pol V conducts translesion synthesis on damaged DNA (84). SSB allows Pol V access to the 3′ end of a DNA gap flanked by RecA filaments and thus facilitates its activity on damaged DNA (83). SbcC is an ssDNA nuclease that is specific for ssDNA palindromic structures; it is localized to the replication fork via its interaction with SSB (15). SSB plays a key role in DNA repair by recruiting, localizing, and stimulating the activity of proteins involved in a variety of repair pathways.

3.5. Other Bacterial SSB/Protein Interfaces

E. coli SSB also interacts with several other proteins with roles outside of canonical genome maintenance pathways. Exonuclease IX, misnamed since it is not an exonuclease and has no known cellular function (85, 86), binds to SSB. Bacteriophage N4 viron RNA polymerase (vRNAP) requires SSB for early transcription (87). N4 phage relies on ecSSB to support vRNAP in displacing nascent RNA from the ssDNA template. By binding to both the DNA template and the RNA product, SSB prevents the formation of a RNA–DNA hybrid and this results in increased template recycling (88). In *B. subtilis* SSB also interacts with XseA and YrrC, localizing them to the replication fork. Neither protein has a well-defined function in the cell (15).

3.6. RPA/DNA Replication Interfaces

As with its bacterial SSB counterparts, RPA interacts with numerous proteins involved in a wide variety of processes: chromosomal and viral DNA replication, DNA repair, DNA recombination, cell division, checkpoint control, DNA damage response, and transcription. RPA/replication protein interactions have been described in a wide variety of species (17, 18). DNA polymerase α-DNA primase interacts with RPA from *Homo sapiens* and *Saccharomyces cerevisiae*. The SV40 large tumor antigen, Bovine papillomavirus E2, and Epstein–Barr virus EBNA-1 all interact with RPA from *H. sapiens*. In fact, RPA was initially identified because it is essential for SV40 replication (89). The large T-antigen of SV-40 directly interacts with RPA, DNA polymerase α/primase, and topoisomerase to coordinate replisome assembly and function (89, 90). RTHI nuclease has been shown to interact with RPA from *S. cerevisiae* (16, 17). In many cases the exact functions of the interaction between RPA and heterologous proteins are not well understood.

RPA is essential for replication in eukaryotes and is involved in both replication initiation and elongation of the replication fork. RPA localizes to origins of replication independent of the origin recognition complex (32). It is thought that RPA binds the small amount of ssDNA exposed at the origin, after which cdc45 unwinds more of the origin and more RPA can bind to the ssDNA (91). After the recognition of the origin by the pre-initiation complex, DNA polymerase α/primase is recruited and bound to the ssDNA via an interaction with RPA. The interaction stabilizes the DNA polymerase α/primase on ssDNA and RPA acts like a fidelity clamp, reducing the rate of nucleotide misincorporation (92, 93). The primase subunit synthesizes ~12 nt RNA primers which are elongated to 20 nt by polymerase α; this protein initiates all of the Okazaki fragments on the lagging strand (94, 95). After initiation, the more processive DNA polymerases ε (leading strand) and δ (lagging strand) replace polymerase α in a replication factor C (RFC) organized switch. RFC competes with polymerase α for binding to RPA causing the polymerase to dissociate (96). RPA helps coordinate the removal of the RNA primers of the Okazaki fragments. RPA recruits Dna2 (helicase/nuclease) to the DNA polymerase δ displaced RNA–DNA flap and stimulates cleavage of the flap resulting in shorter fragments. The short fragment is processed by Fen1 and the resulting nicked duplex DNA is fixed by DNA ligase (97, 98).

3.7. RPA/DNA Repair Interfaces

RPA functions in DNA repair via protein–protein interactions by recruiting and organizing repair proteins at the site of damage. Homology-directed repair (HDR) is employed to repair double-strand DNA breaks, ssDNA gaps, interstrand cross-links, and in recovery of collapsed or stalled replication forks (99, 100). At the site of a double-strand DNA break, the DNA is resected 5′ to 3′ and RPA binds to the ssDNA. RPA is then removed by RAD51, likely via its interaction with RPA70 (101). RAD51 then mediates strand exchange, aided by RPA, and by a second SSB, hSSB1, in humans (18, 32). The loading of RAD51 is facilitated by RAD52, which binds to both RPA70 and RPA32 (102, 103). Additionally, RPA displacement may be facilitated by BRCA2 via a direct interaction. In fact, a common cancer-predisposing mutant of BCRA2 does not interact with RPA (18, 104).

At stalled replication forks, RPA appears to function by exchanging proteins required during different parts of the repair process. RPA interacts with members of the RecQ family of repair helicases, including Bloom syndrome protein (BLM) and Werner syndrome protein (WRN) (105–107). The interaction recruits the proteins to stalled replication forks and may limit RAD51 activity. RPA stimulates the helicase activity of both BLM and WRN (108, 109), which could facilitate their activity in fork repair. A number of the Fanconi anemia (FA) proteins, which are involved in replication fork stabilization and recovery, also associate with RPA (110). One of the FA proteins, FANCD2, also functions in concert with BLM (111). After a replication fork has stalled RPA interacts with FANCJ to increase its processivity and helicase activity (112).

Translesion synthesis (TLS) utilizes polymerases with more open active sites that lack proofreading activity, enabling them to synthesize DNA over lesions (113). In eukaryotes, the ubiquitination of PCNA likely regulates the switch from HDR to TLS. If PCNA is polyubiquitinated, the lesion is repaired by HDR but if it is monoubiquitinated then it is repaired by TLS (114, 115). RPA interacts with Rad18/Rad6 which catalyzes the monoubiquitination of PCNA; thus RPA is involved in regulating the switch between TLS and HDR (116). RPA may also interact with the mutagenic DNA polymerase λ (117, 118). PCNA in conjunction with RPA likely has a role in recruiting and regulating the action of the TLS polymerases (117, 118).

Since DNA damage can occur anywhere in the genome other mechanisms that operate independently of replication structures exist to deal with these fidelity problems and RPA is involved. The nucleotide excision repair (NER) pathway removes a variety of different DNA lesions that occur in response to environmental or endogenous genotoxic stressors throughout the entire genome. RPA plays an essential role in NER, stabilizing ssDNA intermediates and recruiting specific proteins to the site of repair (119–121). As mentioned previously, RPA interacts with UDG, which removes uracil from DNA (122). The RPA34 subunit interacts with XPA nucleases and the RPA–XPA complex fully opens the DNA around the lesion site, with RPA bound to the undamaged strand (123). While the 5′ end of RPA interacts with XPG to open the DNA, the 3′ end recruits and interacts with the nuclease ERCCI-XPF determining the orientation with which the nucleases bind while simultaneously protecting the undamaged DNA from nuclease attack (123, 124).

RPA also functions in the DNA damage checkpoint response pathways, which delays cell cycle entry after DNA damage. RPA interacts with the 9-1-1 complex, specifically directing its loading onto DNA (125). The 9-1-1 complex recruits a variety of proteins that bind to and stimulate the ATR–ATRIP kinase activity (126, 127). ATR is a PI3 family kinase that functions to regulate replication stress caused by replication inhibitors or DNA damage (126). ssDNA bound by RPA may help localize the ATR–ATRIP complex to the site of damaged DNA. ATRIP interacts with RPA70 and then recruits ATR. When the complex binds to RPA-coated ssDNA (127), ATR becomes activated and is able to phosphorylate additional targets in the checkpoint response.

3.8. Archaeal SSB/Protein Interactions

Archaeal SSB interaction discovery is in its beginning stages. *S. solfataricus* SSB interacts with RNA polymerase (RNA pol) via its C-terminal tail, stimulating its activity under limiting conditions (128). Additionally, SSB can melt AT-rich promoter sequences allowing RNA pol access to those genes. Finally, SSB may aid in the formation of the pre-initiation complex at promoter site of archaeal chromatin (128). This highlights a potentially important role of SSB in transcription.

Methanothermobacter thermautotrophicus RPA interacts with the DNA repair helicase Hel308 via its C-terminus resulting in a very modest stimulation of its helicase activity (129). The interaction aids localization and loading of the helicase to its site of activity at blocked replication forks. It is not clear if the interaction is critical for helicase activity. The hetrotrimeric pfRPA co-precipitates with RadA, which functions to resolve Holliday Junction intermediates (24). The interaction between pfRPA and RadA stimulates its strand exchange activity in vitro. In addition, pfRPA may interact with Hjc, a recombination protein, as well as with DNA polymerases and primase (24). The exact functions and consequences of pfRPA's interactions have yet to be determined.

In summary, SSBs from all organisms mediate important genome maintenance processes via protein–protein interactions. Both SSB and RPA are capable of stimulating the activity of enzymes as well as localizing proteins to their sites of action (SSB/ssDNA substrates). Bacterial SSB protein–protein interactions all appear to be mediated by the conserved SSB-Ct. RPA's interactions can involve several different domains in the protein. Archaeal RPA/SSB protein interaction studies have not advanced enough to determine a common binding domain, but the known interactions utilize the C-terminus. Protein–protein interactions allow SSB to function as a central organizer in replication, recombination, and repair.

4. Posttranslational Modifications

Posttranslational modifications are a well-studied biological strategy for altering the function of a protein after it is synthesized by the addition or removal of modifying groups. It has been known for over 10 years that RPA is phosphorylated, mostly on the N-terminus of RPA32, but the exact function remains unclear (17, 18). Identification of the roles of SSB posttranslational modifications in bacteria lags behind eukaryotes. Recently phosphorylation of bsSSB has been described; the modification may result in stronger ssDNA binding (130). A deeper appreciation of the effects of posttranslational modifications of SSB and RPA promises to enhance our understanding of their roles in DNA maintenance.

Phosphorylation of bacterial SSBs has thus far been described in *B. subtilis* (130). The phosphorylation of bsSSB does not correlate with any specific cellular process. BsSSB appears to be phosphorylated by the YwqD kinase and dephosphorylated by the YwqE phosphatase. BsSSB phosphorylation reportedly enhances ssDNA binding ~200-fold over that of unphosphorylated bsSSB. In addition, the level of phosphorylated bsSSB decreases in response to DNA damage, possibly resulting from the protein being removed from damaged ssDNA by proteins involved in DNA repair.

The phosphorylation of RPA appears to be critical for its role in DNA replication and repair. The N-terminus of RPA32 is phosphorylated during the S and G2/M phase of the cell cycle by Cdk2 family kinases (131–133). During mitosis, RPA32 is inactivated by hyperphosphorylation, resulting in RPA disassembly from chromatin, a potentially essential step (18). In addition RPA is hyperphosphorylated in response to DNA damage by the kinases ATM, ATR, and DNA-PK; the exact role of this phosphorylation event remains under debate. Some studies have shown that modification inhibits DNA replication in vivo and in vitro whereas others show no effect on ssDNA binding or in vitro SV40 DNA replication assays (18). Clearly, a significant amount of work is needed to better define the role of RPA phosphorylation in replication and DNA repair.

5. Organisms with Multiple SSBs

SSBs have many important roles in the cell, and in most cases these appear to be dealt with by a single, general protein. However, there are also some situations in which specialized SSBs may be needed and, as a result, alternative SSBs have evolved. Frequently the alternative SSBs are nonessential, have different domain arrangements, or lack certain domains found in the primary cellular SSB. In bacteria, two alternative SSBs have recently been described. Eukaryotes also have a second SSB that is found in the mitochondria and supports its DNA replication and human cells have two additional SSBs, hSSB1 and hSSB2, which have specialized functions in DNA repair.

Bacillus subtilis (and many other naturally competent bacteria) possesses two SSBs. One is similar to the *E. coli* SSB and contains an N-terminal OB fold and a C-terminal tail that mediates its interactions with other proteins. This protein (SsbA) is primarily expressed in log phase in actively replicating cells, with about fourfold lower expression in stationary phase (134). In contrast the second SSB (SsbB), encoded by the *ywpH* gene, contains an OB fold but lacks the C-terminus found in the primary SSB. The single OB fold forms a stable tetramer capable of binding ssDNA (134). SsbB is expressed primarily in stationary phase and aids *B. subtilis* in its natural competence. Transcriptional regulation of *ywpH* is controlled by ComK and its pattern of expression is consistent with late competence genes (134). While SsbB is not essential for cell survival, mutants that lack *ywpH* are 50-fold less competent than wild-type cells (134). The precise function of SsbB in natural competence is not yet known. Many bacteria that are naturally competent possess homologues to SsbB (134).

D. radiodurans also contains at least two SSBs. As described earlier, the primary SSB contains two OB folds per monomer for binding DNA and dimerization and a C-terminal tail that directs the interactions with other proteins (7). Its second SSB, DdrB, is not expressed under normal conditions but is induced to high levels in response to DNA damage caused by ionizing radiation. Similar to the primary SSB, DdrB contains two domains, an N-terminal domain with a unique fold responsible for oligomerization and ssDNA binding and a dynamic C-terminus that may play a role in protein–protein interactions similar to SSB. The C-terminus has a strong similarity to that of the canonical SSB, implying that the two proteins may bind to similar groups of interaction partners. Unlike previously characterized SSBs, DdrB forms a stable pentamer arranged in a ring structure (135, 136). DdrB has been shown to bind ~5 nt per monomer (135). The exact function of DdrB in the cell is not clear; there is some evidence that it may affect RecA binding to ssDNA (136).

All eukaryotic cells have at least two SSBs: RPA, which is found in the nucleus, and mtSSB (Fig. 1c), which is located in the mitochondria. RPA and mtSSB are not evolutionarily related; mtSSB is more structurally similar to tetrameric bacterial SSBs (Fig. 1a, c). The structure and organization of mtSSB are slightly different from ecSSB however; mtSSB has an extended N-terminal domain, followed by an OB fold domain and it lacks the amphipathic C-terminus. Deletion of mtSSB from the mitochondria results in a loss of mitochondrial DNA replication and eventual mitochondrial loss. mtSSB interacts with, and promotes the activity of, two proteins' function in mtDNA replication: DNA polymerase γ and mtDNA helicase (137). The N- and C-terminal regions of mtSSB may play a role in modulating the protein–protein interactions (137).

All eukaryotes have mtSSB; however humans have at least two additional SSB proteins. The structures of hSSB1 and hSSB2 appear to be more closely related to archaeal and bacterial SSBs than to RPA. The domains of the protein are organized into an N-terminal OB fold and a C-terminal tail that may be involved in protein–protein interactions. hSSB2 appears to be expressed preferentially in lymphocytes and testes, hinting that this protein may play a role in meiotic recombination and class switching (18). Unlike hSSB2, hSSB1 is expressed in many cell types and has an important role in the repair of double-strand DNA breaks. hSSB1-depleted cells have increased sensitivity to DNA-damaging agents, greater genomic instability, faulty checkpoint activation, and defective DNA repair (32).

6. Summary

SSBs are remarkably well-adapted structural and organizational proteins that are central factors in genome biological processes in all cells. In addition to their eponymous DNA-binding functions, SSBs play essential roles in establishing critical protein organizational units through multivalent protein complex assembly. From eukaryotes to prokaryotes, SSBs are components in every nucleic acid transaction that requires single-stranded intermediates. This volume assembles the methods that have led to critical discoveries in the SSB field for the past 40 years and that will continue to pave the way to new findings that better define the biological functions of SSBs.

References

1. Flynn RL, Zou L (2010) Oligonucleotide/oligosaccharide-binding fold proteins: a growing family of genome guardians. Crit Rev Biochem Mol Biol 45(4):266–275
2. Theobald DL, Mitton-Fry RM, Wuttke DS (2003) Nucleic acid recognition by OB-fold proteins. Annu Rev Biophys Biomol Struct 32:115–133
3. Murzin AG (1993) OB(oligonucleotide/oligosaccharide binding)-fold: common structural and functional solution for non-homologous sequences. EMBO J 12 (3):861–867
4. Wadsworth RI, White MF (2001) Identification and properties of the crenarchaeal single-stranded DNA binding protein from Sulfolobus solfataricus. Nucleic Acids Res 29 (4):914–920
5. Kelly TJ, Simancek P, Brush GS (1998) Identification and characterization of a single-stranded DNA-binding protein from the archaeon Methanococcus jannaschii. Proc Natl Acad Sci U S A 95(25):14634–14639
6. Raghunathan S et al (2000) Structure of the DNA binding domain of E. coli SSB bound to ssDNA. Nat Struct Biol 7(8):648–652
7. Bernstein DA et al (2004) Crystal structure of the Deinococcus radiodurans single-stranded DNA-binding protein suggests a mechanism for coping with DNA damage. Proc Natl Acad Sci U S A 101(23):8575–8580
8. Bochkarev A et al (1997) Structure of the single-stranded-DNA-binding domain of replication protein A bound to DNA. Nature 385 (6612):176–181
9. Bochkareva E et al (2001) Structure of the major single-stranded DNA-binding domain of replication protein A suggests a dynamic mechanism for DNA binding. EMBO J 20 (3):612–618

10. Shereda RD et al (2008) SSB as an organizer/mobilizer of genome maintenance complexes. Crit Rev Biochem Mol Biol 43(5):289–318
11. Dabrowski S et al (2002) Identification and characterization of single-stranded-DNA-binding proteins from Thermus thermophilus and Thermus aquaticus—new arrangement of binding domains. Microbiology 148(Pt 10):3307–3315
12. Lohman TM, Ferrari ME (1994) Escherichia coli single-stranded DNA-binding protein: multiple DNA-binding modes and cooperativities. Annu Rev Biochem 63:527–570
13. Bujalowski W, Overman LB, Lohman TM (1988) Binding mode transitions of Escherichia coli single strand binding protein-single-stranded DNA complexes. Cation, anion, pH, and binding density effects. J Biol Chem 263 (10):4629–4640
14. Casas-Finet JR et al (1987) Tryptophan 54 and phenylalanine 60 are involved synergistically in the binding of *E. coli* SSB protein to single-stranded polynucleotides. FEBS Lett 220(2):347–352
15. Costes A et al (2010) The C-terminal domain of the bacterial SSB protein acts as a DNA maintenance hub at active chromosome replication forks. PLoS Genet 6(12):e1001238
16. Wold MS (1997) Replication protein A: a heterotrimeric, single-stranded DNA-binding protein required for eukaryotic DNA metabolism. Annu Rev Biochem 66:61–92
17. Fanning E, Klimovich V, Nager AR (2006) A dynamic model for replication protein A (RPA) function in DNA processing pathways. Nucleic Acids Res 34(15):4126–4137
18. Richard DJ, Bolderson E, Khanna KK (2009) Multiple human single-stranded DNA binding proteins function in genome maintenance: structural, biochemical and functional analysis. Crit Rev Biochem Mol Biol 44 (2–3):98–116
19. Bochkareva E et al (2002) Structure of the RPA trimerization core and its role in the multistep DNA-binding mechanism of RPA. EMBO J 21(7):1855–1863
20. Bochkareva E, Korolev S, Bochkarev A (2000) The role for zinc in replication protein A. J Biol Chem 275(35):27332–27338
21. Park JS et al (1999) Zinc finger of replication protein A, a non-DNA binding element, regulates its DNA binding activity through redox. J Biol Chem 274 (41):29075–29080
22. Haseltine CA, Kowalczykowski SC (2002) A distinctive single-strand DNA-binding protein from the Archaeon Sulfolobus solfataricus. Mol Microbiol 43(6):1505–1515
23. Kerr ID et al (2003) Insights into ssDNA recognition by the OB fold from a structural and thermodynamic study of Sulfolobus SSB protein. EMBO J 22(11):2561–2570
24. Komori K, Ishino Y (2001) Replication protein A in Pyrococcus furiosus is involved in homologous DNA recombination. J Biol Chem 276(28):25654–25660
25. Shamoo Y et al (1995) Crystal structure of a replication fork single-stranded DNA binding protein (T4 gp32) complexed to DNA. Nature 376(6538):362–366
26. Hollis T et al (2001) Structure of the gene 2.5 protein, a single-stranded DNA binding protein encoded by bacteriophage T7. Proc Natl Acad Sci U S A 98(17):9557–9562
27. Kim YT, Richardson CC (1994) Acidic carboxyl-terminal domain of gene 2.5 protein of bacteriophage T7 is essential for protein--protein interactions. J Biol Chem 269 (7):5270–5278
28. Ferrari ME, Bujalowski W, Lohman TM (1994) Co-operative binding of Escherichia coli SSB tetramers to single-stranded DNA in the (SSB)35 binding mode. J Mol Biol 236(1):106–123
29. Lohman TM, Bujalowski W (1994) Effects of base composition on the negative cooperativity and binding mode transitions of *Escherichia coli* SSB-single-stranded DNA complexes. Biochemistry 33(20):6167–6176
30. Ferrari ME, Fang J, Lohman TM (1997) A mutation in E. coli SSB protein (W54S) alters intra-tetramer negative cooperativity and inter-tetramer positive cooperativity for single-stranded DNA binding. Biophys Chem 64(1–3):235–251
31. Merrill BM et al (1984) Photochemical cross-linking of the Escherichia coli single-stranded DNA-binding protein to oligodeoxynucleotides. Identification of phenylalanine 60 as the site of cross-linking. J Biol Chem 259(17):10850–10856
32. Richard DJ et al (2008) Single-stranded DNA-binding protein hSSB1 is critical for genomic stability. Nature 453(7195):677–681
33. Wyka IM et al (2003) Replication protein A interactions with DNA: differential binding of the core domains and analysis of the DNA interaction surface. Biochemistry 42 (44):12909–12918
34. Arunkumar AI et al (2003) Independent and coordinated functions of replication protein A tandem high affinity single-stranded DNA

binding domains. J Biol Chem 278 (42):41077–41082

35. Broderick S et al (2010) Eukaryotic single-stranded DNA binding proteins: central factors in genome stability. Subcell Biochem 50:143–163
36. Oakley GG, Patrick SM (2010) Replication protein A: directing traffic at the intersection of replication and repair. Front Biosci 15:883–900
37. Pestryakov PE et al (2004) Human replication protein A (RPA) binds a primer-template junction in the absence of its major ssDNA-binding domains. Nucleic Acids Res 32 (6):1894–1903
38. Williams KR et al (1983) Limited proteolysis studies on the Escherichia coli single-stranded DNA binding protein. Evidence for a functionally homologous domain in both the Escherichia coli and T4 DNA binding proteins. J Biol Chem 258(5):3346–3355
39. Zhou R et al (2011) SSB functions as a sliding platform that migrates on DNA via reptation. Cell 146(2):222–232
40. Chase JW et al (1984) Characterization of the Escherichia coli SSB-113 mutant single-stranded DNA-binding protein. Cloning of the gene, DNA and protein sequence analysis, high pressure liquid chromatography peptide mapping, and DNA-binding studies. J Biol Chem 259(2):805–814
41. Wang TC, Smith KC (1982) Effects of the ssb-1 and ssb-113 mutations on survival and DNA repair in UV-irradiated delta uvrB strains of Escherichia coli K-12. J Bacteriol 151(1):186–192
42. Meyer RR et al (1980) A temperature-sensitive single-stranded DNA-binding protein from Escherichia coli. J Biol Chem 255(7):2897–2901
43. Yuzhakov A, Kelman Z, O'Donnell M (1999) Trading places on DNA—a three-point switch underlies primer handoff from primase to the replicative DNA polymerase. Cell 96 (1):153–163
44. Kelman Z et al (1998) Devoted to the lagging strand-the subunit of DNA polymerase III holoenzyme contacts SSB to promote processive elongation and sliding clamp assembly. EMBO J 17(8):2436–2449
45. Greenberg J, Donch J (1974) Sensitivity to elevated temperatures in exrB strains of *Escherichia coli*. Mutat Res 25(3):403–405
46. Breier AM (2005) Independence of replisomes in *Escherichia coli* chromosomal replication. Proc Natl Acad Sci 102 (11):3942–3947
47. Gulbis JM et al (2004) Crystal structure of the chi:psi subassembly of the *Escherichia coli* DNA polymerase clamp-loader complex. Eur J Biochem 271(2):439–449
48. Kelman Z, O'Donnell M (1995) DNA polymerase III holoenzyme: structure and function of a chromosomal replicating machine. Annu Rev Biochem 64:171–200
49. Simonetta KR et al (2009) The mechanism of ATP-dependent primer-template recognition by a clamp loader complex. Cell 137 (4):659–671
50. Glover BP, McHenry CS (2001) The DNA polymerase III holoenzyme: an asymmetric dimeric replicative complex with leading and lagging strand polymerases. Cell 105 (7):925–934
51. Glover BP, McHenry CS (1998) The chi psi subunits of DNA polymerase III holoenzyme bind to single-stranded DNA-binding protein (SSB) and facilitate replication of an SSB-coated template. J Biol Chem 273 (36):23476–23484
52. Marceau AH et al (2011) Structure of the SSB–DNA polymerase III interface and its role in DNA replication. EMBO J. 30(20): 4236–4247
53. Wu CA, Zechner EL, Marians KJ (1992) Coordinated leading- and lagging-strand synthesis at the Escherichia coli DNA replication fork. I. Multiple effectors act to modulate Okazaki fragment size. J Biol Chem 267 (6):4030–4044
54. Rowen L, Kornberg A (1978) Primase, the dnaG protein of *Escherichia coli*. An enzyme which starts DNA chains. J Biol Chem 253 (3):758–764
55. Cadman CJ, McGlynn P (2004) PriA helicase and SSB interact physically and functionally. Nucleic Acids Res 32(21):6378–6387
56. Kozlov AG et al (2010) Binding specificity of Escherichia coli single-stranded DNA binding protein for the χ subunit of DNA pol III holoenzyme and PriA helicase. Biochemistry 49(17):3555–3566
57. Lecointe F et al (2007) Anticipating chromosomal replication fork arrest: SSB targets repair DNA helicases to active forks. EMBO J 26(19):4239–4251
58. Arai K, Low RL, Kornberg A (1981) Movement and site selection for priming by the primosome in phage phi X174 DNA replication. Proc Natl Acad Sci USA 78 (2):707–711
59. Shereda RD, Bernstein DA, Keck JL (2007) A central role for SSB in Escherichia coli RecQ

DNA helicase function. J Biol Chem 282 (26):19247–19258

60. Lovett ST, Kolodner RD (1989) Identification and purification of a single-stranded-DNA-specific exonuclease encoded by the recJ gene of *Escherichia coli*. Proc Natl Acad Sci U S A 86(8):2627–2631
61. Courcelle J, Carswell-Crumpton C, Hanawalt PC (1997) recF and recR are required for the resumption of replication at DNA replication forks in *Escherichia coli*. Proc Natl Acad Sci U S A 94(8):3714–3719
62. Courcelle J, Crowley DJ, Hanawalt PC (1999) Recovery of DNA replication in UV-irradiated Escherichia coli requires both excision repair and recF protein function. J Bacteriol 181(3):916–922
63. Han ES et al (2006) RecJ exonuclease: substrates, products and interaction with SSB. Nucleic Acids Res 34(4):1084–1091
64. Hobbs MD, Sakai A, Cox MM (2007) SSB protein limits RecOR binding onto single-stranded DNA. J Biol Chem 282 (15):11058–11067
65. Inoue J et al (2008) The process of displacing the single-stranded DNA-binding protein from single-stranded DNA by RecO and RecR proteins. Nucleic Acids Res 36 (1):94–109
66. Ryzhikov M et al (2011) Mechanism of RecO recruitment to DNA by single-stranded DNA binding protein. Nucleic Acids Res 39 (14):6305–6314
67. Luisi-DeLuca C, Kolodner R (1994) Purification and characterization of the Escherichia coli RecO protein. Renaturation of complementary single-stranded DNA molecules catalyzed by the RecO protein. J Mol Biol 236(1):124–138
68. Luisi-DeLuca C (1995) Homologous pairing of single-stranded DNA and superhelical double-stranded DNA catalyzed by RecO protein from Escherichia coli. J Bacteriol 177 (3):566–572
69. Page AN et al (2011) Structure and biochemical activities of Escherichia coli MgsA. J Biol Chem 286(14):12075–12085
70. McGlynn P, Mahdi AA, Lloyd RG (2000) Characterisation of the catalytically active form of RecG helicase. Nucleic Acids Res 28 (12):2324–2332
71. McGlynn P, Lloyd RG (2002) Genome stability and the processing of damaged replication forks by RecG. Trends Genet 18 (8):413–419
72. Slocum SL et al (2007) Characterization of the ATPase activity of the Escherichia coli RecG protein reveals that the preferred cofactor is negatively supercoiled DNA. J Mol Biol 367(3):647–664
73. Lu D, Keck JL (2008) Structural basis of Escherichia coli single-stranded DNA-binding protein stimulation of exonuclease I. Proc Natl Acad Sci USA 105(27):9169–9174
74. Genschel J, Curth U, Urbanke C (2000) Interaction of E. coli single-stranded DNA binding protein (SSB) with exonuclease I. The carboxy-terminus of SSB is the recognition site for the nuclease. Biol Chem 381 (3):183–192
75. Dianov G, Lindahl T (1994) Reconstitution of the DNA base excision-repair pathway. Curr Biol 4(12):1069–1076
76. Purnapatre K et al (1999) Differential effects of single-stranded DNA binding proteins (SSBs) on uracil DNA glycosylases (UDGs) from Escherichia coli and mycobacteria. Nucleic Acids Res 27(17):3487–3492
77. Iwasaki H et al (1990) The Escherichia coli polB gene, which encodes DNA polymerase II, is regulated by the SOS system. J Bacteriol 172(11):6268–6273
78. Pages V, Janel-Bintz R, Fuchs RP (2005) Pol III proofreading activity prevents lesion bypass as evidenced by its molecular signature within E.coli cells. J Mol Biol 352 (3):501–509
79. Weiner JH, Bertsch LL, Kornberg A (1975) The deoxyribonucleic acid unwinding protein of Escherichia coli. Properties and functions in replication. J Biol Chem 250 (6):1972–1980
80. Molineux IJ, Gefter ML (1974) Properties of the Escherichia coli in DNA binding (unwinding) protein: interaction with DNA polymerase and DNA. Proc Natl Acad Sci U S A 71 (10):3858–3862
81. Bonner CA et al (1992) Processive DNA synthesis by DNA polymerase II mediated by DNA polymerase III accessory proteins. J Biol Chem 267(16):11431–11438
82. Dalrymple BP et al (2001) A universal protein-protein interaction motif in the eubacterial DNA replication and repair systems. Proc Natl Acad Sci USA 98 (20):11627–11632
83. Arad G et al (2008) Single-stranded DNA-binding protein recruits DNA polymerase V to primer termini on RecA-coated DNA. J Biol Chem 283(13):8274–8282

84. Goodman MF (2000) Coping with replication "train wrecks" in Escherichia coli using Pol V, Pol II and RecA proteins. Trends Biochem Sci 25(4):189–195
85. Shafritz KM, Sandigursky M, Franklin WA (1998) Exonuclease IX of Escherichia coli. Nucleic Acids Res 26(11):2593–2597
86. Hodskinson MR et al (2007) Molecular interactions of Escherichia coli ExoIX and identification of its associated 3'-5' exonuclease activity. Nucleic Acids Res 35 (12):4094–4102
87. Glucksmann-Kuis MA et al (1996) *E. coli* SSB activates N4 virion RNA polymerase promoters by stabilizing a DNA hairpin required for promoter recognition. Cell 84(1):147–154
88. Davydova EK, Rothman-Denes LB (2003) Escherichia coli single-stranded DNA-binding protein mediates template recycling during transcription by bacteriophage N4 virion RNA polymerase. Proc Natl Acad Sci U S A 100(16):9250–9255
89. Borowiec JA et al (1990) Binding and unwinding—how T antigen engages the SV40 origin of DNA replication. Cell 60 (2):181–184
90. Hurwitz J et al (1990) The in vitro replication of DNA containing the SV40 origin. J Biol Chem 265(30):18043–18046
91. Mimura S et al (2000) Central role for cdc45 in establishing an initiation complex of DNA replication in Xenopus egg extracts. Genes Cells 5(6):439–452
92. Frick DN, Richardson CC (2001) DNA primases. Annu Rev Biochem 70:39–80
93. Maga G et al (2001) Replication protein A as a "fidelity clamp" for DNA polymerase alpha. J Biol Chem 276(21):18235–18242
94. Conaway RC, Lehman IR (1982) Synthesis by the DNA primase of Drosophila melanogaster of a primer with a unique chain length. Proc Natl Acad Sci U S A 79(15):4585–4588
95. Conaway RC, Lehman IR (1982) A DNA primase activity associated with DNA polymerase alpha from Drosophila melanogaster embryos. Proc Natl Acad Sci U S A 79 (8):2523–2527
96. Maga G, Hubscher U (1996) DNA replication machinery: functional characterization of a complex containing DNA polymerase alpha, DNA polymerase delta, and replication factor C suggests an asymmetric DNA polymerase dimer. Biochemistry 35(18):5764–5777
97. Bae SH et al (2001) RPA governs endonuclease switching during processing of Okazaki fragments in eukaryotes. Nature 412 (6845):456–461
98. Bae SH, Seo YS (2000) Characterization of the enzymatic properties of the yeast dna2 Helicase/endonuclease suggests a new model for Okazaki fragment processing. J Biol Chem 275(48):38022–38031
99. Paques F, Haber JE (1999) Multiple pathways of recombination induced by double-strand breaks in Saccharomyces cerevisiae. Microbiol Mol Biol Rev 63(2):349–404
100. Michel B et al (2004) Multiple pathways process stalled replication forks. Proc Natl Acad Sci U S A 101(35):12783–12788
101. Stauffer ME, Chazin WJ (2004) Physical interaction between replication protein A and Rad51 promotes exchange on single-stranded DNA. J Biol Chem 279 (24):25638–25645
102. Butland G et al (2005) Interaction network containing conserved and essential protein complexes in Escherichia coli. Nature 433 (7025):531–537
103. Mer G et al (2000) Structural basis for the recognition of DNA repair proteins UNG2, XPA, and RAD52 by replication factor RPA. Cell 103(3):449–456
104. Wong JM, Ionescu D, Ingles CJ (2003) Interaction between BRCA2 and replication protein A is compromised by a cancer-predisposing mutation in BRCA2. Oncogene 22(1):28–33
105. Doherty KM et al (2005) Physical and functional mapping of the replication protein a interaction domain of the werner and bloom syndrome helicases. J Biol Chem 280 (33):29494–29505
106. Shen JC et al (2003) The N-terminal domain of the large subunit of human replication protein A binds to Werner syndrome protein and stimulates helicase activity. Mech Ageing Dev 124(8–9):921–930
107. Constantinou A et al (2000) Werner's syndrome protein (WRN) migrates Holliday junctions and co-localizes with RPA upon replication arrest. EMBO Rep 1(1):80–84
108. Wu L (2008) Wrestling off RAD51: a novel role for RecQ helicases. Bioessays 30 (4):291–295
109. Bugreev DV et al (2007) Novel pro- and anti-recombination activities of the Bloom's syndrome helicase. Genes Dev 21(23):3085–3094
110. Wang LC et al (2008) Fanconi anemia proteins stabilize replication forks. DNA Repair (Amst) 7(12):1973–1981
111. Pichierri P, Franchitto A, Rosselli F (2004) BLM and the FANC proteins collaborate in a common pathway in response to stalled replication forks. EMBO J 23(15):3154–3163

112. Gupta R et al (2007) FANCJ (BACH1) helicase forms DNA damage inducible foci with replication protein A and interacts physically and functionally with the single-stranded DNA-binding protein. Blood 110(7):2390–2398
113. Kunkel TA, Pavlov YI, Bebenek K (2003) Functions of human DNA polymerases eta, kappa and iota suggested by their properties, including fidelity with undamaged DNA templates. DNA Repair (Amst) 2(2):135–149
114. Kannouche PL, Lehmann AR (2004) Ubiquitination of PCNA and the polymerase switch in human cells. Cell Cycle 3(8):1011–1013
115. Kannouche PL, Wing J, Lehmann AR (2004) Interaction of human DNA polymerase eta with monoubiquitinated PCNA: a possible mechanism for the polymerase switch in response to DNA damage. Mol Cell 14 (4):491–500
116. Davies AA et al (2008) Activation of ubiquitin-dependent DNA damage bypass is mediated by replication protein a. Mol Cell 29(5):625–636
117. Crespan E et al (2007) Expanding the repertoire of DNA polymerase substrates: template-instructed incorporation of non-nucleoside triphosphate analogues by DNA polymerases beta and lambda. Nucleic Acids Res 35(1):45–57
118. Krasikova YS et al (2008) Interaction between DNA Polymerase lambda and RPA during translesion synthesis. Biochemistry (Mosc) 73(9):1042–1046
119. Guzder SN et al (1996) Nucleotide excision repair in yeast is mediated by sequential assembly of repair factors and not by a pre-assembled repairosome. J Biol Chem 271 (15):8903–8910
120. Coverley D et al (1992) A role for the human single-stranded DNA binding protein HSSB/RPA in an early stage of nucleotide excision repair. Nucleic Acids Res 20(15):3873–3880
121. Coverley D et al (1991) Requirement for the replication protein SSB in human DNA excision repair. Nature 349(6309):538–541
122. Nagelhus TA et al (1997) A sequence in the N-terminal region of human uracil-DNA glycosylase with homology to XPA interacts with the C-terminal part of the 34-kDa subunit of replication protein A. J Biol Chem 272 (10):6561–6566
123. He Z et al (1995) RPA involvement in the damage-recognition and incision steps of nucleotide excision repair. Nature 374 (6522):566–569
124. Matsunaga T et al (1996) Replication protein A confers structure-specific endonuclease activities to the XPF-ERCC1 and XPG subunits of human DNA repair excision nuclease. J Biol Chem 271(19):11047–11050
125. Kumagai A et al (2006) TopBP1 activates the ATR-ATRIP complex. Cell 124(5): 943–955
126. Zou L, Liu D, Elledge SJ (2003) Replication protein A-mediated recruitment and activation of Rad17 complexes. Proc Natl Acad Sci U S A 100(24):13827–13832
127. Zou Y et al (2006) Functions of human replication protein A (RPA): from DNA replication to DNA damage and stress responses. J Cell Physiol 208(2):267–273
128. Richard DJ, Bell SD, White MF (2004) Physical and functional interaction of the archaeal single-stranded DNA-binding protein SSB with RNA polymerase. Nucleic Acids Res 32 (3):1065–1074
129. Woodman IL, Brammer K, Bolt EL (2011) Physical interaction between archaeal DNA repair helicase Hel308 and replication protein A (RPA). DNA Repair (Amst) 10(3):306–313
130. Mijakovic I et al (2006) Bacterial single-stranded DNA-binding proteins are phosphorylated on tyrosine. Nucleic Acids Res 34 (5):1588–1596
131. Dutta A et al (1991) Phosphorylation of replication protein A: a role for cdc2 kinase in G1/S regulation. Cold Spring Harb Symp Quant Biol 56:315–324
132. Dutta A, Stillman B (1992) cdc2 family kinases phosphorylate a human cell DNA replication factor, RPA, and activate DNA replication. EMBO J 11(6):2189–2199
133. Din S et al (1990) Cell-cycle-regulated phosphorylation of DNA replication factor A from human and yeast cells. Genes Dev 4 (6):968–977
134. Lindner C et al (2004) Differential expression of two paralogous genes of Bacillus subtilis encoding single-stranded DNA binding protein. J Bacteriol 186(4):1097–1105
135. Sugiman-Marangos S, Junop MS (2010) The structure of DdrB from deinococcus: a new fold for single-stranded DNA binding proteins. Nucleic Acids Res 38 (10):3432–3440
136. Norais CA et al (2009) DdrB protein, an alternative deinococcus radiodurans SSB induced by ionizing radiation. J Biol Chem 284(32):21402–21411
137. Oliveira MT, Kaguni LS (2010) Functional roles of the N- and C-terminal regions of the human mitochondrial single-stranded DNA-binding protein. PLoS One 5(10): e15379

Chapter 2

Structural Diversity Based on Variability in Quaternary Association. A Case Study Involving Eubacterial and Related SSBs

S.M. Arif and M. Vijayan

Abstract

Eubacterial and related single-stranded DNA-binding proteins (SSBs) exhibit considerable variability in their quaternary association in spite of their having the same tertiary fold. The variability involves differences in the orientation of dimers in the tetrameric molecule (or of two-domain subunits in the dimeric molecule) and that of monomers in each dimer. The presence of an additional strand in mycobacterial and related SSBs, which clamps the dimers together, is a major determinant of the mode of quaternary association in them. The variability in quaternary structure has implications to the stability of the protein and possibly to its mode of DNA binding.

Key words: Eubacterial SSBs, Mycobacterial SSBs, OB fold, Quaternary structure, Protein stability, DNA binding

1. Introduction

Proteins employ a variety of strategies to produce structural diversity based on the same folding motif. One of them is variability in quaternary association. The first major systematic attempt to explore this variability, mainly using X-ray crystallography, was made in relation to legume lectins. It was established that legume lectins are a family of proteins in which small alterations in essentially the same tertiary structure lead to large variations in quaternary association (1–7). Since then such variability has been observed and explored in several other proteins, including other classes of plant lectins (6, 8–11). Eubacterial and related single-stranded DNA-binding proteins (SSBs) constitute an important family of proteins which exhibit substantial variation in quaternary association among its members while retaining the basic fold of the subunit.

James L. Keck (ed.), *Single-Stranded DNA Binding Proteins: Methods and Protocols*, Methods in Molecular Biology, vol. 922, DOI 10.1007/978-1-62703-032-8_2,

DNA transactions such as replication, recombination, and repair involve unwinding of duplex DNA into single-stranded DNA (ssDNA). SSBs are involved in protecting vulnerable ssDNA from chemical and nuclease attacks and formation of aberrant secondary structures (12). They bind ssDNA with high affinity and in a sequence-independent manner. They have been found in all classes of organisms performing similar function, but displaying little sequence similarity. They exhibit some common features in three-dimensional structure. The commonality is most pronounced among the subunit structures of eubacterial and related SSBs. Among them, *E.coli* SSB (*Ec*SSB) (13) and human mitochondrial SSB (HMtSSB) (14) were the first to be structurally characterized using X-ray crystallography. Like other similar SSBs, both consist of a globular N-terminal domain and a disordered C-terminal domain. While the N-terminal domain binds DNA, the C-terminal domain is believed to be involved in interacting with other proteins (12). The DNA-binding domain has the OB fold (15) in *Ec*SSB as well as HMtSSB. Both of them form similar tetramers with 222 (D2) symmetry. The structure of the SSB from *Mycobacterium tuberculosis* (*Mt*SSB) subsequently became available (16). *Mt*SSB has the same tertiary structure as in *Ec*SSB and HMtSSB, but with a somewhat different quaternary arrangement. It is then that the variability in the quaternary association of SSBs came into focus. The subsequent structure determination of SSBs from other sources showed this variability to be fairly extensive, as indeed has been noted in earlier reports. The present review attempts a comprehensive treatment of this variability and related issues.

2. Methods

The crystal structures reported in the literature and/or the coordinates of which have been deposited in the Protein Data Bank (PDB) (17) form the data used in the present analysis. They have been determined by the well-known methods of macromolecular crystallography. Several computer programs such as ALIGN (18), NACCESS (19), COOT (20), CALCOM (21), CHIMERA (22), and routines in Collaborative Computation Project, Number 4 (CCP4) (23) were used for analysis of the available structures. PYMOL (24) and CHIMERA were used for preparing figures. Sequence alignment was carried out using CLUSTALW (25).

3. Database

The crystal structures of eubacterial and related SSBs from the following sources have been reported:

1. *E. coli* (*Ec*SSB) (13).
2. Human mitochondria (HMtSSB) (14).
3. *Mycobacterium tuberculosis* (*Mt*SSB) (16).
4. *Deinococcus radiodurans* (*Dr*SSB) (26).
5. *Mycobacterium smegmatis* (*Ms*SSB) (27).
6. *Thermus aquaticus* (*Ta*SSB) (28, 29).
7. *Thermotoga maritima* (*Tm*SSB) (30).
8. *Mycoplasma pneumoniae* (*Mp*SSB) (31).
9. *Streptomyces coelicolor* (*Sc*SSB) (32).
10. *Mycobacterium leprae* (*Ml*SSB) (33).

DNA complexes of SSBs from the following sources have been published:

1. *E. coli* (*Ec*SSB + DNA) (34).
2. *Helicobacter pylori* (*Hp*SSB + DNA) (35).

Coordinates of the structures of SSBs from the following sources are available in the PDB, but the results are yet to be published:

1. *Thermus thermophilus* (*Tt*SSB) (PDB code 2cwa).
2. *Bartonella henselae* (*Bh*SSB) (3pgz).
3. *Synechococcus sp* (*Ss*SSB) (3koj).

The coordinates of the structure of a complex of *Ms*SSB with DNA are available in the PDB (3a5u), but the results have not been published.

4. Results

4.1. Structural Features

Among the SSBs listed above, all except four are tetramers with 222 symmetry (Fig. 1). Each subunit is less than 200 amino acid residues long (177 in *E. coli*). The C-terminal stretch of varying length (about 60 residues in *E. coli*) involved in interacting with other proteins is disordered in all the crystal structures analyzed so far. Parts of it are believed to be intrinsically disordered. The larger N-terminal domain, which binds DNA, is characterized by the OB

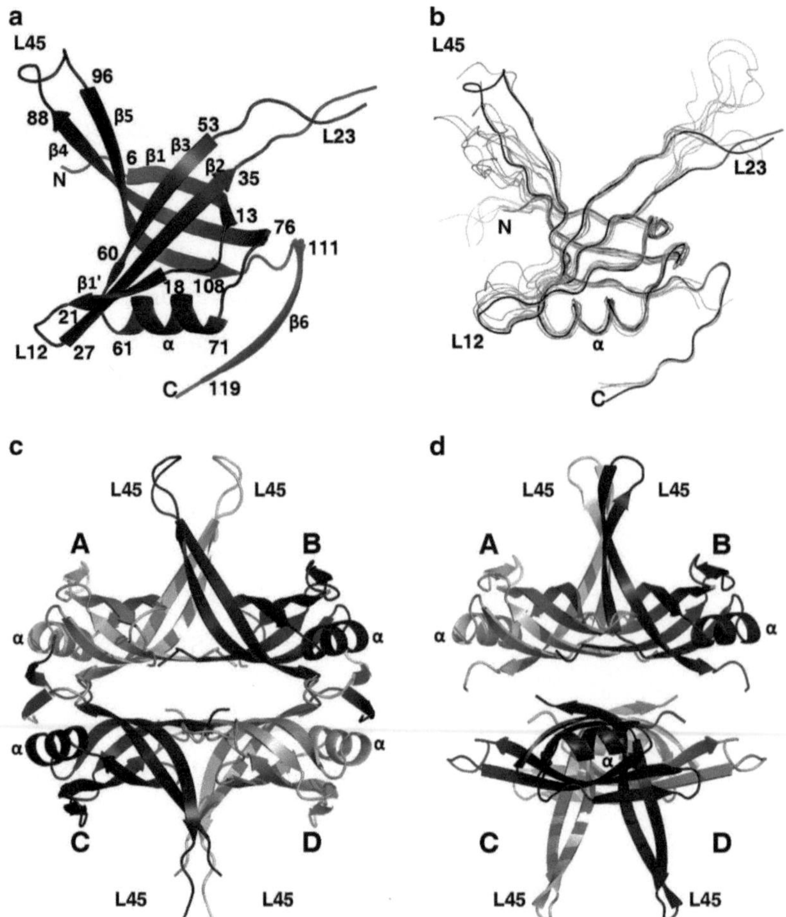

Fig. 1. (**a**) The structure of the N-terminal domain of *Mt*SSB, (**b**) superposition of the N-domain of tetrameric SSBs (the line corresponding to *Mt*SSB is thick), and the quaternary structure of (**c**) *Mt*SSB and (**d**) *Hp*SSB. In (**c**) and (**d**), the AB dimer has the same orientation.

fold in all cases (Fig. 1a). The DNA-binding domain has nearly the same structure in all the concerned SSBs, except in the flexible loops (Fig. 1b). The SSB tetramer can be described as a dimer of dimers (Fig. 1c, d). Partly for convenience and partly on the basis of intersubunit interactions in a majority of the SSBs considered here, subunits A and B can be treated as the dimer. The interactions between A and B include inter-subunit hydrogen bonds of the type, which occur in antiparallel β-sheets. They are between the two N-terminal stretches (as in Fig. 2b).

Mycobacterial SSBs and *Sc*SSB have some distinctive additional structural features. The ordered DNA-binding domain in them contains an additional stretch (β6 in Fig. 1a), which is absent in other SSBs. This additional feature results in the clamping of the two dimers together at two extremities (Fig. 1c), which lends

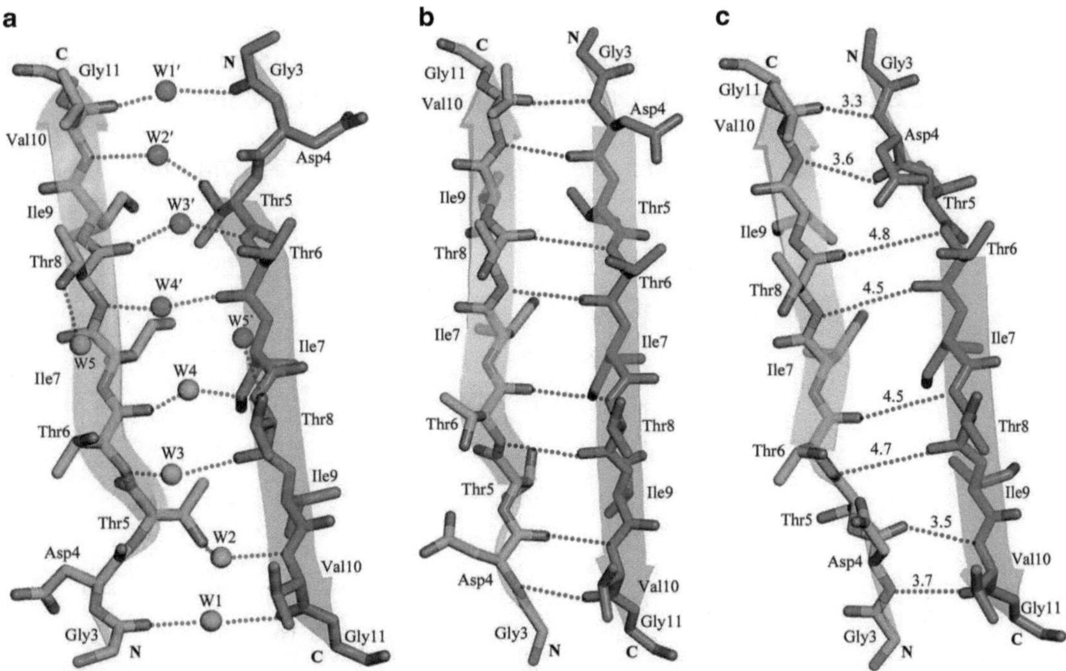

Fig. 2. Interactions between monomers (**a**) in form I *Ml*SSB, (**b**) in form II *Mt*SSB, and (**c**) at the AB interface in form II *Ml*SSB. The distances in (**c**) indicate that the four inner interactions are water bridges. Reproduced with permission from Acta Cryst (2010) D66, 1048–1058.

additional stability to the tetramer. Furthermore, in these SSBs, the interactions between the N-terminal stretches in the AB dimer could be direct hydrogen bonds or water bridges (Fig. 2). Crystal structures of two forms of *Mt*SSB, one form of *Ms*SSB, and two forms of *Ml*SSB are available. The two subunits are interconnected by water bridges in form I of *Mt*SSB, *Ms*SSB, and form I of *Ml*SSB (Fig. 2a). Direct hydrogen bonds exist between them in form II *Mt*SSB (Fig. 2b), while an intermediate situation is seen in form II of *Ml*SSB (Fig. 2c). The interconnection is through water bridges in *Sc*SSB as well. Direct hydrogen bonds exist in all other tetrameric SSBs. The similarity between mycobacterial SSBs and *Sc*SSB is reflected in amino acid sequences as well (Table 1). The N-domains of the three mycobacterial SSBs have a sequence identity of 94–95 % among themselves. The sequence identity between these SSBs and *Sc*SSB ranges between 73 and 76 %.

*Dr*SSB, *Ta*SSB, and *Tt*SSB form another group. They are dimers, but each subunit is made up of two OB domains in such a way that the structure of the subunit is similar to the AB dimer in tetrameric SSBs. Therefore, the overall structure of these dimeric SSBs is similar to that of the more common tetrameric SSBs. The sequence identity between the two domains in each subunit of the dimeric SSBs is typically a little over 30 %, which is roughly comparable to the sequence identity of each such domain with the

Table 1
Alignment scores among N-terminal domains of SSBs with known structure. N and C refer to the N-terminal and C-terminal domains of dimeric SSBs

	Ml	*Mt*	*Ms*	*Sc*	*Ec*	*Tm*	*Bh*	*Hp*	HMt	*Ss*	*Mp*	*Dr*N	*Ta*N	*Tt*N	*Dr*C	*Ta*C	*Tt*C
Ml		95	95	73	30	30	23	22	16	17	7	30	28	28	31	35	35
Mt			94	75	33	30	25	24	15	16	2	29	28	29	32	35	37
Ms				76	33	31	25	25	17	16	2	30	29	29	33	35	37
Sc					30	26	26	24	18	19	2	28	22	25	33	29	32
Ec						39	54	37	31	20	10	35	30	31	36	41	40
Tm							33	36	25	24	9	34	25	29	28	33	33
Bh								27	33	14	8	26	25	23	31	26	27
Hp									23	10	8	27	28	31	32	33	33
HMt										5	5	23	16	16	21	24	22
Ss											8	22	16	13	23	22	19
Mp												5	4	5	3	1	3
*Dr*N													46	48	31	37	40
*Ta*N														75	28	33	32
*Tt*N															23	30	31
*Dr*C																52	53
*Ta*C																	91
*Tt*C																	

domains of tetrameric SSBs (Table 1). Among the SSBs considered here, *Mp*SSB is unique in that it is a dimer with each subunit made up of a single OB domain. Each single domain subunit shares a alignment score of only 2–10 with the subunits of tetrameric SSBs. Despite this very low sequence identity, the structure of dimeric *Mp*SSB is very similar to that of the AB dimer in tetrameric SSBs.

4.2. Variability in Quaternary Association

The variability in the quaternary association of tetrameric SSBs arises from the change in the mutual orientation of the two dimers and of the two monomers within a dimer. The changes are of course consistent with the 222 symmetry of the whole tetramer. In *Mp*SSB, only the orientation between the two monomers is relevant. On the other hand, in dimeric SSBs with each subunit containing two OB domains, only the orientation similar to that between the two dimers in tetrameric SSBs is relevant.

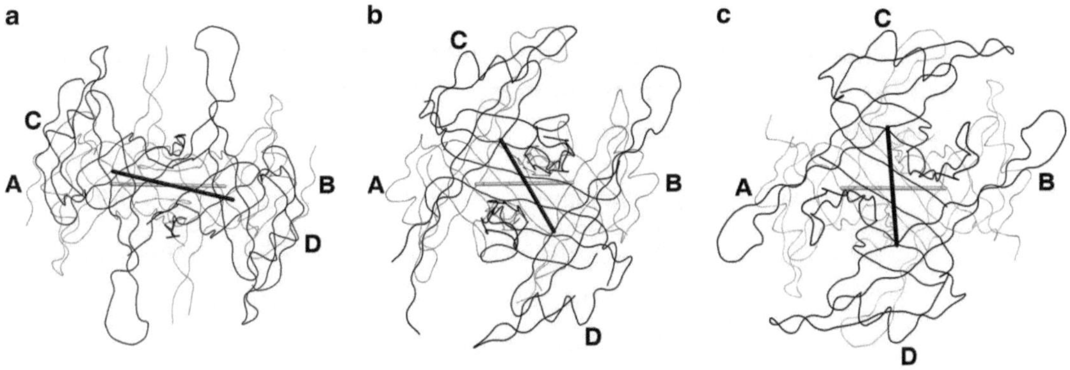

Fig. 3. Mutual orientation of the CD dimer (*thick line*) with respect to the AB dimer (*thin line*) in the tetramer in (**a**) *Mt*SSB, (**b**) *Ec*SSB, and (**c**) *Hp*SSB. Broad straight lines join the centroids of A and B (*gray*) and C and D (*black*).

The variability in the mutual orientation of the dimers (or the two monomers in the case of SSBs with two domains in each subunit) is extensive (Fig. 3, Table 2). In mycobacterial SSBs and *Sc*SSB, the two dimers almost eclipse each other with their longest dimensions pointing in the same direction. On the other extreme, the longest dimensions are nearly perpendicular to each other in *Hp*SSB. In terms of orientation between the two dimers, *Ec*SSB, HMtSSB, and *Bh*SSB cluster together closer to *Hp*SSB. *Tm*SSB exhibits an orientation still closer to *Hp*SSB, while the orientation in *Ss*SSB involves a movement in the opposite direction. The three dimeric SSBs cluster together with an orientation closer to that in mycobacterial SSBs and *Sc*SSB. The variability in the mutual orientation of the two monomers within a dimer is much less extensive and is best described in terms of the distance between the centers of masses of subunits A and B and the angle the two centers subtend at the center of mass of the tetramer (Fig. 4, Table 2). Mycobacterial SSBs and to some extent *Sc*SSB again cluster together in terms of these parameters. *Ec*SSB, HMtSSB, and *Bh*SSB also cluster together as in the case of the orientation between the tetramers.

The variability in quaternary association outlined above leads to interesting differences in the surface area buried on oligomerization (Table 2). The monomer–monomer interface in the dimer involves a long loop which is not only highly flexible, but also disordered and hence undefined to different extents in different structures. Therefore, the estimate of surface area buried on dimerization could have some error. Yet it is clear from the table that the surface area buried on dimerization is somewhat lower in mycobacterial SSBs and *Sc*SSB than in other SSBs. This could partly be on account of the comparatively high separation between the two monomers and partly because of the occurrence of water molecules in water bridges in the interface in some of the structures. The two dimers almost eclipse each other in the first group and the surface area buried at the AC (and equivalent BD) interface on oligomerization

Table 2
(1) Mutual orientation (°) of dimers with respect to that in mycobacterial SSBs, (2) distance (Å) between the centers of mass of subunits A and B, and surface areas (Å²) buried on oligomerization

Source	PDB code	(1)	(2)	Tetramer	AB	AC	AD
MtSSB	1uel	0	23.7	8,966(5,554)	1,708(1,021)	2,771(1,740)	83(22)
MlSSB	3afp	0	23.9	9,104(5,300)	1,799(985)	2,800(1,724)	17(1)
MsSSB	1x3e	0	24.3	8,932(5,446)	1,573(895)	2,770(1,735)	35(20)
ScSSB	3eiv	4	27.0	9,312(5,661)	1,654(917)	2,838(1,777)	17(13)
SsSSB	3koj	40	23.5	6,895(4,525)	2,138(1,457)	580(373)	906(595)
EcSSB	3kaw	44	20.3	6,542(3,553)	2,319(1,135)	458(389)	637(424)
HMtSSB	3ull	45	18.3	7,132(4,239)	2,777(1,482)	588(391)	772(417)
BhSSB	3pgz	49	18.4	6,905(3,757)	2,440(1,209)	453(407)	815(487)
TmSSB	1z9f	59	21.4	7,935(5,143)	2,594(1,599)	387(368)	1,257(832)
HpSSB + DNA	2vw9	71	20.0	5,690(3,774)	2,367(1,601)	94(93)	500(360)
TaSSB	2fxq	21		2,413(1,430)[a]			
DrSSB	1se8	23		2,189(1,429)[a]			
TtSSB	2cwa	24		3,290(1,709)[a]			
MpSSB	2hql		18.3		3,178(2,084)		

The nonpolar component is given in parentheses. In case of more than one crystal form of the same SSB, only one is used

[a]The values represent the area buried on dimerization of SSBs with two-domain subunits

is substantial. The buried area at this interface is, as expected, much lower in the group constituted by the rest of the tetrameric SSBs. The situation is the opposite at the AD (and BC) interface, although the difference between the two sets of values is less pronounced. The observed variability in the quaternary association of SSBs does not exhibit any obvious linear correlation with the differences in amino acid sequence. However, some broad correlation is discernible. In terms of sequence similarity, mycobacterial SSBs and *Sc*SSB form one group. They form one group also in terms of the different components of surface area buried on oligomerization. The rest of the tetrameric SSBs form another group with similar surface areas buried at different interfaces in the oligomer. The sequence identity among them is not in general higher than that between them and mycobacterial SSBs; yet they form a group distinct from mycobacterial proteins in terms of oligomerization. Perhaps, the additional β strand (β6) which clamps the two

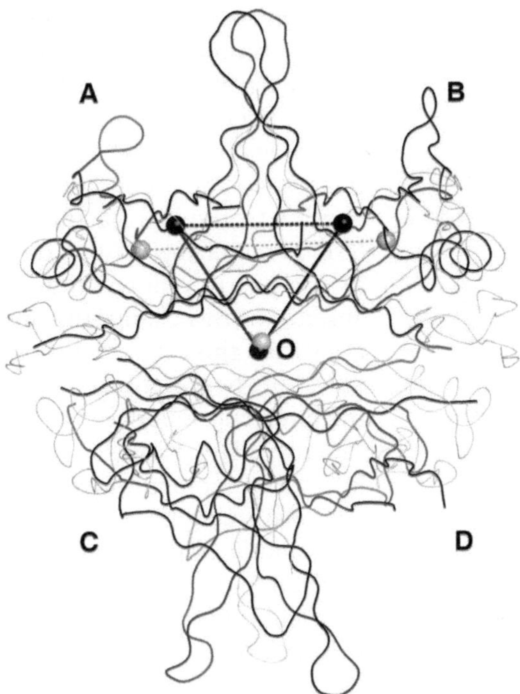

Fig. 4. Mutual orientation of A and B in HMtSSB (*black*) and *Sc*SSB (*gray*). The balls represent centers of mass of A and B and the tetramer. The distance between the center of mass of A and B in HMtSSB is 18.3 Å and they subtend an angle of 66.5° at the center of mass of the tetramer. The corresponding values in *Sc*SSB are 27.0 Å and 102.7°, respectively. Intermediate values are found in other SSBs of known structure.

dimers together in mycobacterial SSBs and *Sc*SSB is a substantial contributor to this difference. It is also interesting that the total area buried on tetramerization is higher in mycobacterial SSBs and *Sc*SSB than in other tetrameric SSBs. This observation is in consonance with the higher stability of *Mt*SSB compared to *Ec*SSB in the presence of guanidine hydrochloride (36).

5. Discussion

Despite their similar function and almost identical tertiary structures based on the OB fold, eubacterial and related SSBs exhibit considerable variability in quaternary association as evidenced by the crystal structures of their DNA-binding domains from different sources. The major source of this variability is changes in the mutual orientation of the two dimers in the tetrameric molecules with 222 symmetry. The mutual orientation of the two subunits in dimeric SSBs, in which each subunit is made of two OB domains, is equivalent to that between the two dimers in tetrameric SSBs. A secondary source of variability in oligomerization is changes in the mutual orientation of the two monomers in the dimer.

In terms of quaternary structure, tetrameric SSBs of known three-dimensional structure can be broadly classified into two categories; mycobacterial SSBs and *Sc*SSB belong to one category and the rest to the second category. The difference between the two appears to result from the presence of β6 in the first category of SSBs. They clamp the two dimers together in the tetramer resulting in the nearly eclipsed mutual orientation of the dimers. In the absence of this clamp, the molecules in the second category move away from this eclipsed disposition; the mutual orientations in them vary, but the variation is within about 30°. The clamps also loosen the interactions between the two monomers in the dimer in the first category. This allows in some instances water molecules to enter into the monomer–monomer interface. These observations are also consistent with the plasticity of the molecules in terms of relatively rigid and flexible regions, estimated earlier taking advantage of the availability of the structures of several crystallographically independent subunits of the same SSB. For example, in *Ml*SSB, which belongs to the first category, much of the dimer–dimer interface is relatively rigid, but the monomer–monomer interface in the dimer is relatively flexible (33). The opposite is true in the case of *Ec*SSB, which belongs to the second category.

The difference in the mode of oligomerization in the two categories appears to have implications to the stability of the tetramer, as evidenced by the area buried on oligomerization in the two cases and the experimental observation of the higher stability of *Mt*SSB than that of *Ec*SSB in the presence of guanidine hydrochloride (36). *M. tuberculosis* has to withstand considerable environmental stress during dormancy and growth in macrophages and the higher stability of *Mt*SSB could be of advantage in coping with and protecting genomic integrity from the stress (16). ssDNA with higher G + C content tends to form larger and more stable secondary structures and this might also dictate the requirement of more stable SSB in mycobacteria which have G + C-rich genomes. The structural results along with a comparative analysis of sequences of SSBs from eubacterial and mitochondrial SSBs had earlier indicated that the insertion of the critical β6 tends to occur in all high G + C Gram-positive bacteria (27). Thus the sturdy quaternary association of SSBs found in the first category could well occur in such bacteria.

It has been demonstrated through complementation studies that *Mt*SSB cannot perform the function of *Ec*SSB and vice versa (36). The barrier could not be overcome even when chimeras constructed by swapping the C-terminal domains of the two SSBs were used. Therefore species specificity cannot be attributed to the C-terminal region of the protein. The difference in the quaternary structures of the two SSBs could also contribute to the species barrier (16).

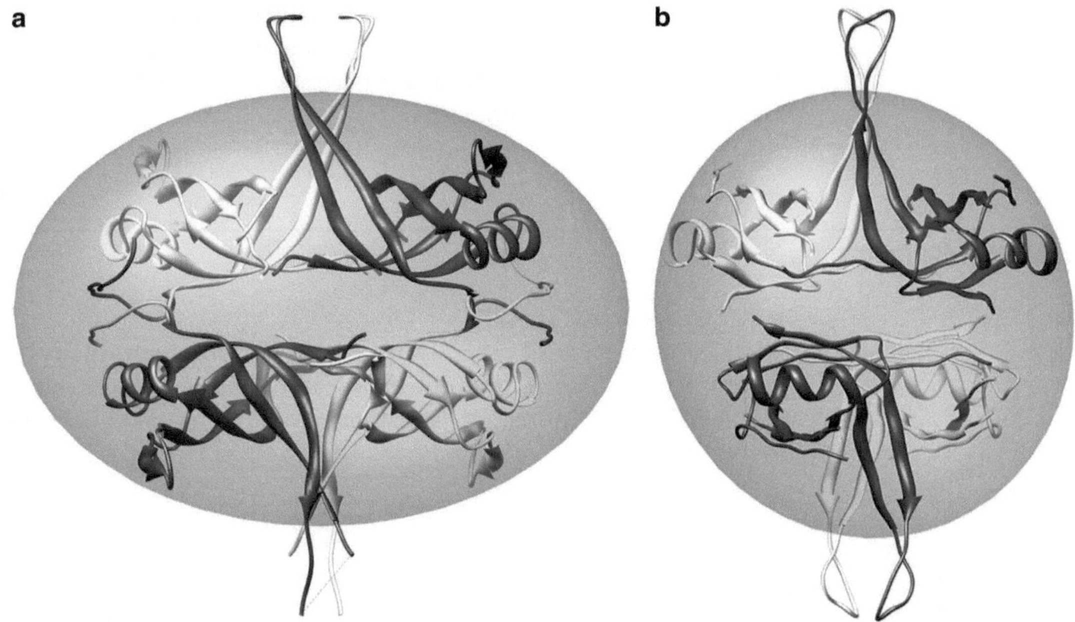

Fig. 5. Ellipsoidal representation of the core of the tetramer in (**a**) *Mt*SSB and (**b**) *Ec*SSB.

The most obvious overall difference between the SSBs of the two categories is in their shapes. For example, *Mt*SSB is approximately ellipsoidal whereas *Ec*SSB is approximately spherical (Fig. 5). Modeling based on this difference suggested that the length of DNA required to wrap around an *Mt*SSB tetramer is lower than that required to wrap around an *Ec*SSB tetramer (16). The length of DNA which binds to tetrameric *Ec*SSB has been established as 65 ± 3 nucleotides (37), but the corresponding length with respect to *Mt*SSB is yet to be determined. Crystal structures of the complexes of shorter DNA fragments with *Ec*SSB (34) and *Hp*SSB (35) have been reported. Coordinates of a similar complex with *Ms*SSB (3a5u) are available in the PDB. The three structures together provide an interesting picture(s) (38). In all the cases, the electron density for DNA was discontinuous, perhaps reflecting disorder; partial occupancy was also indicated. Therefore, the structure of DNA in the complexes was partly modeled. Again, the DNA molecules do not obey the symmetry of the tetrameric protein. The paths followed by DNA in the three cases exhibit substantial differences, probably reflecting the differences in the quaternary structure of the proteins. There are notable commonalities as well in some regions of the path of DNA and the amino acid residues involved in interactions with DNA. The structural features of eubacterial SSBs, the commonalities among them, and the variations in their quaternary structure are very well characterized. The same cannot be said about SSB–DNA interactions. Further work is needed to fully structurally characterize them, to discern commonalities among them, and to elucidate the effect of variability in the quaternary association of SSBs on the features of their DNA binding.

Acknowledgements

Financial support of the Department of Biotechnology is acknowledged. S.M.A. is a Junior Research Fellow of the Council of Scientific and Industrial Research and M.V. is a DAE Homi Bhabha Professor.

References

1. Banerjee R et al (1994) Crystal structure of peanut lectin, a protein with an unusual quaternary structure. Proc Natl Acad Sci U S A 91:227–231
2. Drickamer K (1995) Multiplicity of lectin-carbohydrate interactions. Nature Struct Biol 2:437–439
3. Loris R et al (1998) Legume lectin structure. Biochim Biophys Acta 1383:9–36
4. Prabu MM et al (1998) Carbohydrate specificity and quaternary association in basic winged bean lectin: X-ray analysis of the lectin at 2.5 Å resolution. J Mol Biol 276:787–796
5. Prabu MM, Suguna K, Vijayan M (1999) Variability in quaternary association of proteins with the same tertiary fold. A case study and rationalization involving legume lectins. Proteins 35:58–69
6. Vijayan M, Chandra N (1999) Lectins. Curr Opin Struct Biol 9:707–714
7. Kulkarni KA et al (2004) Effect of glycosylation on the structure of *Erythrina corallodendron* lectin. Proteins 56:821–827
8. Hester G et al (1995) Structure of mannose-specific snowdrop (*Galanthus nivalis*) lectin is representative of a new plant lectin family. Nat Struct Biol 2:472–9
9. Chandra N et al (1999) Crystal structure of a dimeric mannose-specific agglutinin from garlic: quaternary association and carbohydrate specificity. J Mol Biol 285:1157–1168
10. Singh DD et al (2005) Unusual sugar specificity of banana lectin from *Musa paradisiaca* and its probable evolutionary origin. Crystallographic and modelling studies. Glycobiol 15:1025–1032
11. Meagher JL et al (2005) Crystal structure of banana lectin reveals a novel second sugar binding site. Glycobiol 15:1033–1042
12. Shereda RD et al (2008) SSB as an organizer/mobilizer of genome maintenance complexes. Crit Rev Biochem Mol Biol 43:289–318
13. Raghunathan S et al (1997) Crystal structure of the homo-tetrameric DNA binding domain of *Escherichia coli* single-stranded DNA-binding protein determined by multiwavelength x-ray diffraction on the selenomethionyl protein at 2.9-Å resolution. Proc Natl Acad Sci U S A 94:6652–6657
14. Yang C et al (1997) Crystal structure of human mitochondrial single-stranded DNA binding protein at 2.4 Å resolution. Nat Struct Biol 4:153–157
15. Murzin AG (1993) OB(oligonucleotide/oligosaccharide binding)-fold: common structural and functional solution for non-homologous sequences. EMBO J 12:861–867
16. Saikrishnan K et al (2003) Structure of *Mycobacterium tuberculosis* single-stranded DNA-binding protein. Variability in quaternary structure and its implications. J Mol Biol 331:385–393
17. Berman HM et al (2000) The protein data bank. Nucleic Acids Res 28:235–242
18. Cohen GE (1997) ALIGN: a program to superimpose protein coordinates, accounting for insertions and deletions. J Appl Cryst 30:1160–1161
19. Hubbard SJ, Thornton JM (1993) NACCESS computer program. Department of Biochemistry and Molecular Biology, University College London

20. Emsley P, Cowtan K (2004) Coot: model-building tools for molecular graphics. Acta Crystallogr D Biol Crystallogr 60:2126–2132
21. Costantini S, Paladino A, Facchiano MA (2008) CALCOM: a software for calculating the center of mass of proteins. Bioinformation 2:271–272
22. Pettersen EF et al (2004) UCSF Chimera—a visualization system for exploratory research and analysis. J Comput Chem 25:1605–1612
23. Collaborative Computational Project, Number 4 (1994) The CCP4 suite: programs for protein crystallography. Acta Crystallogr D Biol Crystallogr 50:760–763
24. DeLano WL (2002) PyMOL http://www.pymol.org.
25. Larkin MA et al (2007) Clustal W and Clustal X version 2.0. Bioinformatics 23:2947–2948
26. Bernstein DA et al (2004) Crystal structure of the *Deinococcus radiodurans* single-stranded DNA-binding protein suggests a mechanism for coping with DNA damage. Proc Natl Acad Sci U S A 101:8575–8580
27. Saikrishnan K et al (2005) Structure of *Mycobacterium smegmatis* single-stranded DNA-binding protein and a comparative study involving homologus SSBs: biological implications of structural plasticity and variability in quaternary association. Acta Crystallogr D Biol Crystallogr 61:1140–1148
28. Jedrzejczak R et al (2006) Structure of the single-stranded DNA-binding protein SSB from *Thermus aquaticus*. Acta Crystallogr D Biol Crystallogr 62:1407–1412
29. Fedorov R et al (2006) 3D structure of *Thermus aquaticus* single-stranded DNA-binding protein gives insight into the functioning of SSB proteins. Nucleic Acids Res 34:6708–6717
30. DiDonato M et al (2006) Crystal structure of a single-stranded DNA-binding protein (TM0604) from *Thermotoga maritima* at 2.60 Å resolution. Proteins 63:256–260
31. Das D et al (2007) Crystal structure of a novel single-stranded DNA-binding protein from *Mycoplasma pneumoniae*. Proteins 67:776–782
32. Stefanić Z, Vujaklija D, Luić M (2009) Structure of the single-stranded DNA-binding protein from *Streptomyces coelicolor*. Acta Crystallogr D Biol Crystallogr 65:974–979
33. Kaushal PS et al (2010) X-ray and molecular-dynamics studies on *Mycobacterium leprae* single-stranded DNA-binding protein and comparison with other eubacterial SSB structures. Acta Crystallogr D Biol Crystallogr 66:1048–1058
34. Raghunathan S et al (2000) Structure of the DNA-binding domain of *E. coli* SSB bound to ssDNA. Nat Struct Biol 7:648–652
35. Chan KW et al (2009) Single-stranded DNA-binding protein complex from *Helicobacter pylori* suggests an ssDNA-binding surface. J Mol Biol 388:508–519
36. Handa P et al (2000) Distinct properties of *Mycobacterium tuberculosis* single-stranded DNA-binding protein and its functional characterization in *Escherichia coli*. Nucleic Acids Res 28:3823–3829
37. Lohman TM, Ferrari ME (1994) *Escherichia coli* single-stranded DNA-binding protein: multiple DNA-binding modes and cooperativities. Annu Rev Biochem 63:527–570
38. Kaushal PS (2010) Structural studies of mycobacterial uracil-DNA glycosylase (Ung) and single-stranded DNA binding proteins. Ph.D. thesis, Indian Institute of Science, Bangalore, India.

Chapter 3

SSB Binding to ssDNA Using Isothermal Titration Calorimetry

Alexander G. Kozlov and Timothy M. Lohman

Abstract

Isothermal titration calorimetry (ITC) is a powerful method for studying protein–DNA interactions in solution. As long as binding is accompanied by an appreciable enthalpy change, ITC studies can yield quantitative information on stoichiometries, binding energetics (affinity, binding enthalpy and entropy) and potential site–site interactions (cooperativity). This can provide a full thermodynamic description of an interacting system which is necessary to understand the stability and specificity of protein–DNA interactions and to correlate the activities or functions of different species. Here we describe procedures to perform and analyze ITC studies using as examples, the *E. coli* SSB (homotetramer with 4 OB-folds) and *D. radiodurans* SSB (homodimer with 4 OB-folds). For oligomeric protein systems such as these, we emphasize the need to be aware of the likelihood that solution conditions will influence not only the affinity and enthalpy of binding but also the mode by which the SSB oligomer binds ssDNA.

Key words: EcoSSB, DrSSB, ssDNA binding, ITC, SSB-ssDNA thermodynamics

1. Introduction

Isothermal titration calorimetry (ITC) is the only method that allows for simultaneous determination of binding affinity (K_{eq}), and binding enthalpy (ΔH^o) for an interacting system from a single equilibrium titration experiment. As a result one ITC experiment can provide a nearly complete thermodynamic characterization of an interacting system under one set of conditions, i.e., standard state binding free energy (ΔG^o), binding enthalpy (ΔH), and binding entropy (ΔS^o), since these functions are related through Eq. 1,

$$\Delta G^o = -RT \ln K_{eq} = \Delta H - T\Delta S^o \qquad (1)$$

where R is the gas constant (cal/mol deg) and T is the temperature (K^o). Additional experiments performed at different temperatures allow for the determination of another

James L. Keck (ed.), *Single-Stranded DNA Binding Proteins: Methods and Protocols*, Methods in Molecular Biology, vol. 922, DOI 10.1007/978-1-62703-032-8_3, © Springer Science+Business Media, LLC 2012

important thermodynamic quantity, the heat capacity change $\left(\Delta C_P = \left(\frac{\partial \Delta H}{\partial T}\right)_p = T\left(\frac{\partial \Delta S}{\partial T}\right)_p\right)$. Further experiments performed as a function of pH, salt concentration and type or another second ligand that may bind any of the primary species under study (protein, DNA, etc.) will also provide information on any heterotropic linkage that affects binding (1). Such thermodynamic information is necessary to understand stability and specificity of protein–DNA complexes and their mechanisms of interactions. The thermodynamics of *Eco*SSB binding to ssDNA has been extensively studied using ITC (2–7). More limited examples of ITC experiments are also available for *Dr*SSB (8) and yeast *sc*RPA (9).

The *Escherichia coli* ssDNA binding protein (*Eco*SSB) is the most studied representative of the large class of bacterial SSB proteins (10). It is a stable homotetramer in solution containing four OB folds (one per subunit) that each provide a site for binding ssDNA (11). Due to the presence of these four potential ssDNA binding sites, the *Eco*SSB tetramer can bind to long, polymeric ssDNA in multiple binding modes that differ by the amount of ssDNA that is wrapped around the tetramer. Two major binding modes, denoted $(SSB)_{35}$ and $(SSB)_{65}$, differ by the number of subunits, either two or four, respectively, that interact with ssDNA with corresponding occluded site sizes of ~35 and 65 nucleotides (10). The stabilities of these binding modes are influenced by salt concentration, type and valence and protein binding density (10, 12). The four C-termini of the SSB tetramer (residues 112-177) are unstructured even when SSB is bound to ssDNA (13). The last 8-10 amino acids in these C-termini serve as the primary sites to which more than a dozen other proteins bind to SSB (14). The binding of *Eco*SSB tetramer to oligodeoxythymidylates that are close in size to the occluded site size (65 nucleotides) is of such high affinity (i.e., stoichiometric) under most solution conditions, even at very high salt concentrations, that the affinity is difficult to measure directly. Binding of *Eco*SSB to $(dT)_{70}$ is characterized by an extremely large and negative enthalpy change (up to −160 kcal/mol at low salt conditions (3)), and a large and negative heat capacity change, $\Delta C_{p,obs}$ (up to −1.2 kcal/mol °K (6)). In fact, this is the largest binding enthalpy yet reported for a protein–DNA interaction. The thermodynamics of *Eco*SSB-ssDNA binding has contributions from the binding of multiple other small molecule species that accompany SSB–DNA binding. These so-called "linked equilibria" or linkage effects include the binding of cations and anions (2, 3, 6, 7), and protonation (5, 15). Conformational transitions within the DNA (4) and SSB (6, 7) have also been documented and quantified.

The *Deinococcus radiodurans* SSB (*Dr*SSB) belongs to the Deinococcus-Thermus group and functions as a dimer; however, each monomer contains two OB-folds linked by a conserved spacer sequence (16). Therefore, the functional form of this protein also is composed of four OB-folds, although the sequence differences between the two OB-folds within each dimer impose an asymmetry that affects its DNA binding properties (8). The occluded site size of this protein is also dependent on salt concentration, although not as dramatically as for the *Eco*SSB tetramer, changing from 50 to 53 nucleotides at moderate and high salt concentrations to 45 nucleotides at lower salt concentrations (8). ITC measurements (8) showed that *Dr*SSB binding to oligo(dT)s with lengths close to the occluded site size (45-55 nucleotides) is stoichiometric at low and moderate salt concentrations (0.02–0.2 M NaCl), although in contrast to *Eco*SSB the binding is weaker at higher salt concentrations (1 M NaCl), such that an affinity can be directly measured. The $\Delta H_{obs} = -94$ kcal/mol determined at 0.2 M NaCl salt conditions is somewhat smaller than observed for *Eco*SSB (approximately −130 kcal/mol) under the same conditions. However, the observed binding enthalpy shows a large sensitivity to NaCl concentration, similar to that observed for *Eco*SSB, but with a smaller $\Delta C_{p,obs}$.

In the following sections we describe the procedures for ITC titration experiments performed using VP-ITC instrument from MicroCal (GE Healthcare) for *Eco*SSB and *Dr*SSB binding to oligo (dT)s of different lengths that exemplify the effects of salt concentration on both stoichiometry, affinity and binding enthalpy.

2. Materials

2.1. Buffer Solutions

All solutions were prepared with reagent grade chemicals and glass distilled water that was subsequently treated with a Milli Q (Millipore, Bedford, MA) water purification system. Buffer T is 10 mM Tris (tris(hydroxymethyl) aminomethane), pH 8.1 and buffer C is 10 mM Cacodylate, pH 7.0, 25 % (V/V) glycerol. Both buffers also contained NaCl at the indicated concentration (either 0.20 and 1.00 M for Buffer T or 0.02 and 0.20 M for Buffer C) and 0.1 mM Na_3EDTA (ethylenediaminetetraacetic acid sodium salt). We emphasize that these buffers do not contain additional reducing agents (e.g., 2-mercaptoethanol or tris(2-carboxyethyl)phosphine (TCEP)) (see Note 1).

1. Buffer T (1 l). Dissolve 1.211 g of Trizma-base and either 11.67 g or 58.44 g of NaCl (for 0.2 and 1 M final concentration of NaCl, respectively) in ~900–950 mL of MilliQ deionized water in a 1 l glass beaker under constant stirring with the magnetic stirrer bar on a stirrer plate (see Note 2). Add 200 μl of 0.5 M EDTA stock solution (pH 8.1–8.3). Adjust the pH to 8.10 (25 °C) with ~5 M HCl stock solution (2–2.2 ml will be required) as monitored with a glass electrode pH meter. Pour

the solution into a volumetric flask (1 l) being careful to rinse several times with a few mL of MiliQ water to ensure all components are transferred to the volumetric flask, then add sufficient MilliQ water to adjust the volume to 1 l. Be sure that the contents are mixed thoroughly by wrapping with Parafilm around the top of the volumetric flask to make a leak proof seal. Filter the solution through a disposable NALGENE filter unit (0.22 μm nitrocellulose) using a standard laboratory bench vacuum outlet and collect the filtrate in the plastic bottle that comes with the unit. Cap and store at 4 °C.

2. Buffer C (1 l). Dissolve 1.38 g of cacodylic acid and the required amount of NaCl as above in ~500 ml of MilliQ deionized water in a 1 l glass beaker and then add 250 ml of glycerol (see Note 3). Add 200 μl of 0.5 M EDTA, mix and add water again up to ~950 ml total volume, mix and adjust the pH to 7.00 with ~5 M NaOH (3–3.5 ml will be required). Proceed with filtering and storage as described above.

2.2. Proteins and ssDNA

1. The *Eco*SSB and *Dr*SSB proteins used in the ITC experiments were expressed and purified as described ((17) and (18), respectively). Protein stock solutions in storage buffer (20 mM Tris (pH 8.3), 50 % (V/V) glycerol, 0.50 M NaCl, 1 mM EDTA, and 1 mM BME) are usually (4–6) mg/ml (50–80 μM *Eco*SSB tetramer) and 10 mg/ml (~160 μM *Dr*SSB dimer). Proteins are stored in 0.5–1.0 mL aliquots in 2 mL cryogenic vials in storage buffer at −80 °C (see Note 4). Protein concentrations are determined spectrophotometrically after dialysis in buffer T containing 0.20 M NaCl using the extinction coefficients, $\in_{280} = 1.13\times10^5$ M^{-1} (tetramer) cm^{-1} for *Eco*SSB and $\in_{280} = 8.2\times10^4$ M^{-1} (dimer) cm^{-1} for *Dr*SSB (see Note 5).

2. The single stranded oligodeoxythymidylates, $(dT)_{70}$ and $(dT)_{45}$, used in the ITC experiments were synthesized in our lab using an Applied Biosystems DNA synthesizer and purified by gel electrophoresis as described (19). The DNA is ≥98% pure as judged by denaturing gel electrophoresis and autoradiography of a sample that were 5′ end-labeled with ^{32}P using polynucleotide kinase (20). The aliquots (~0.5 ml) of synthesized oligodeoxynucleotides (0.5–0.7 mM (molecule concentration)) are usually stored in water or Tris buffer at −20 °C. Concentrations of oligo(dT) are determined spectrophotometrically by UV absorbance in buffer T (pH 8.1), 0.1 M NaCl using the extinction coefficient, $\in_{260} = 8.1 \times 10^3$ M^{-1} (nucleotide) cm^{-1} or 5.67×10^5 M^{-1} ($(dT)_{70}$ molecule) cm^{-1} and 3.65×10^5 M^{-1} ($(dT)_{45}$ molecule) cm^{-1} (see Note 5).

3. Methods

3.1. Estimates of Amounts of SSB and DNA Required for ITC Experiment

An estimate of the amount of SSB and DNA needed for an ITC experiment is facilitated if some initial information on the system is known. For example, in the experiments presented in Figs. 1a and 2a, b the affinity of the SSB-$(dT)_L$ complexes is so high that binding is "stoichiometric" and the reaction is complete at total DNA to protein ratio $R = 1$ (mole $(dT)_L$/mole SSB). As used here, the term "stoichiometric" means that essentially all of the titrant being added (ligand) becomes bound to the species in the cell (macromolecule), so that there is essentially no free ligand in solution up until the "stoichiometric point" of $R = 1$. The results in the experiment presented in Fig. 1b indicates that this titration is not stoichiometric due to the weaker binding of $(dT)_{45}$ to *Dr*SSB under these conditions, hence saturation of the SSB requires a ratio, $R = [(dT)_{45}]_{tot}/[SSB]_{tot} \approx 4$. In any case, required concentrations of species along with the volume and number of experimental injections can be roughly calculated using simple formula in Eq. 2:

$$R \cdot V_{cell} \cdot C_{cell} = n_{inj} \cdot V_{inj} \cdot C_{syr} \qquad (2)$$

Where R is the desirable total DNA to total protein ratio in the calorimetric cell at the end of the titration, where $V_{cell} \approx 1.45$ ml is the volume of calorimetric cell, C_{cell} is the concentration of the SSB in the cell, C_{syr} is concentration of DNA in the syringe, n_{inj} is number of injections, and V_{inj} is injection volume. Note that the two latter parameters are constrained by the syringe volume $V_{syr} = n_{inj} \cdot v_{inj} < 300$ μl. Therefore, for example, to reach $R = 4$ (as in Fig. 1b) by performing 19 injections (15 μl each) at the concentration of SSB in the cell 0.6 μM, one needs a $(dT)_{45}$ concentration in the syringe near 12 μM. The following procedures describe the preparation of SSB and DNA samples used in performing 2–3 ITC experiments (including reference titrations) for 0.5–1.0 μM of SSB in the cell and 10–20 μM of DNA in the syringe (see Note 6).

3.2. Dialysis of Proteins and ssDNA

Extensive dialysis of the protein and DNA solutions is required to exactly match buffer conditions for titrant in the syringe ($(dT)_L$) and sample in the calorimetric cell (SSB). This is particularly important for interactions that have small binding enthalpies, where the reference heats of dilution are comparable to the experimental heats (for example, see Fig. 1b).

1. Prepare ~3–4 mL of an approximately 2 μM solution of either *Dr*SSB or *Eco*SSB by adding some of the concentrated SSB stocks to dialysis buffer.

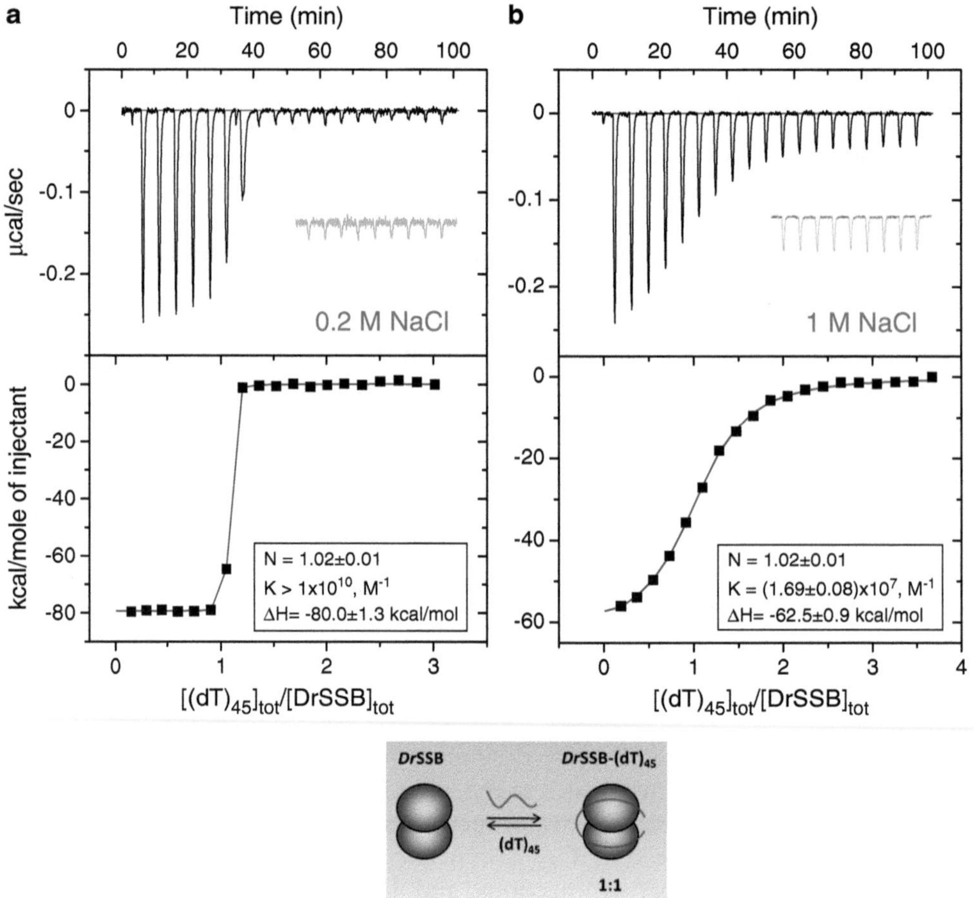

Fig. 1. ITC isotherms for the binding of $(dT)_{45}$ to *Dr*SSB to form a 1:1 molar complex at two different NaCl concentrations. (**a**) Calorimetric titration of $(dT)_{45}$ into *Dr*SSB in the presence of 0.2 M NaCl (Tris buffer (pH 8.1), 0.2 M NaCl, 25 °C). *Upper panel*: Raw titration data, plotted as the heat signal (microcalories per second) versus time (minutes), obtained for 19 injections (15 μl each) of $(dT)_{45}$ (8.6 μM, syringe) into a solution containing *Dr*SSB protein (0.6 μM dimer, cell). *Lower panel*: the ITC data form the upper panel in the form of titration isotherm, where the integrated heat responses per injection, after subtraction of the heats of dilution obtained from a blank titration of $(dT)_{45}$ into buffer (shown in insert), were normalized to the moles of injected ligand ($(dT)_{45}$) and plotted versus total DNA to protein ratio. The continuous curve shows the best fit of the data to a 1:1 binding model with $N = 1.02 \pm 0.01$, $\Delta H_{obs} = -80.0 \pm 1.3$ kcal/mol. The affinity of $(dT)_{45}$ is too high to measure under these "stoichiometric" conditions. A minimum estimate of $K_{obs} = 2\times10^{10}\ M^{-1}$ was used for the simulation. (**b**) Calorimetric titration of $(dT)_{45}$ (10.5 μM) into *Dr*SSB (0.6 μM dimer) in the presence of 1.0 M NaCl (experimental setup and buffer conditions are the same as in Fig. 1a). The raw data and corresponding titration isotherm are presented in *upper* and *lower panels*, respectively. The continuous curve in the lower panel shows the best fit of the data to a 1:1 binding model with $N = 1.02 \pm 0.01$, $K_{obs} = (1.69 \pm 0.08)\times10^7\ M^{-1}$, $\Delta H_{obs} = -62.5 \pm 0.9$ kcal/mol.

2. Prepare ~1 mL of an approximately 50 μM $(dT)_L$ solution by adding calculated aliquot of concentrated DNA stock (usually 70–100 μl) to dialysis buffer.
3. Place the SSB and DNA solutions in dialysis bags of corresponding size (MWCO: 10,000 Spectra/Por) and secure the contents with plastic clamps (see Note 7).

4. Dialyze protein and DNA in separate beakers containing 300 ml of dialysis buffer at 4 °C with constant stirring of the buffer. Change dialysis buffer two times every 3–4 h.
5. Remove SSB solution after dialysis, distribute in Eppendorf tubes and centrifuge at 10,000 rpm (~7000 *g*) for 15 min in a cold room. Pipette out the protein solution from each Eppendorf tube avoiding 5–10 μl at the bottom (to avoid any possible precipitated protein) and combine them in a larger 5–10 mL plastic tube. Remove DNA from dialysis bag and place in Eppendorf tube for further use.

3.3. Preparing for ITC Experiment and Performing the Titrations

Here we describe the experimental procedure for a titration of SSB in the cell with DNA in the syringe, followed by a reference titration of buffer in the cell with DNA in the syringe. We assume that the experimenter is familiar with the basics of instrument operation and software. Otherwise, we recommend reading the VP-ITC Microcalorimeter user's manual first.

1. Turn on the instrument and leave it to equilibrate for 1 h.
2. While waiting determine the concentrations of dialyzed samples (see Note 5).
3. Use dialysis buffer to dilute the dialyzed SSB and DNA samples to the concentrations calculated previously using Eq. 2 for any experiment of your choice (see Fig. 1 or Fig. 2). Prepare 2.6 mL of protein solution and 1 mL of DNA solution to fill the calorimetric cell and the syringe (see Note 8).
4. Prepare the cell. The cell should have been cleaned after the previous run and filled with distilled water. Remove the water and fill the cell with dialysis buffer using a disposable plastic syringe attached to a long stainless steel needle (see Note 9). Remove the buffer and using a fresh syringe fill the cell with protein solution. It is important to avoid introducing bubbles into the cell. Save the rest of the protein solution (~500 μl) for a redetermination of its concentration.
5. While waiting for the instrument equilibration to be completed (15–20 min) (see Note 10), fill the syringe with the DNA solution using a plastic syringe connected to the filling port (~500 μl required) and carefully introduce syringe assembly into the cell.
6. Allow for equilibration (another 10–15 min). Using the instrument computer software set all parameters for the titration (number and volume of the injections, injections intervals, reference power, filter period, etc., see manual for details) and start the run (see Note 11).
7. After completion of the run carefully remove the syringe assembly to avoid bending the syringe needle and place it in the

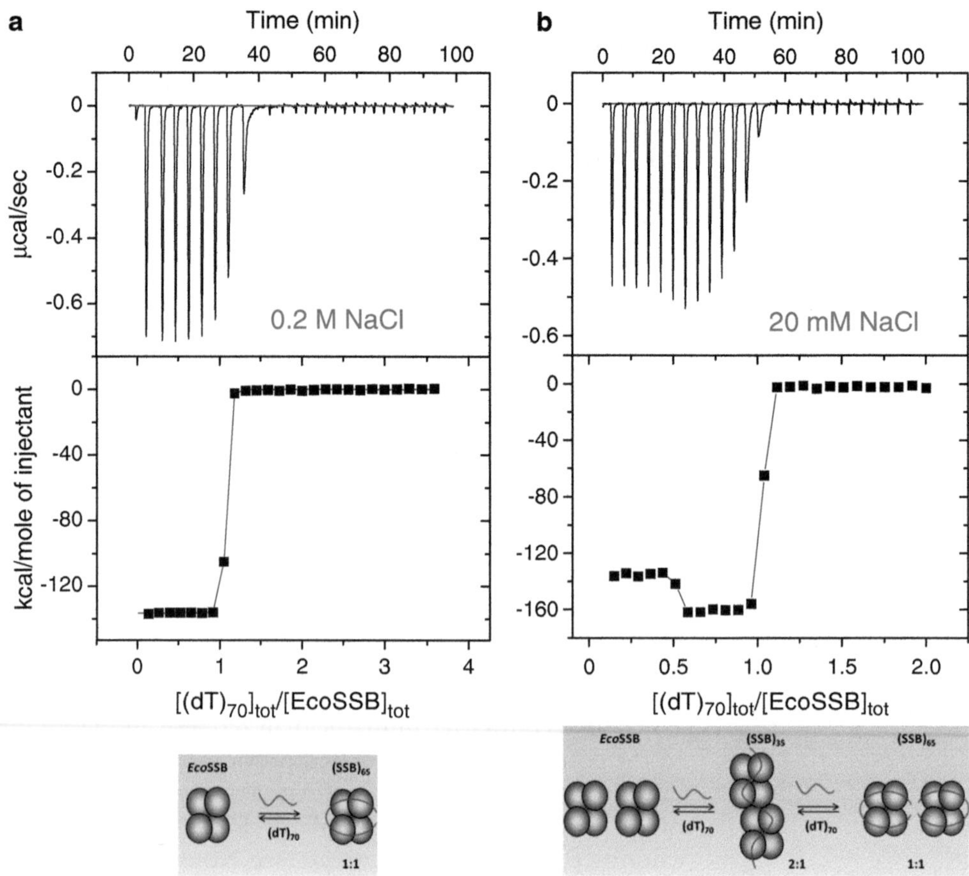

Fig. 2. The stoichiometry of *Eco*SSB:$(dT)_{70}$ complexes depends on salt and protein concentration reflecting its salt-dependent binding mode transition. (**a**) ITC titration of $(dT)_{70}$ into *Eco*SSB at moderate salt (0.2 M NaCl) reveals the formation of only the fully wrapped 1:1 molar complex under high affinity "stoichiometric" conditions. *Upper panel*: The raw titration data, plotted as heat signal (microcalories per second) versus time (minutes), was obtained for 26 injections (10 μl each) of $(dT)_{70}$ (17 μM) into a solution containing *Eco*SSB protein (1.0 μM tetramer) (Cacodylate buffer (pH 7.0), 25 % Glycerol, 0.2 M NaCl, 25 °C). *Lower panel*: standard titration isotherm of normalized heats versus total DNA to protein ratio calculated as described in Fig. 1a. The continuous curve shows the best fit of the data to a 1:1 binding model with $N = 1.03 \pm 0.01$, $\Delta H_{obs} = -136.5 \pm 0.7$ kcal/mol. The affinity of $(dT)_{70}$ is too high to measure under these conditions. A minimum estimate of $K_{obs} = 1 \times 10^{10}\ M^{-1}$ was used for the simulation. (**b**) ITC titration of $(dT)_{70}$ (10.0 μM, syringe) into *Eco*SSB (1.0 μM, cell) at low salt (0.02 M NaCl) reveals a transition from a 2:1 (2 SSB per DNA) complex to a 1:1 complex as the DNA to protein ratio increases (experimental setup and buffer conditions are the same as in Fig. 2a). This data represents an example of a reverse titration (macromolecule ($(dT)_{70}$) in the syringe and ligand (*Eco*SSB) in the cell). The raw data and corresponding titration isotherm are presented in the *upper* and *lower panels*, respectively. In the DNA to protein ratio range from 0 to ½ (from very large to twofold excess of *Eco*SSB over DNA) only the 2:1 *Eco*SSB:$(dT)_{70}$ complex is formed, whereas as more free ssDNA is added the transition to the 1:1 complex occurs, which is completed at $[(dT)_{70}]_{tot}$:$[EcoSSB]_{tot} = 1$. The fitting of these data to obtain binding parameters (affinities, enthalpies, cooperativity) will depend on the type of the model (see text), but in any case will be difficult due to the very high affinities of SSB to DNA for both complexes reflected by the very sharp transitions within the titration isotherm. Estimates of the enthalpies of formation of these complexes can be obtained from analysis of the flat portions of the titration isotherm (see text and Note 21).

holder, carefully rinse the needle with distilled water and wipe it with "Kimwipes."

8. To prepare for reference titration, remove the contents of the cell using a disposable syringe, then rinse the cell with distilled water a few times and once with dialysis buffer (see Note 12). Fill the cell with dialysis buffer and allow it to equilibrate as before.
9. Fill the syringe with the rest of the DNA solution manually manipulating the plunger position in the Auto-Pipette window. This is different from what is described in step 5 above and serves to save up to 200 μl of ligand solution (see Note 13).
10. Put the syringe assembly back into the cell, allow equilibration, and run a reference titration keeping all the settings identical to those used in the actual titration performed above.
11. While performing the titrations redetermine the concentrations of SSB and ssDNA used in the experiment as exactly as possible (see Notes 5 and 14).

3.4. Data Analysis

The analysis of ITC titration data and the subsequent fit of the data to an appropriate binding model can be performed using the software provided by the instrument manufacturer or using MicroMath "Scientist" (St. Louis, MO), a software package specifically designed for data fitting (see Note 15). The results of ITC titrations of *Dr*SSB with $(dT)_{45}$ and *Eco*SSB with $(dT)_{70}$ at different NaCl concentrations are presented in Figs. 1 and 2, respectively. The solution conditions and concentrations of SSB and ssDNA for each particular experiment are specified in the figure legends along with the experimental design and fitted parameters. The upper panels show raw titration data representing the heat signal (microcalories per sec) versus time (minutes). Each injection causes a negative deviation of the signal from the baseline indicating the release of heat (exothermic reaction, $\Delta H < 0$) which returns back to baseline as the injected amount of DNA reaches equilibrium with the protein in the cell. As the titration progresses the SSB in the cell becomes saturated with DNA and at the end of the titration the only heats observed reflect those due to DNA dilution (reference heats). Reference titrations of the DNA into the buffer (see inserts in Fig. 1a, b) are required (see Note 16) to be certain that saturation is attained and to correct experimental heats for use in subsequent analyses.

Standard ITC titration isotherms (known as Wiseman isotherms (21)) are shown in the lower panels of Figs. 1 and 2. These are constructed from the raw data in the upper panels. In these isotherms the integrated heat responses are normalized per amount of injected DNA (after subtraction of corresponding heats of dilution) and plotted versus the mole ratio of total DNA to total protein. This so called differential representation of the ITC data is more informative than the integral representation (see Note 17), particularly for the case of very tight (stoichiometric) binding since it provides a visual estimate of the stoichiometry and binding enthalpy (from the flat portions of the isotherms) for complex formation (see Figs. 1a and 2a, b). For a system with weaker binding affinity (nonstoichiometric) (see Fig. 1b) it is necessary to fit these data to a model to obtain estimates of the stoichiometry, binding enthalpy, and equilibrium binding constant(s).

The data in Figs. 1a, b and 2a can be fit to an n-independent and identical sites model using the instrument software and described elsewhere (21). This model describes binding of a ligand (X) to a macromolecule (M) possessing n independent and identical sites for the ligand with macroscopic affinity, $K = [MX]/[X][M]$, and enthalpy, ΔH per binding site. Here we provide basic equations for analysis and simulation of the ITC data for this model.

The integral or cumulative heat (see Note 17) is defined by Eq. 3

$$Q = V_{\mathrm{cell}} \Delta H [M]_{\mathrm{tot}} <X> = V_{\mathrm{cell}} \Delta H [M]_{\mathrm{tot}} \frac{nK[X]}{1 + K[X]} \quad (3)$$

It is assumed to be directly proportional (see Note 18) to the binding density $<X> = nK[X]/(1 + K[X])$, with the calorimetric cell volume (V_{cell}), enthalpy change (ΔH), and total macromolecule concentration in the cell ($[M]_{\mathrm{tot}}$) being proportionality constants. An expression for free ligand concentration ($[X]$) can be found by solving the mass conservation Eq. 4

$$[X]_{\mathrm{tot}} = [X] + <X>[M]_{\mathrm{tot}} = [X] + \frac{nK[X]}{1 + K[X]}[M]_{\mathrm{tot}} \quad (4)$$

where $[X]_{\mathrm{tot}}$ is total concentration of the ligand in the cell. The analytical expression for the differential heat, $q_{\mathrm{n}} = 1/V_{\mathrm{cell}}\, \mathrm{d}Q/\mathrm{d}[X]_{\mathrm{tot}}$ (heat of injection normalized per amount of injected ligand), can be obtained differentiating Eqs. 3 and 4 with respect to the free ligand concentration to yield Eq. 5

$$\begin{aligned} q_{\mathrm{n}} &= \frac{1}{V_{\mathrm{cell}}} \frac{\mathrm{d}Q}{\mathrm{d}[X]_{\mathrm{tot}}} = \frac{1}{V_{\mathrm{cell}}} \frac{\mathrm{d}Q/[\mathrm{d}[X]}{[\mathrm{d}X]_{\mathrm{tot}}/\mathrm{d}[X]} \\ &= \Delta H \frac{nK[M]_{\mathrm{tot}}}{(1 + K[X])^2 + nK[M]_{\mathrm{tot}}} \end{aligned} \quad (5)$$

Where $[X]$ is determined from Eq. 4. Equations 3 for the integral heat and 5 for the differential heat are functions of the total ligand and total macromolecule concentrations and can be used to predict and fit the experimental ITC isotherms (see Note 19).

Equations 4 and 5 can be rewritten in the form of the Wiseman isotherm (21) introducing the variable $R = [X]_{tot}/[M]_{tot}$ and a new parameter $r = 1/C$ (where $C = K[M]_{tot}$) as shown in Eq. 6

$$q_n = \frac{1}{V_{cell}} \frac{dQ}{d[X]_{tot}} = \Delta H \left(\frac{1}{2} + \frac{1 - (1 + r)/2 - R/2}{\left(R^2 - 2R(1 - r) + (1 + r)^2\right)^{1/2}} \right) \quad (6)$$

This equation predicts that the optimal range for performing an ITC experiment in which the binding constant and enthalpy change can both be determined accurately is when the product of the total macromolecule concentration and binding constant ($C = K[M]_{tot}$) is in the range between 1 and 1,000 (21). If C exceeds 1,000, then stoichiometric binding is observed and only ΔH, but not K can be determined accurately. When $C < 1$ the observed isotherm becomes too shallow to allow determinations of either K or ΔH reliably. If C is outside the appropriate range, one can increase or decrease the concentration of macromolecule ($[M]_{tot}$) to shift the value of C into the appropriate range. One can also modify the solution conditions (e.g., salt, pH, or temperature) to change the value of K (see below). This should be taken into account when planning the experiment.

The ITC titrations of *Dr*SSB with $(dT)_{45}$ at two NaCl concentrations shown in Fig. 1 illustrate two cases of strong "stoichiometric" binding at 0.2 M NaCl, when the value of C exceeds 1,000 (Fig. 1a), and weaker binding ($C = K[M]_{tot} = 10$) at the much higher NaCl concentration of 1 M (Fig. 1b). In the first case, one cannot estimate K with any accuracy, only a lower limit of $K > 1{,}000/[M]_{tot} = 1.7 \times 10^9\ M^{-1}$ can be determined. However, a very precise estimate of $\Delta H = -79.9 \pm 1.3$ kcal/mol can be obtained from these data from the flat portion of the isotherm in the R range from 0 to 1, since for this conditions $q_n \rightarrow \Delta H$ when $[X] \rightarrow 0$ in Eq. 5 or when $r = 1/C$ is very small in Eq. 6. As the salt concentration increases to 1 M NaCl both K and ΔH can be determined from a fit of the data to the n-independent and identical sites model described above (see Fig. 2b).

The results of two ITC titrations of *Eco*SSB with$(dT)_{70}$ at moderate (0.2 M NaCl) and low (20 mM NaCl) salt concentrations are presented in Fig. 2a, b, respectively, and demonstrate differences between two proteins when bound to oligonucleotides of approximately the length of the occluded site size. At 0.2 M NaCl *Eco*SSB binds $(dT)_{70}$ stoichiometrically, similarly to *Dr*SSB (Fig. 1a), but the binding enthalpies differ significantly (1.7 fold more negative (favorable) for *Eco*SSB). As the salt concentration increases the affinity of DrSSB to ssDNA decreases and the binding enthalpy becomes less favorable (see Fig. 1b). The same trend in the binding enthalpy is observed for *Eco*SSB (3) (data not shown), and is due to the linkage of a release of anions upon formation of the SSB–DNA complex, although the binding of *Eco*SSB to ssDNA remains stoichiometric up to 3.0 M NaCl (22).

Upon decreasing the salt concentration to 20 mM NaCl the *Eco*SSB switches its mode of interaction with $(dT)_{70}$ from its $(SSB)_{65}$ mode (a 1:1 complex in which 65 nucleotides of ssDNA are fully wrapped around the tetramer) to its $(SSB)_{35}$ binding mode in which two SSB tetramers bind to $(dT)_{70}$, both in a partially wrapped state in which only two subunits of each tetramer (on average) interact with ssDNA). In contrast, *Dr*SSB still forms a 1:1 complex with ssDNA at this lower NaCl concentration (8). The data in Fig. 2b for the titration of *Eco*SSB with $(dT)_{70}$ at low salt demonstrate that although $(dT)_{70}$ binding to *Eco*SSB remains stoichiometric, it undergoes the transition from the 2:1 $(SSB)_{35}$ mode to 1:1 $(SSB)_{65}$ mode upon increasing the $(dT)_{70}$ concentration so that the protein to DNA ratio decreases (see Note 20).

At the beginning of the titration when a small amount of $(dT)_{70}$ is added to a large excess of *Eco*SSB the 1:2 complex forms stoichiometrically and the overall heat per injection corresponds to overall binding enthalpy for formation of a 2 SSB per $(dT)_{70}$ complex (≈ -135 kcal/mol). This complex becomes fully populated at $R = [(dT)_{70}]_{tot}/[EcoSSB]_{tot} = 1/2$. Further addition of $(dT)_{70}$ causes a redistribution of the 2:1 complexes to form two fully wrapped 1:1 $(SSB)_{65}$ complexes (see Scheme in Fig. 2b). This process is accompanied by a more favorable overall heat $q_n \approx -160$ kcal/mol. Finally, at $R = 1$ the transition is complete and only the 1:1 $(SSB)_{65}$ complex exists after this point (see Note 21).

4. Notes

1. Neither DrSSB nor EcoSSB contains cysteines and, therefore, do not require reducing agents to prevent inter and intra protein disulfide bonds formation. Otherwise, 1–2 mM concentrations of BME may be necessary. TCEP proved to be a more effective reducing agent being odorless, more powerful, hydrophilic, and more resistant to oxidation in air. It also can be used at lower concentrations—0.5–1.0 mM. However, it is more expensive and should be added to the buffer solution before pH adjustment (in Subheading 1, step 1 or Subheading 1, step 2), since it is highly acidic and will affect the pH of the buffer even at low concentrations. DTT (dithiothreitol) is not recommended for ITC experiments since it can affect the stability of the baseline, presumably due to oxidation of the DTT. Due to the fact that these reducing agents (particularly 2-ME) are oxidized rapidly,

buffers containing these should be used within a day after preparation.

2. Stock solution of 5 M NaCl in water can also be used for buffer preparations. For buffer T it is required to add 40 and 200 mL of, such stock to obtain final concentrations of 0.2 and 1 M, respectively. The precise concentration of the stock buffer can be checked using refractive index $n = 1.3770$ (20 °C) (23). This becomes particular important when using such hydroscopic salts as $MgCl_2$ (usually exists as a hexahydrate in solid state). In the latter case preparation of approximately 2 M $MgCl_2 \times 6H_2O$ stock solution in water with subsequent correction of concentration using refractive index for 2 M concentration ($n = 1.3766$ (20 °C)) (23) is highly recommended for further use.

3. We recommend to use ≥99.5 % spectrophotometric grade glycerol from "Sigma-Aldrich." If the glycerol is first placed in an intermediate vessel (e.g., a graduated cylinder), it is essential that after pouring the glycerol into the beaker, the graduated cylinder is rinsed several times with MiliQ water and the rinse added to the beaker so that all of the glycerol is transferred. If this is not done, then it will be impossible to ensure reproducibility of the glycerol content in future Buffers.

4. EcoSSB protein can be stored at −20 °C in storage buffer up to a few years. Due to the presence of the 50 % (V/V) glycerol in the storage buffer it does not freeze at this temperature, which allows one to remove aliquots and avoid freeze–thaw cycles.

5. Precise values of extinction coefficients for proteins and ssDNA are required for determination of their concentrations. An accurate value of the extinction coefficient (usually at 280 nm) for an unfolded protein in 6.0 M Guanidine HCl can be calculated based on its amino acid composition (24). Various calculators are available on the Internet to assist this (for example, one is available in "Sednterp" software: http://www.jphilo.mailway.com/download.htm). One can obtain the extinction coefficient of the fully folded protein in the buffer of interest by measuring absorbance spectra (230–320 nm) of the same amount of the protein in 6.0 M Guanidine HCl and in the particular buffer (X) of interest. The value of the extinction coefficient in the buffer is obtained from the calculated value in 6 M Guanidine HCl, multiplied by the ratio of the absorbance values in the native buffer and in 6 M Guanidine HCL ($\in_{\lambda,X} = \in_{\lambda,GHCl} \times (A_{\lambda,X}/A_{\lambda,GHCl})$). Free calculators are also available for calculating extinction coefficients of ssDNA of any sequence and length (see for example from "Integrated DNA Technologies" (IDT) Web site: http://www.idtdna.com/analyzer/Applications/OligoAnalyzer/). These calculations are based on the nearest-neighbor model (25, 26). Reliable estimates of extinction

coefficients can also be obtained by degrading the DNA to its mononucleotides using Phosphodiesterase I digestion and calculating the extinction coefficient from the known mononucleotide extinction coefficients and the DNA composition (27). For both protein and DNA a full UV spectrum must be taken from 240 to 350 nm and at least 3 determinations at different concentrations should be made and a plot of the absorbance versus concentration should follow the linear Beer's law relationship with a zero intercept. Make sure that the absorbance between 310 and 350 is zero after subtraction of the absorbance of the buffer. Any signal in this range above 10 % of the maximum absorbance at 280 (or 260 nm for DNA) is due to light scattering and is likely the result of some protein insolubility problems (usually not the case for DNA) and this sample may need to be discarded.

6. A 10–30-fold difference between the concentrations of the species in the cell and in the syringe is typical for ITC experiments.
7. Prior to use wash the dialysis membrane with distilled water and let it soak for 5–10 min in dialysis buffer. Put one plastic clamp on one end of the bag first, then fill the bag with sample and then secure the other end with the second clamp.
8. The minimum volume to fill the cell is 2.0–2.1 mL. The rest of the solution is used to redetermine the protein concentration. The remaining 500 μl of DNA will be enough to refill the syringe for the reference titration and for a redetermination of the DNA concentration. We have not found it necessary to degas our SSB and DNA solutions, since controls with degassed samples showed no difference.
9. We recommend to use disposable plastic syringes (slip tip) of 3 or 5 mL with the set of 3 attachable long stainless steel needles (9 in. long from "Hamilton") instead of the syringes provided with the instrument since it facilitate handling and reduce time required for cleaning.
10. Equilibration can be considered as completed when the DP (Differential Power) signal is about 0 ± 0.2 μcal/s. We have noticed that equilibration of the system with the syringe introduced immediately after filling the cell requires much longer times.
11. Remember, when you press the "START" button, the syringe starts to rotate (we usually use 300 rpm) and some time is required for the DP signal to reach the preset reference plateau value (we usually use 2 μcal/s). We do not recommend using the automatic option to start the titration but rather waiting to make sure that the baseline is stable, indicating that the protein solution remains stable during rotation of the

injection syringe. Do not leave until the end of the titration, so that you can check the progress of the run from time to time. Make sure that injection intervals are long enough for heat signal to return to the baseline (not waiting for baseline return is often a mistake made in ITC experiments). Peak equilibration times can increase during the course of the titration (see Fig. 1a for example) and may not fit within the injection time intervals that were initially set. Note that the parameters of the titration, such as injection intervals and number, duration and volume of the injections can be corrected during the run.

12. It is desirable to run the reference titration (if necessary) immediately after the experimental one. Since SSB–ssDNA complexes are quite soluble and show no aggregation or precipitation, there is no need to subject the ITC cell to thorough cleaning with the detergent after each run. However, if you study a new protein–DNA system for which you have not performed previous ITC experiments, we recommend running a reference titration(s) first.

13. After the first titration the syringe can be refilled fully or partially with the same solution without opening the filling port by using buttons "distance," "up," and "down" in the Auto-Pipette window. To make sure there is no air in the needle, first, set short distance (~0.02 in.) and move plunger down until solution appears at the end of the needle. Place the syringe in the tube containing the DNA solution, set appropriate distance and move the plunger up until the syringe is filled to the desired level. Do Purge/Refill.

14. It is important to know the precise values of the concentrations of SSB and DNA used in the experiment for meaningful quantitative analyses of the data. Mistakes can be made when diluting the samples to the experimental concentrations. Up to 5–7 % of SSB can be lost if the protein solution is degassed. An accurate determination of the concentration of the species in the syringe (DNA) is of particular importance under stoichiometric binding conditions (see Figs. 1a and 2) since it is used when the observed heats are normalized.

15. Recent upgrades of the Origin 5 ITC analysis software provide the following models for data analysis: "One set of sites," "Two set of sites," "Sequential binding sites," "Competitive binding," and "Dissociation." As an alternative we also use the data analysis software "Scientist" (St. Louis, MO), particularly when a more complex equilibrium model is needed that is not available in Origin.

16. Titration of the buffer in the cell with ligand in the syringe is the most effective reference titration, since this corresponds to

more than a 100-fold dilution of the ligand into the cell, resulting in the largest effects due to heat of dilution. Another useful control is to have buffer in the syringe and titrate this into the protein in the cell. This is useful to determine whether the protein solution in the cell is stable to rotation of the injection syringe. A mismatch of the heats at the end of titration to those obtained from the reference titration should not be ignored, since this may result from the following: (1) The reaction is not completed. (2) Syringe and cell buffers do not match. (3) Oligomerization state of the species in the syringe changes upon dilution in the cell.

17. The data can also be presented in integral form, where at each titration point the heat is calculated as a sum of the heats for the current and all previous injections and is plotted versus the total titrant concentration. This form corresponds to a general representation of titration data where the integral response of the system is measured as a function of saturation of the macromolecule with the ligand (for example, absorbance or fluorescence).

18. As discussed previously (28) for a macromolecule that binds multiple ligands, the change in the experimentally observed property of the system that is being monitored to follow binding (e.g., absorbance or fluorescence or heat) is not always directly proportional to the average extent of ligand binding to the macromolecule (binding density). If this is not the case then incorrect isotherms will be obtained if direct proportionality is assumed. Rather than making this assumption a model independent analysis of a number of experimental titrations performed at different macromolecule concentrations should be performed and analyzed as described to obtain the actual relationship between the signal change (ΔQ) and the average extent of ligand binding (28).

19. When analyzing and fitting experimental data the corrections for heat displacement effects and ligand and macromolecule dilutions in the calorimetric cell during the titration should be applied as described in the ITC Data Analysis in Origin: Tutorial Guide (MicroCal, LLC) and in ref. (3).

20. This titration is an example of a so-called "reverse titration," in which the ligand in the cell (*Eco*SSB) is titrated with the macromolecule in the syringe $(dT)_{70}$. (The macromolecule is defined as the species that binds multiple ligands.) When existing models incorporated in the ITC software, such as "two independent nonidentical sites" or "two sequential sites" models are used, you should switch options in the main menu from "macromolecule in the cell" to "ligand in the cell" to analyze these data properly.

21. A more detailed interpretation of the data shown in Fig. 2b and fitting of this data to an appropriate model is complicated by the fact that in both complexes *Eco*SSB binds to $(dT)_{70}$ with very high affinity (stoichiometrically), which makes it difficult to distinguish between different models, as for example, a simple model of two independent sites, from models that include cooperativity between sites or in more complicated cases, when intermediate species are populated significantly.

Acknowledgments

We thank Dr. Edwin Antony and Dr. Binh Nguyen for careful reading of the manuscript and useful suggestions. This work was supported in part by grants to T.M.L. from NIH (GM030498 and GM045948).

References

1. Wyman J, Gill SJ (1990) Bindng and linkage. University Science Books, Mill Valley, CA
2. Lohman TM, Overman LB, Ferrari ME, Kozlov AG (1996) A highly salt-dependent enthalpy change for Escherichia coli SSB protein-nucleic acid binding due to ion–protein interactions. Biochemistry 35:5272–5279
3. Kozlov AG, Lohman TM (1998) Calorimetric studies of E-coli SSB protein single-stranded DNA interactions. Effects of monovalent salts on binding enthalpy. J Mol Biol 278:999–1014
4. Kozlov AG, Lohman TM (1999) Adenine base unstacking dominates the observed enthalpy and heat capacity changes for the Escherichia coli SSB tetramer binding to single-stranded oligoadenylates. Biochemistry 38:7388–7397
5. Kozlov AG, Lohman TM (2000) Large contributions of coupled protonation equilibria to the observed enthalpy and heat capacity changes for ssDNA binding to Escherichia coli SSB protein. Proteins Suppl 4:8–22
6. Kozlov AG, Lohman TM (2006) Effects of monovalent anions on a temperature-dependent heat capacity change for Escherichia coli SSB tetramer binding to single-stranded DNA. Biochemistry 45:5190–5205
7. Kozlov AG, Lohman TM (2011) E. coli SSB tetramer binds the first and second molecules of (dT)(35) with heat capacities of opposite sign. Biophys Chem 159:48–57
8. Kozlov AG, Eggington JM, Cox MM, Lohman TM (2010) Binding of the dimeric Deinococcus radiodurans single-stranded DNA binding protein to single-stranded DNA. Biochemistry 49:8266–8275
9. Kumaran S, Kozlov AG, Lohman TM (2006) Saccharomyces cerevisiae replication protein A binds to single-stranded DNA in multiple salt-dependent modes. Biochemistry 45:11958–11973
10. Lohman TM, Ferrari ME (1994) Escherichia coli single-stranded DNA-binding protein: multiple DNA-binding modes and cooperativities. Annu Rev Biochem 63:527–570
11. Raghunathan S, Ricard CS, Lohman TM, Waksman G (1997) Crystal structure of the homo-tetrameric DNA binding domain of Escherichia coli single-stranded DNA-binding protein determined by multiwavelength X-ray diffraction on the selenomethionyl protein at 2.9-A resolution. Proc Natl Acad Sci U S A 94:6652–6657
12. Bujalowski W, Overman LB, Lohman TM (1988) Binding mode transitions of Escherichia coli single strand binding protein-single-stranded DNA complexes. Cation, anion, pH, and binding density effects. J Biol Chem 263:4629–4640
13. Savvides SN, Raghunathan S, Futterer K, Kozlov AG, Lohman TM, Waksman G (2004)

The C-terminal domain of full-length E-coli SSB is disordered even when bound to DNA. Protein Sci 13:1942–1947

14. Shereda RD, Kozlov AG, Lohman TM, Cox MM, Keck JL (2008) SSB as an organizer/ mobilizer of genome maintenance complexes. Crit Rev Biochem Mol Biol 43:289–318
15. Overman LB, Lohman TM (1994) Linkage of pH, anion and cation effects in protein-nucleic acid equilibria. Escherichia coli SSB protein-single stranded nucleic acid interactions. J Mol Biol 236:165–178
16. Bernstein DA, Eggington JM, Killoran MP, Misic AM, Cox MM, Keck JL (2004) Crystal structure of the Deinococcus radiodurans single-stranded DNA-binding protein suggests a mechanism for coping with DNA damage. Proc Natl Acad Sci U S A 101:8575–8580
17. Lohman TM, Green JM, Beyer RS (1986) Large-scale overproduction and rapid purification of the Escherichia coli ssb gene product. Expression of the ssb gene under lambda PL control. Biochemistry 25:21–25
18. Eggington JM, Haruta N, Wood EA, Cox MM (2004) The single-stranded DNA-binding protein of Deinococcus radiodurans. BMC Microbiol 4:2
19. Ferrari ME, Bujalowski W, Lohman TM (1994) Co-operative binding of Escherichia coli SSB tetramers to single-stranded DNA in the (SSB) 35 binding mode. J Mol Biol 236:106–123
20. Wong I, Chao KL, Bujalowski W, Lohman TM (1992) DNA-induced dimerization of the Escherichia coli rep helicase, Allosteric effects of single-stranded and duplex DNA. J Biol Chem 267:7596–7610
21. Wiseman T, Williston S, Brandts JF, Lin LN (1989) Rapid measurement of binding constants and heats of binding using a new titration calorimeter. Anal Biochem 179:131–137
22. Bujalowski W, Lohman TM (1989) Negative co-operativity in Escherichia coli single strand binding protein–oligonucleotide interactions. I. Evidence and a quantitative model. J Mol Biol 207:249–268
23. Weast RC (ed) (1974–1975) Handbook of chemistry and physics, 55 edn. CRC Press.
24. Lohman TM, Chao K, Green JM, Sage S, Runyon GT (1989) Large-scale purification and characterization of the Escherichia coli rep gene product. J Biol Chem 264:10139–10147
25. Cantor CR, Warshaw MM, Shapiro H (1970) Oligonucleotide interactions. 3. Circular dichroism studies of the conformation of deoxyoligonucleotides. Biopolymers 9:1059–1077
26. Fasman GD (ed) (1975) Handbook of biochemistry and molecular biology, vol 1: nucleic acids, 3rd edn., CRC Press.
27. Holbrook JA, Capp MW, Saecker RM, Record MT Jr (1999) Enthalpy and heat capacity changes for formation of an oligomeric DNA duplex: interpretation in terms of coupled processes of formation and association of single-stranded helices. Biochemistry 38:8409–8422
28. Lohman TM, Bujalowski W (1991) Thermodynamic methods for model-independent determination of equilibrium binding isotherms for protein–DNA interactions: spectroscopic approaches to monitor binding. Methods Enzymol 208:258–290

Chapter 4

SSB–DNA Binding Monitored by Fluorescence Intensity and Anisotropy

Alexander G. Kozlov, Roberto Galletto, and Timothy M. Lohman

Abstract

Fluorescence methods have proven to be extremely useful tools for quantitative studies of the equilibria and kinetics of protein–DNA interactions. If the protein contains tryptophan (Trp), as is often the case, and there is a change in intrinsic Trp fluorescence of the protein, one can use this change in signal (quenching/enhancement) to monitor binding. One can also attach an extrinsic fluorophore to either the protein or the DNA and monitor binding due to a change in fluorescence intensity or a change in fluorescence anisotropy. Such equilibrium studies can provide important quantitative information on stoichiometries (occluded site size, number of binding sites) and energetics (affinities and cooperativities) of the interactions. This information is needed to understand the mechanisms of protein–DNA interactions. A critical aspect of such approaches for systems that have non-unity stoichiometries (e.g., a protein that binds multiple ligands) is knowledge of the relationship between the change in fluorescence signal (intensity or anisotropy) and the average extent of binding. Here we describe procedures for using fluorescence approaches to examine the stoichiometries and equilibrium binding affinities of *Escherichia coli* single-stranded DNA-binding protein (SSB) and *Deinococcus radiodurans* SSB with long polymeric ssDNA to determine an occluded site size. We also provide examples of studies of SSB binding to shorter oligonucleotides to demonstrate analysis and fitting of the data to an appropriate model (monitoring fluorescence intensity or anisotropy) to obtain quantitative estimates of equilibrium binding parameters. We emphasize that the solution conditions (especially salt concentration and type) can influence not only the binding affinity, but also the mode by which an SSB oligomer binds ssDNA.

Key words: EcoSSB, DrSSB, ssDNA binding, Fluorescence equilibrium titrations, Intrinsic tryptophan quenching, SSB–ssDNA thermodynamics, Anisotropy

1. Introduction

If a change in the fluorescence properties (e.g., intensity or anisotropy) accompanies the binding of two species, this can be used to quantitatively monitor the binding interaction and determine both the stoichiometry and energetics (affinity and thermodynamics) of binding using equilibrium titration methods. Such approaches

James L. Keck (ed.), *Single-Stranded DNA Binding Proteins: Methods and Protocols*, Methods in Molecular Biology, vol. 922, DOI 10.1007/978-1-62703-032-8_4, © Springer Science+Business Media, LLC 2012

have been used extensively to examine numerous protein–DNA interactions in solution including many SSB–ssDNA interacting systems. Generally, if the molecule undergoes a change in its intrinsic fluorescence (e.g., tryptophan (Trp) in a protein), then this species is titrated with the nonfluorescent molecule (e.g., DNA) and the change in fluorescence signal (intensity or anisotropy) is monitored as a function of total DNA concentration. Alternatively, if the changes in the intrinsic fluorescence of the protein are not sufficient to reliably monitor the interaction, an extrinsic fluorophore can be attached to one species (either protein or DNA) to study the interaction. However in this case, it is often observed that the modification affects the energetics of the interaction. Independent of the nature of the fluorescence signal that is monitored (intensity or anisotropy), one needs to know (or determine) how the change in fluorescence signal relates to $\langle X \rangle$, namely, the average number of ligands (in this case, DNA) that bind to the macromolecule (in this case, protein). We emphasize this point since for macromolecules that bind multiple ligands, the average fractional fluorescence signal change is often not directly proportional to $\langle X \rangle$. However, once this relationship is established, then one can obtain an equilibrium binding isotherm that can be analyzed to determine stoichiometries (number of binding sites), equilibrium binding constants (K), and cooperativities of protein–DNA interactions. Equilibrium experiments performed at different temperatures and different solution conditions (salt, pH) allow one to examine how the binding is thermodynamically linked to these variables (1). Such linkage analysis can yield information on the binding enthalpy (linkage to temperature) and whether ions or protons bind preferentially to the protein–DNA complex or the free protein and/or DNA (2–6). This provides information about the thermodynamic forces that drive the formation of the complex and more practically how solution conditions (pH, salt concentration and type, etc.) can influence these interactions. This information is needed to understand the energetics of the protein–DNA interaction and allows predictions of their behavior under conditions that either have not been examined or are not as easily accessed experimentally.

Fluorescence equilibrium titrations have been used extensively to study interactions of single-stranded DNA-binding proteins (SSBs) with ssDNA. The *Escherichia coli* SSB (*Eco*SSB) and the *Deinococcus radiodurans* SSB (*Dr*SSB) are used as examples in this chapter and are also discussed in the chapter describing isothermal titration calorimetry (ITC) methods; hence the reader should refer to that chapter for more introductory comments about these protein–DNA systems.

Here we briefly review the equilibrium fluorescence titration studies that have been performed to examine the interactions of *Eco*SSB and *Dr*SSB with ssDNA. We first remind the reader that in solution, *Eco*SSB and *Dr*SSB are stable homotetramers and homodimers, respectively. As a result of the multiple ssDNA-binding sites (4 OB folds each), these SSB proteins can bind to long polymeric ssDNA in different binding modes in vitro that depend on solution conditions, particularly salt concentration and type (7). The *Eco*SSB tetramer shows dramatic changes in its mode of binding to ssDNA as a function of salt concentration, type, and valence. On poly(dT) *Eco*SSB binds in its fully wrapped $(SSB)_{65}$ mode at [NaCl] >0.2 M in which ssDNA interacts with all four subunits of the SSB tetramer with an occluded site size of 65 nts. However, at lower [NaCl] ($\leq$10 mM) it binds in its highly cooperative $(SSB)_{35}$ mode, in which ssDNA binds to an average of only two subunits of the tetramer with an occluded site size of ~35 nts (2, 3, 7). At intermediate salts (40–100 mM NaCl), *Eco*SSB forms its $(SSB)_{56}$ binding mode on poly(dT). Equilibrium fluorescence titrations have been used extensively to study the effects of various solution conditions on its different binding modes with polynucleotides and on the energetics of ssDNA binding. Effects of salt concentration (2, 3), anion and cation type (4–6), pH (4, 6), polyamines (8), binding density (4), and base composition (5, 9) have been examined. Statistical thermodynamic models describing the binding of SSB tetramers to long ssDNA in its highly cooperative $(SSB)_{35}$ mode at low salt conditions (<0.02 M NaCl) (10) and in its low cooperativity $(SSB)_{65}$ mode at higher salt concentrations (>0.2 M NaCl) (11) have also been developed.

Oligodeoxynucleotides of different lengths have also been used to examine binding of ssDNA to the individual tetramer. Four molecules of $(dX)_{16}$, two molecules of $(dX)_{35}$, or one molecule of $(dX)_{70}$ can bind per SSB tetramer. However, the shorter oligodeoxynucleotides, $(dX)_{16}$ or $(dX)_{35}$, display negative cooperative binding to an individual SSB tetramer such that ssDNA binds with high affinity to the first two subunits of the tetramer and with much lower affinity to the second two subunits. In addition, this negative cooperativity is highly salt dependent, decreasing with increasing [NaCl] or [$MgCl_2$], although high salt does not eliminate the negative cooperativity (12). In fact, this salt-dependent negative cooperativity provides part of the explanation for the salt-dependent change in the SSB–ssDNA binding modes observed with poly(dT) (12–14). Studies of the *Eco*SSB binding to oligodeoxynucleotides of lengths close to the occluded site size ($n = 65$) have been used to estimate the binding enthalpies and have shown that the unusually high negative heat capacity change associated with the binding of $(dA)_{70}$ is due to the temperature-dependent stacking of adenine within the ssDNA and the loss of these stacking interactions upon binding SSB (15, 16). The energetics of cooperative binding of two SSB tetramers to $(dA)_{70}$ has also been examined at low salt conditions (17). More limited fluorescence binding studies have been performed with the *Dr*SSB dimer (18) and yeast *sc*RPA hetero-trimer (19).

We note that in all of the studies mentioned above the intrinsic Trp fluorescence quenching of the protein has been used to monitor binding, although recent studies (20) have used *Eco*SSB that has been extrinsically labeled with fluorophores that undergo fluorescence intensity changes upon binding ssDNA. Generally these extrinsic fluorophores also change the binding affinities and properties and thus care should be taken to characterize these labeled proteins rather than assume that their behavior mirrors the unlabeled protein.

The energetics and kinetics of *Eco*SSB binding to single-stranded oligodeoxynucleotides have also been studied by monitoring the change in fluorescence intensity of fluorescently labeled nucleic acid (21–23) as well as SSB (20, 24). Although changes in the fluorescence anisotropy of a labeled ssDNA can also be utilized to monitor binding, these have not been extensively used to study SSB–ssDNA interactions. Changes in the fluorescence anisotropy of *Eco*SSB have been used to study the self-assembly of a His-55 to Tyr mutant of SSB (SSB-1) that destabilizes the tetramer to form monomers (25, 26).

In the following sections we describe procedures for performing and analyzing titration experiments monitoring changes in either fluorescence intensity or anisotropy. These experiments were performed using a PTI QM-2000 spectrofluorometer (Photon Technologies, Inc., Lawrenceville, NJ) in which the quenching of the intrinsic Trp fluorescence of the protein or the change in fluorescence intensity or anisotropy of an ssDNA labeled with an extrinsic fluorophore (fluorescein) was used to monitor binding.

2. Materials

2.1. Buffer Solutions

All solutions were prepared with reagent grade chemicals and glass distilled water that was subsequently treated with a Milli Q (Millipore, Bedford, MA) water purification system. The buffer used is Buffer T (10 mM Tris (tris(hydroxymethyl) aminomethane), pH 8.1, 0.1 mM EDTA) with different NaCl concentrations as indicated in the text for each particular experiment. Buffers were prepared as described in the ITC Chapter.

2.2. Proteins and ssDNA

1. *Eco*SSB (27) and *Dr*SSB (28) proteins were expressed, purified, and stored as described in the ITC Chapter. Protein concentrations were determined spectrophotometrically in buffer T containing 0.20 M NaCl using the extinction coefficients, $\varepsilon_{280} = 1.13\times10^5$ M^{-1} (tetramer) cm^{-1} for *Eco*SSB and $\varepsilon_{280} = 8.2\times10^4$ M^{-1} (dimer) cm^{-1} for *Dr*SSB (see Note 1).
2. The single-stranded oligodeoxythymidylates, $(dT)_{70}$, $(dT)_{50}$, $(dT)_{35}$, and F-$(dT)_{50}$ (labeled with fluorescein at the 5′ end),

were synthesized and stored as described in the ITC Chapter of this book. Poly(dT) has an average length ~1,100 nucleotides and was purchased from Midland Certified Reagents Co. (Midland, TX). The concentration of poly(dT) is expressed in nucleotides and was determined spectrophotometrically in buffer T (pH 8.1), 0.1 M NaCl using the extinction coefficient, $\varepsilon_{260,\mathrm{dT}} = 8.1\times10^3\ \mathrm{M}^{-1}\ \mathrm{cm}^{-1}$. Concentrations of oligodeoxythymidylates were determined using extinction coefficients (per mole of ssDNA molecule) at 260 nm calculated according to the expression $\varepsilon_{260} = N\times\varepsilon_{260,\mathrm{dT}} + \varepsilon_{260,\mathrm{F}}$, where N is the number of nucleotides and $\varepsilon_{260,\mathrm{F}} = 2.096\times10^4\ \mathrm{M}^{-1}\ \mathrm{cm}^{-1}$ is the extinction coefficient for fluorescein (Glen Research) (see Note 1).

3. Methods

3.1. Equilibrium Binding Models Used for Analysis of Titration Data

For the binding of a ligand (X) to a one-site macromolecule (M), the equilibrium can be defined as in Scheme 1,

$$M + X \underset{}{\overset{K_{obs}}{\rightleftarrows}} MX$$

Scheme 1

where $K_{\mathrm{obs}} = [MX]/[X][M]$ is the equilibrium association constant and $[MX]$, $[X]$, $[M]$ are the equilibrium concentrations of each species in solution. The expression for the average degree of binding $<X>$ (average number of ligands bound per macromolecule) can be obtained using the mass conservation equation $[M]_{\mathrm{tot}} = [M] + [MX]$ and is given in Eq. 1

$$<X> = \frac{[MX]}{[M]_{\mathrm{tot}}} = \frac{K_{\mathrm{obs}}[X]}{1 + K_{\mathrm{obs}}[X]} \tag{1}$$

If the macromolecule has n identical, independent sites the expression for $<X>$ is given in Eq. 1a (1)

$$<X> = \frac{[MX]}{[M]_{\mathrm{tot}}} = \frac{nK_{\mathrm{obs}}[X]}{1 + K_{\mathrm{obs}}[X]} \tag{1a}$$

The equilibrium binding model in Scheme 1 was used to fit the titration curves for *Dr*SSB binding to $(\mathrm{dT})_{50}$ shown in Figs. 2a and 3a, b (see Subheading 10).

Scheme 2 defines the equilibrium for the binding of two molecules of ligand to a macromolecule,

$$M + X \overset{K_{1,obs}}{\rightleftharpoons} MX + X \overset{K_{2,obs}}{\rightleftharpoons} MX_2$$

Scheme 2

where $K_{1,\text{obs}} = [MX]/[X][M]$ and $K_{2,\text{obs}} = [MX_2]/[X][MX]$ are the macroscopic (see Note 2) equilibrium binding constants for binding of the first and second ligands, respectively. The expression for $<X>$ for this system is given in Eq. 2 (1):

$$<X> = \frac{d\ln P}{d\ln X} = \frac{K_{1,\text{obs}}[X] + 2K_{1,\text{obs}}K_{2,\text{obs}}[X]^2}{1 + K_{1,\text{obs}}[X] + K_{1,\text{obs}}K_{2,\text{obs}}[X]^2} \quad (2)$$

This model was used to fit the data in Fig. 2b for the binding of two molecules of $(dT)_{35}$ to *Eco*SSB (see Subheading 10).

3.2. Monitoring Binding by a Fluorescence Intensity Change

The total fluorescence intensity of a mixture of "i" fluorescent species can be defined as in Eq. 3:

$$F = \sum F_i[X]_i \quad (3)$$

where F_i is the molar fluorescence intensity of species i and $[X]_i$ is its concentration. For a macromolecule (M) with fluorescence intensity, $F_M = F_0$, that upon formation of a 1:1 complex (MX) undergoes a quenching of its fluorescence intensity, $F_{MX} = F_{\min}$, the observed change in relative fluorescence quenching is described by Eq. 4:

$$Q_{\text{obs}} = Q_{\max}\frac{K_{\text{obs}}[X]}{1 + K_{\text{obs}}[X]} \quad (4)$$

where $Q_{\text{obs}} = (F_0 - F_{\text{obs}})/F_0$ is the observed fluorescence quenching and $Q_{\max} = (F_0 - F_{\min})/F_0$ is the fluorescence quenching at saturation of the macromolecule with ligand, where $<X>$ and K_{obs} are as defined above. The concentration of free ligand, $[X]$, can be determined from the mass conservation Eq. 4a:

$$[X]_{\text{tot}} = [X] + <X>[M]_{\text{tot}} = [X] + \frac{K_{\text{obs}}[X]}{1 + K_{\text{obs}}[X]}[M]_{\text{tot}} \quad (4a)$$

For a macromolecule with n independent and identical sites, Eqs. 4 and 4a transform into Eqs. 5 and 5a:

$$Q_{\text{obs}} = Q_{\max}\frac{nK_{\text{obs}}[X]}{1 + K_{\text{obs}}[X]} \quad (5)$$

$$[X]_{\text{tot}} = [X] + <X>[M]_{\text{tot}} = [X] + \frac{nK_{\text{obs}}[X]}{1 + K_{\text{obs}}[X]}[M]_{\text{tot}} \quad (5a)$$

The expression describing the binding of two ligands to a macromolecule according to Scheme 2 is given by Eq. 6:

$$Q_{obs} = \frac{Q_1 K_{1,obs}[X] + Q_2 K_{1,obs} K_{2,obs}[X]^2}{1 + K_{obs}[X] + K_{1,obs} K_{2,obs}[X]^2} \tag{6}$$

where $K_{1,obs}$ and $K_{2,obs}$ are the macroscopic equilibrium binding constants (see Note 2) for binding of the first and the second ligands, X, to the macromolecule M, and Q_1 and Q_2 are the fluorescence quenching upon binding of one and two molecules of ligand, respectively. In Eq. 6 the concentration of free ligand ($[X]$) can be determined from the mass conservation Eq. 6a:

$$[X]_{tot} = [X] + \langle X \rangle [M]_{tot} \tag{6a}$$

where $\langle X \rangle$ is defined in Eq. 2.

3.3. Monitoring Binding by Fluorescence Anisotropy Changes

SSB–DNA binding can also be monitored by the change in fluorescence anisotropy of an extrinsic fluorophore attached to either the SSB or the DNA. Fluorescence anisotropy (r) reflects the extent of depolarization of the fluorescence emission of a species in solution when excited with polarized light. It is directly related to the rotational diffusion of the species and is low for small and flexible molecules (higher depolarization), and increases when larger complexes are formed (slower rotational movement, smaller depolarization). Therefore, the change in fluorescence anisotropy can generally be used to monitor a binding event (see Note 3). This is particularly convenient for systems where no change in fluorescence intensity is observed between the free and bound states. For one-channel spectrofluorometers (L-format) the anisotropy is defined in Eq. 7 (29):

$$r = \frac{I_{VV} - GI_{VH}}{I_{VV} + 2GI_{VH}} \tag{7}$$

where I_{VV} is the fluorescence emission intensity measured for vertically polarized excitation and vertically polarized emission. I_{VH} is the intensity measured for vertically polarized excitation and horizontally polarized emission, and G is the correction factor (see Note 4). We note that in many studies polarization (p) changes have been used directly to quantitatively analyze interacting systems rather than anisotropy (r). This will result in an incorrect binding isotherm and incorrect binding constant and should be avoided (see Note 5).

The anisotropy of a mixture of "i" fluorescent species is given by Eq. 8:

$$\bar{r} = \sum r_i f_i \tag{8}$$

where $f_i = X_i F_i / \sum X_i F_i$ is the fractional contribution to the observed fluorescence intensity of component i. It is important to note the differences in the weighting factors in Eq. 3 versus Eq. 8. In Eq. 3, the observed fluorescence intensity is the sum of the fluorescence intensities of each species weighted by their concentrations, whereas in Eq. 8, the observed anisotropy is the sum of the anisotropies of each species weighted by the fractional contribution of each species to the total fluorescence. Based on Eq. 8 the expression for r for a 1:1 binding system is given by Eq. 9:

$$r = r_0 + (r_{max} - r_0)\frac{K_{obs}[X]c}{1 + K_{obs}[X]c} \tag{9}$$

where r_0 and r_{max} are the limiting values of the anisotropy for (M) and (MX), respectively, $[X]$ is the free protein concentration, and K_{obs} is as defined above. The factor $c = F_{bound}/F_{free} = F_{MX}/F_M$ is the ratio of the fluorescence of the bound and free macromolecule (DNA) (assuming that it is labeled with fluorophore).

3.4. Amounts of SSB and ssDNA Required for Fluorescence Titrations and the Design of the Experiment

In general, preliminary information on the system (such as the oligomerization state of the protein and its stability) should be obtained prior to performing titration experiments (see Note 6). *Eco*SSB (7) is a stable tetramer and *Dr*SSB (28) is a stable dimer under the conditions of the experiments presented here. Any preliminary information on the value of binding affinities (if available from literature) can also be useful in designing the experiments and to estimate the amounts of SSB and DNA needed. For example, for most of the experiments presented in Fig. 1 (except the titrations with *Dr*SSB in 1.0 M NaCl) the affinity of the SSB for ssDNA is so high that binding is "stoichiometric" (i.e., all DNA added is bound until the "stoichiometric" point n is achieved) and the reaction is completed at a total DNA-to-protein ratio, $R = 1$ (mole $(dT)_L$/ mole SSB) in the case of oligonucleotides (Fig. 1c, d, 0.2 M NaCl) or $R \approx n$ for poly(dT) (Fig. 1a, b, 0.2 M NaCl), where n is the occluded site size in nucleotides (poly(dT) concentration is expressed in moles of nucleotides and not molecules). For such experiments the spectroscopic signal changes linearly with titrant concentration until the stoichiometric point is reached and does not change further upon addition of more titrant. A two- to three-fold excess of $[DNA]_{tot}$ over $[SSB]_{tot}$ is usually sufficient for these titrations while generally obtaining 20–25 points in the titration. The amounts of SSB and DNA required for a single titration can be estimated using the formula in Eq. 10:

$$R \cdot V_{SSB,cell} \cdot C_{SSB,cell} = V_{DNA} \cdot C_{DNA} \tag{10}$$

where R is the final total DNA-to-total protein ratio desired in the spectrophotometric cell at the end of titration, $V_{SSB,cell}$ is the minimum volume of protein solution in the cell (usually ~2 mL), $C_{SSB,cell}$ is the concentration of SSB in the cell (usually varying from 0.1 to 1 μM), and V_{DNA} is the volume of the titrant (DNA solution,

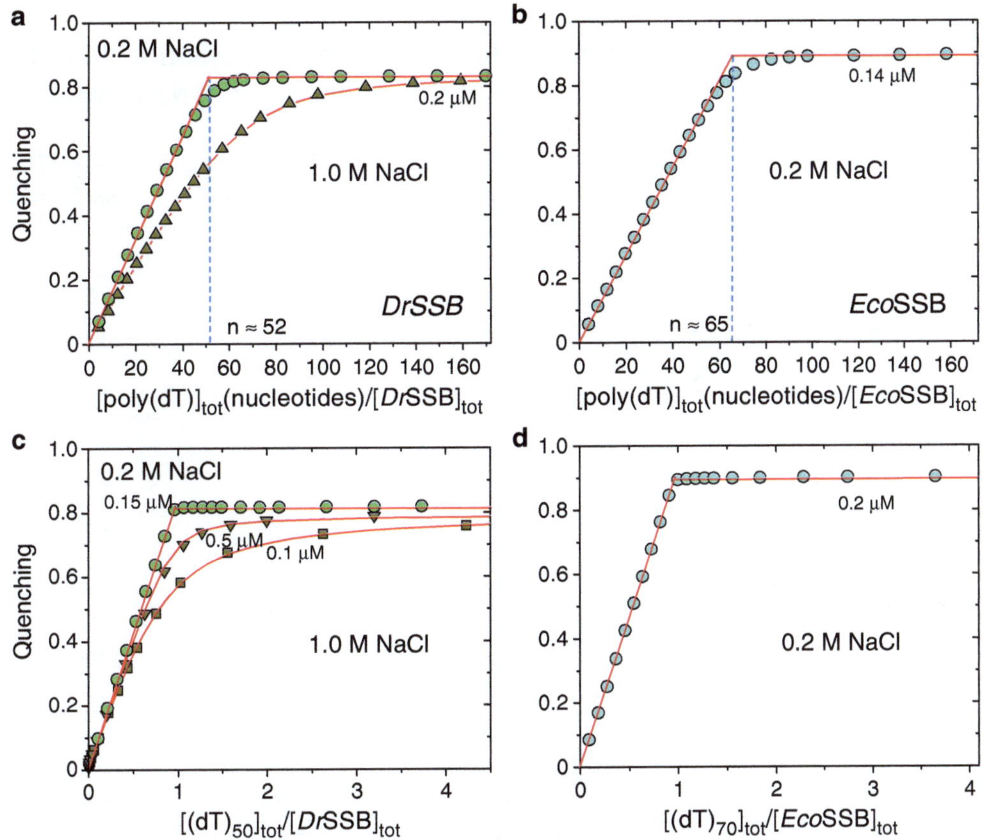

Fig. 1. Determination of the occluded site size of *Dr*SSB (**a**) and *Eco*SSB (**b**) on poly(dT) under stoichiometric conditions monitoring change in intrinsic Trp fluorescence. (**a**) Fluorescence titrations of *Dr*SSB (0.2 μM) with poly(dT) (829 μM (nucleotides)) under stoichiometric conditions (0.2 M NaCl, buffer T, pH 8.1, 25°C) are shown in *green circles*. The results are presented as a change in the extent of fluorescent quenching relative to free protein fluorescence versus the ratio of total poly(dT) concentration (nucleotides) to total concentration of the *Dr*SSB dimer in the cell. The linear increase in relative quenching indicates stoichiometric binding of the protein to poly(dT) until the point of saturation (plateau value) is reached (~83% quenching). The occluded site size (≈52 nucleotides per *Dr*SSB dimer) was determined by extrapolation of the linear part of the titration curve to the point of intersection with the corresponding plateau value after saturation (shown in *red*). For 1.0 M NaCl (*yellow triangles*) the titration is not stoichiometric, so a determination of occluded site size is not possible (in this case the concentrations used in the experiment were 0.2 μM *Dr*SSB and 817 μM of poly(dT)). (**b**) Fluorescence titration of *Eco*SSB (0.14 μM) with Poly(dT) (495 μM (nucleotides)) under stoichiometric conditions (0.2 M NaCl, buffer T, pH 8.1, 25°C). For this titration the occluded site size determined from the intersection point (as described in *panel A*) is ~65 and maximum quenching is ~90%. (**c**) Fluorescence titrations of *Dr*SSB with $(dT)_{50}$ in buffer T (pH 8.1, 25°C) under stoichiometric (0.2 M NaCl, *green circles*) and nonstoichiometric conditions (1.0 M NaCl, *yellow squares* and *triangles*). The titration isotherms are presented in the form of the dependence of relative fluorescence quenching versus the ratio of total $(dT)_{50}$ to total protein concentration. The concentrations of *Dr*SSB (dimer) used in the titrations are shown under corresponding plots. (**d**) Fluorescence titration of *Eco*SSB (0.1 μM tetramer) with $(dT)_{70}$ in buffer T (0.2 M NaCl, pH 8.1, 25°C) is indicative of 1:1 stoichiometric binding of *Eco*SSB tetramer to ssDNA.

usually ~100 μl), which generally does not exceed 5% of the solution volume in the cell (see Subheading 7). The concentration of the titrant (C_{DNA}) required for the experiment can then be determined from Eq. 10. For example, in the stoichiometric titrations

Table 1
Experimental Excel spreadsheet design for the titration of *Eco*SSB (0.14 μM, 1,800 μL total volume) with 495 M of poly(dT) in buffer T, 0.2 M NaCl (see Fig. 1b)

#	Titrant conc. (μM)	Titrant aliquot (μL)	Total titrant volume (μL)	Titrant conc. in cell (μM)	$[Poly(dT)]_{tot}$ per $[EcoSSB]_{tot}$
1	495	2	2	0.55	4
2	495	2	4	1.10	8
3	495	2	6	1.64	12
4	495	2	8	2.19	16
5	495	2	10	2.73	20
6	495	2	12	3.28	24
7	495	2	14	3.82	28
8	495	2	16	4.36	31
9	495	2	18	4.90	35
10	495	2	20	5.44	39
11	495	2	22	5.98	43
12	495	2	24	6.51	47
13	495	2	26	7.05	51
14	495	2	28	7.58	55
15	495	2	30	8.11	59
16	495	2	32	8.65	63
17	495	2	34	9.18	67
18	495	4	38	10.23	75
19	495	4	42	11.29	83
20	495	4	46	12.33	90
21	495	4	50	13.38	98
22	495	10	60	15.97	118
23	495	10	70	18.53	138
24	495	10	80	21.06	157

shown in Fig. 1b ($C_{SSB,cell} = 0.14$ μM, $R \approx 160$) and in Fig. 1d ($C_{SSB,cell} = 0.10$ μM, $R \approx 4$) the suggested titrant concentrations of poly(dT) and $(dT)_{70}$ are ~450 μM and ~8μM, respectively. These experiments are performed by adding equal-volume aliquots of the titrant to the protein solution as shown in Table 1 (see Note 7) for the titration of *Eco*SSB (0.14 μM) with poly(dT) (Fig. 1b).

Since the only information that can be obtained from such a titration is the occluded site size on poly(dT) (or the stoichiometry of SSB binding to a $(dT)_L$) and the maximum fluorescence change at saturation (Q_{max}) these data are generally plotted in the form of relative fluorescence quenching ($Q = (F_0 - F_i)/F_0$) versus the ratio of total DNA concentration to total protein concentration as shown in Fig. 1, where F_0 is the initial protein fluorescence before the addition of DNA and F_i is the fluorescence for the *i*th addition of DNA (see Subheading 5 for more details).

The results presented in Fig. 1a, c for *Dr*SSB binding to poly (dT) and $(dT)_{50}$ in the presence of 1.0 M NaCl indicate that the binding affinity is not high enough to be stoichiometric under these conditions (see Note 8) and therefore larger ratios of DNA to protein are required to achieve saturation, $R = [\text{poly(dT)}]_{tot}/[Dr\text{SSB}]_{tot} > n$ (Fig. 1a) and $R = [(dT)_{50}]_{tot}/[Dr\text{SSB}]_{tot} \gg 1$ (Fig. 1c). In this case, it is better to plot the data as ($Q = (F_0 - F_i)/F_0$) versus log [DNA] (see Fig. 2) since for accurate analysis (see Subheadings 9 and 10) more experimental points should be obtained at low and high DNA concentrations. For example, in the titrations presented in Fig. 2a the concentration range of $[(dT)_{50}]_{tot}$ is varied from 1 nM to ~4 μM. In designing these experiments four different titrant stock concentrations with varying volumes of added titrant were used to achieve an even distribution of the experimental points over a logarithmic scale of titrant concentration in the cell. This is exemplified in Table 2 (see Note 7) for the particular titration of 0.5 μM of *Dr*SSB with $(dT)_{50}$ (see Fig. 2a). In general, for this type of experiment we recommend preparing a concentrated stock of the titrant (~200 μM) (see Note 9), which can be diluted to the desired concentrations to perform the experiments.

The following procedures describe the preparation of SSB and DNA samples used in performing 2–10 fluorescent titration experiments with 0.1–0.5 μM of SSB or F-$(dT)_{50}$ in the cell.

3.5. Dialysis of Proteins and ssDNA

We recommend dialyzing protein and DNA samples prior to performing titration experiments since the storage conditions for DNA (water or Tris buffer) and SSB (20 mM Tris (pH 8.3), 50% (V/V) glycerol, 0.50 M NaCl, 1 mM EDTA, and 1 mM BME) are very different from the typical experimental conditions.

1. Prepare ~1–1.5 mL of an approximately 2 μM solution of either *Dr*SSB or *Eco*SSB by diluting an aliquot of the concentrated SSB stocks into the buffer to be used for dialysis. For titrations of fluorescein-labeled DNA (e.g., F-$(dT)_{50}$) make 500 μL of an ~15 μM *Dr*SSB solution.
2. Prepare ~100–200 μL of an approximately 200 μM solution of $(dT)_L$ or a 500–800 μM solution of poly(dT) by adding the

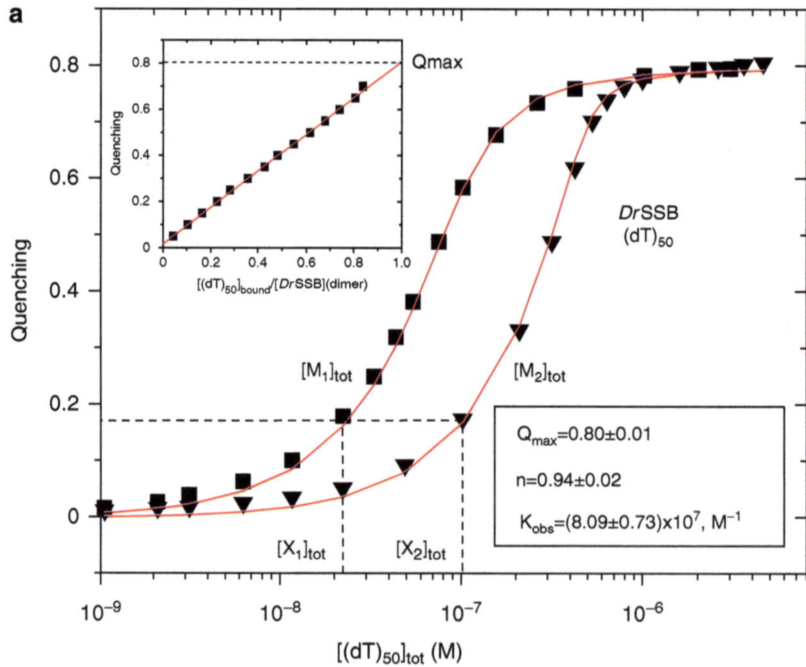

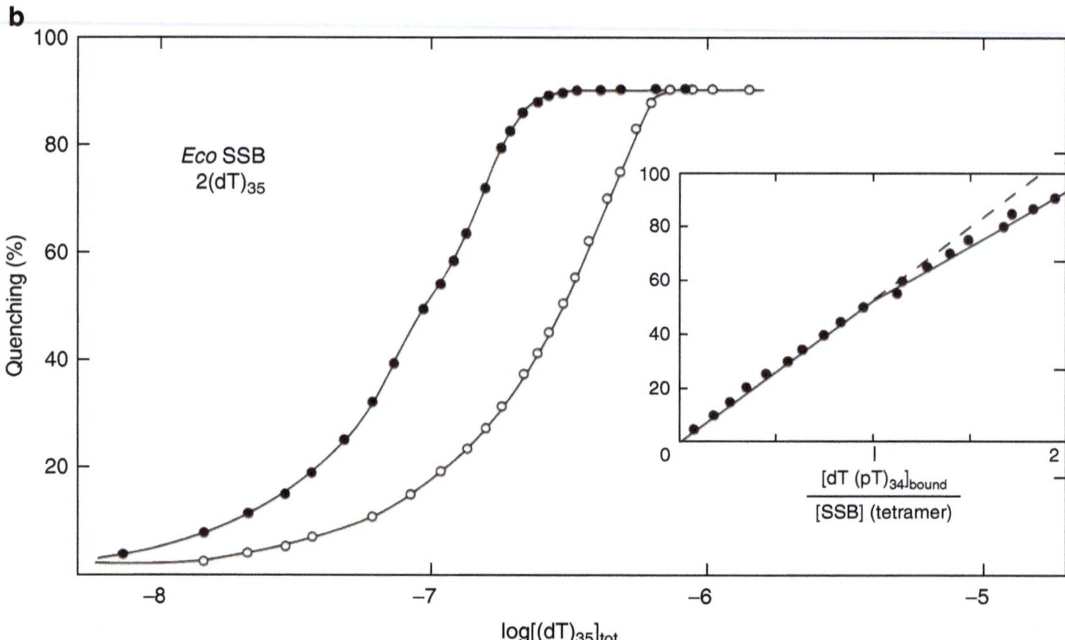

Fig. 2. The analysis of titration isotherms of SSB binding to oligo(dT) using model-independent approach and fitting the data to a particular model. (**a**) The titration isotherms from Fig. 1c for $(dT)_{50}$ binding to *Dr*SSB (0.2 M NaCl, pH 8.1, 25°C) at two protein concentrations, 0.1 μM dimer (*squares*) and 0.5 μM dimer (*triangles*), shown in the form of relative fluorescence quenching versus total concentration of the $(dT)_{50}$. The insert shows the results of MBDF analysis of the data (see text for details) indicative that relative change in fluorescence is directly proportional to the extent of DNA binding (binding density) and that the stoichiometry of the formed complex is 1 mol of $(dT)_{50}$ per mole of *Dr*SSB dimer (the maximum quenching ~80% is reached when the binding density $<X> = 1$). Smooth curves through the data points

appropriate volume of a concentrated DNA stock solution to dialysis buffer.

3. Place the SSB and DNA solutions in dialysis bags of appropriate size (MWCO: 10,000 Spectra/Por) and proceed with the dialysis as described in ITC Chapter.
4. After dialysis equilibrium has been reached (3 changes of 300 mL volume), remove the contents from the dialysis bag and determine the concentrations of protein and ssDNA (see Note 1).

3.6. Titration Protocol

Here we describe the experimental procedure for titration of the fluorescent species in the cell (SSB) with nonfluorescent ssDNA in which binding is monitored by quenching of the intrinsic Trp fluorescence of SSB. The procedure is essentially the same if the fluorescent species in the cell is DNA (e.g., $(dT)_{50}$ labeled with fluorescein) that is titrated with SSB while monitoring the change in fluorescein intensity or anisotropy (see Subheadings 10 and 11). We assume that the experimenter is familiar with the principles of operation of the fluorometer.

1. After starting the instrument allow the lamp to warm up for at least 15 min. Fluorescence intensity is temperature sensitive, as are the equilibrium binding constants for protein–DNA interactions (see Note 10). Hence, the temperature of the sample cell should always be controlled precisely when performing fluorescence experiments. Turn on the water bath or peltier block that controls the temperature of the cuvette holder, and set to the desired temperature, in this case, 25°C.
2. Place a magnetic stir bar in a 3 mL (10 mm pathway) quartz cuvette and dilute SSB directly in the cell to the desired concentration with dialysis buffer such that the total volume is 1.9–2 mL (see Note 11). Make certain that no bubbles are stuck to the stir bar. Prepare a reference cell by adding the same volume of dialysis buffer into another cuvette (also with a stir bar). Prepare solutions of the appropriate concentrations and volumes of the titrant by diluting the DNA stock with dialysis buffer.
3. Place cuvettes with protein and buffer into the cuvette holder of the instrument (see Note 12) and open the "Time base"

Fig. 2. (continued) represent a global fit of the two experimental data sets to the model of n-independent and identical sites (see Eqs. 5 and 5a and text for details) with the following parameters: $Q_{max} = 0.80 \pm 0.01$; $n = 0.94 \pm 0.02$; and $K_{obs} = (8.09 \pm 0.73) \times 10^7\ M^{-1}$. (**b**) Titration curves representing fluorescence titrations of the *Eco*SSB tetramer with $(dT)_{35}$ in buffer T (0.2 M NaCl, pH 8.1, 25°C) at two protein concentrations, 0.1 μM tetramer (*closed circles*) and 0.32 μM tetramer (*open circles*) (14). The insert shows the results of an MBDF analysis of the data (as described for Fig. 1a), which indicates that two molecules of $(dT)_{35}$ bind to *Eco*SSB tetramer upon saturation. The binding of the first one is characterized by the fluorescence quenching of ~50% and the second one by fluorescence quenching of ~40% (~90% quenching for the final complex having 2 mol of $(dT)_{35}$ bound per mole of *Eco*SSB tetramer).

Table 2
Experimental Excel spreadsheet design for the titration of *Dr*SSB (0.5 μM, 1,900 μL total volume) with $(dT)_{50}$ in buffer T, 1.0 M NaCl (see Figs. 1c and 2a)

#	Titrant conc. (μM)[a]	Titrant aliquot (μL)[b]	Total titrant volume (μL)	Titrant conc. in cell (M)	$[(dT)_{50}]_{tot}$ per $[DrSSB]_{tot}$
1	1	2	2	1.1×10^{-9}	0.002
2	1	2	4	2.1×10^{-9}	0.004
1	1	2	6	3.1×10^{-9}	0.006
4	1	6	12	6.3×10^{-9}	0.013
5	10	1	13	1.2×10^{-8}	0.023
6	10	2	15	2.2×10^{-8}	0.044
7	10	5	20	4.8×10^{-8}	0.097
8	100	1	21	1.0×10^{-7}	0.202
9	100	2	23	2.0×10^{-7}	0.413
10	100	2	25	3.1×10^{-7}	0.623
11	100	2	27	4.1×10^{-7}	0.834
12	100	2	29	5.1×10^{-7}	1.044
13	100	2	31	6.2×10^{-7}	1.255
14	100	3	34	7.7×10^{-7}	1.571
15	190	2	36	9.7×10^{-7}	1.971
16	190	6	42	1.6×10^{-6}	3.171
17	190	10	52	2.5×10^{-6}	5.171
18	190	10	62	3.5×10^{-6}	7.171
19	190	10	72	4.4×10^{-6}	9.171

[a]Concentrations and [b]aliquots of titrant ($[(dT)_{50}]_{tot}$) are varied in a way that results in even distribution of experimental points over the logarithm of concentrations of the titrant in the cell

mode in the software, which allows one to monitor the fluorescence emission at a particular wavelength (for a given excitation wavelength) as a function of time. Set the excitation (296 nm) and emission (350 nm) wavelengths (see Note 13) and slit widths (usually corresponding to a 2 nm bandpass for excitation and a 4–8 nm bandpass for emission, depending on SSB concentration (see Note 14)). Specify a number of acquisition points and the averaging time for each point (we usually set 8 points with an averaging time of 1 s and then take average of these 8 points to obtain a single data point for the fluorescence intensity, $F_{obs,i}$).

4. Take the first measurement for the protein solution ($F_{obs,0}$) and then for the buffer ($F_{ref,0}$). Repeat the measurements a few times waiting 2–3 min between measurements to make sure that the signal is stable (see Note 15) and record the observed intensity values for the sample and the buffer.
5. Begin the titration by adding the same-volume aliquots of the titrant (DNA) into the sample cell and the reference cell containing buffer using a micropipette (see Note 16). Wait for 2–3 min for equilibration and take measurements ($F_{obs,1}$ and $F_{ref,1}$). Repeat acquisition after 2–3 min to make sure that the signal intensity is no longer changing indicating that the system has reached equilibrium (if a slow change in fluorescence is observed a separate kinetic experiment should be performed to determine the equilibration times required between additions of titrant). Record the fluorescence intensity values for both cuvettes.
6. Proceed with the serial addition of aliquots of the titrant as described in the previous step while recording $F_{obs,i}$ and $F_{ref,i}$ for each addition until saturation is reached (i.e., the fluorescence intensity is no longer changing). Further additions of the titrant (DNA) may still be needed (see Note 17).
7. After finishing the titration redetermine the concentration of the DNA solution used as the titrant (see Note 18).

3.7. Correction of Experimental Data for Background Signal, Dilution, Inner Filter Effects, and Photobleaching

Prior to analysis of the data it is necessary to correct the observed fluorescence intensities and concentrations for dilution, photobleaching, and inner filter effects (30, 31).

1. First the concentrations of the species in the cell need to be corrected for dilution. The total concentration of the SSB in the cell (P) after addition of the *i*th aliquot of titrant DNA (D) solution can be calculated using Eq. 11:

$$P_{tot,i} = \frac{P_0 \cdot V_0}{(V_0 + V_i)} \tag{11}$$

where P_0 is the initial concentration in the cell, V_0 is the initial sample volume, and V_i is the total volume of titrant added up to the point *i* in titration. The total concentration of the titrant in the cell at each point *i* in the titration is calculated using Eq. 12:

$$D_{tot,i} = \frac{D_0 \cdot V_i}{(V_0 + V_i)} \tag{12}$$

where D_0 is the concentration of the titrant (see Note 19).

2. The first correction to obtain the true fluorescence intensities, F_i, is to subtract the background contribution to the signal due to any light scattering or impurities in the DNA or buffer. This background signal, $F_{ref,i}$, is obtained from the titration with

DNA of the reference cuvette containing buffer, yielding $F_i = F_{obs,i} - F_{ref,i}$. Next, the experimentally observed intensities, F_i, must be further corrected for dilution, photobleaching, and inner filter effects (31). Photobleaching is the excitation-dependent loss of fluorescence intensity, whereas inner filter effects can result when the absorbance spectrum of the sample overlaps either the fluorescence excitation or emission wavelength, such that some of the excitation or emission intensity is re-absorbed ("filtered") by the sample, resulting in attenuation of the fluorescence signal (30). These factors are multiplicative, so the corrected fluorescence for each titration point can be expressed as in Eq. 13:

$$F_{i,\mathrm{corr}} = \left(\frac{(V_0 + V_i)}{(V_0)}\right)\left(\frac{1}{C_i}\right)\left(\frac{f_0}{f_i}\right) \tag{13}$$

where the first term accounts for dilution, the second term is for inner filter effects (see Note 20), and the third term is for photobleaching, where f_0 and f_i are the fluorescence intensities measured before and during the experiment for a sample which was not titrated, but exposed to the excitation light for the same periods of time as a titrated sample (see Note 15).

3. The actual fluorescence signal depends on the concentration of the fluorescent species and will vary in different experiments. Hence, the change in fluorescence is usually expressed as a quenching (Q_i) or an enhancement (En_i), which normalizes the signal relative to the initial fluorescence of the sample before the start of the titration, i.e., $Q_i = (F_{\mathrm{corr},0} - F_{\mathrm{corr},i})/F_{\mathrm{corr},0}$ or $En_i = (F_{\mathrm{corr},i} - F_{\mathrm{corr},0})/F_{\mathrm{corr},0}$.

3.8. Determination of the Occluded Site Size of SSB on Poly(dT) Under Stoichiometric Conditions

Titration of SSB proteins with polynucleotides under stoichiometric conditions can be used to determine the occluded site size of the protein on the DNA. The occluded site size is defined as the average number of nucleotides made inaccessible to other proteins by the binding of one protein (32). This differs from the number of nucleotides that are physically contacted by the protein, which is usually a lower number. Figure 1a, b shows the results of titrations of *Dr*SSB and *Eco*SSB with poly(dT) at 0.2 M NaCl, a condition under which binding of these proteins to ssDNA is stoichiometric. Under these conditions, an accurate determination of the equilibrium binding constant (affinity) is not possible; however, this allows for an accurate determination of the occluded site size. In these titrations the intrinsic Trp fluorescence of the protein in the cell is monitored as a function of total added ssDNA. The data are plotted as the relative fluorescence change versus the ratio of the total poly(dT) concentration (in nucleotides) to total protein concentration. Such a plot allows one to determine the occluded site size as the point of intersection of a linear extrapolation of the linear part of the titration curve with the plateau value of the fluorescence at saturation. The occluded site sizes, n, and maximum fluorescence quenching, Q_{max}, differ for *Dr*SSB ($n \sim 52$ nucleotides, $Q_{max} = 82\%$) and *Eco*SSB ($n \sim 65$ nucleotides, $Q_{max} = 90\%$). To ensure that binding is stoichiometric, multiple titrations should be performed at a few different SSB concentrations. If binding is stoichiometric, then plots of the type shown in Fig. 1a obtained for different SSB concentrations should overlay.

Upon increasing the [NaCl], the binding isotherm for *Dr*SSB becomes non-stoichiometric (less sharp) (see Fig. 1a) indicating a decrease in affinity (lowering of the equilibrium binding constant, K_{obs}). In contrast, the binding of *Eco*SSB to poly(dT) remains stoichiometric up to 3 M NaCl (14). This does not mean that increasing [NaCl] does not reduce the value of K_{obs} for the *Eco*SSB–DNA interaction, but rather that K_{obs} remains above the value (generally $\sim 10^9\ M^{-1}$) needed for binding to appear "stoichiometric". In each case an increase in [NaCl] does not affect the maximum quenching suggesting no change in the mode of DNA binding. On the other hand, as the [NaCl] decreases to ~0.010 M NaCl, the occluded site size and maximum fluorescence quenching change for both *Eco*SSB (2, 7) and *Dr*SSB (18). For *Eco*SSB this change is more dramatic ($n = 35$ and $Q = 50\%$) and indicates a transition from its $(SSB)_{65}$ mode to its $(SSB)_{35}$ mode of DNA binding in which on average only two instead of four subunits of SSB tetramer are bound to ssDNA (7, 33). For *Dr*SSB the change is less dramatic ($n = 45$ and $Q = 76\%$) (18) but still represents a change in ssDNA binding mode (18). For this reason one should always examine the effects of solution conditions on the occluded site size, rather than assume that a site size measured under one condition is applicable to all conditions.

3.9. Model-Independent Analysis of Titration Data Using the Macromolecule Binding Density Function Method

A model-independent analysis of two or more titrations performed at different macromolecule concentrations can be used to determine the stoichiometry of binding and to construct a true binding isotherm (34–36). This is most useful for systems in which one species binds multiple copies of the other. For this discussion, the macromolecule is defined as the species that binds multiple ligands. This analysis is also needed when the relationship between the average extent of ligand binding ($<X>$ = ligands bound per total macromolecule) and the spectroscopic signal change (in this case fluorescence intensity or anisotropy) is not known a priori. Application of the following analysis to two or more titrations performed at different SSB concentrations as shown in Fig. 2a, b results in a true isotherm, i.e., a plot of $<X>$ versus free or total ligand concentration (34–36). The two titrations need to be performed under conditions such that the total protein concentration is high so that the free ligand and total ligand concentrations differ significantly throughout the titrations.

The basis for this analysis is the thermodynamic constraint that for a macromolecule that does not undergo self-association, the average number of ligands bound per macromolecule, $<X>$, is determined solely by the free ligand concentration (strictly the chemical potential of the ligand) (34–36). Moreover, it is important to recognize that for a given set of solution conditions the relative fluorescence change ($Q_i = (F_{corr,0} - F_{corr,i})/F_{corr,0}$) is an intrinsic property of the system and is determined by the average degree of saturation of the binding sites, which in turn is determined solely by the free ligand concentration. Therefore, if two titrations are performed at two different total macromolecule (SSB) concentrations, $[M_1]_{tot}$ and $[M_2]_{tot}$, and plotted as a function of the total ligand concentration $[X]_{tot}$ (DNA), then whenever the fluorescence quenching is the same for the two titrations (e.g., for the horizontal line in Fig. 2a, where $Q = 0.18$), it follows that the free ligand concentration must be the same at that same fluorescence quenching value (e.g., $Q = 0.18$) (34–36). As such, one can calculate the value of $<X>$ and the free ligand concentration, $[X]$, from knowing the pair of $[M_1]_{tot}$ and $[M_2]_{tot}$ and $[X_1]_{tot}$ and $[X_2]_{tot}$ that are needed to attain the same value of Q as indicated in Eqs. 14 and 15:

$$<X> = \frac{[X_2]_{tot} - [X_1]_{tot}}{[M_2]_{tot} - [M_1]_{tot}} \tag{14}$$

$$[X] = \frac{[M_1]_{tot}[X_2]_{tot} - [M_2]_{tot}[X_1]_{tot}}{[M_1]_{tot} - [M_2]_{tot}} \tag{15}$$

By repeating this calculation at multiple values of the observed Q for the two titrations, the titration curves can be analyzed as shown in Fig. 2a to obtain the dependence of Q_{obs} on $<X>$ ($[(dT)_{50}]_{bound}/[Dr\text{SSB}]_{tot}$ or $[(dT)_{35}]_{bound}/[Eco\text{SSB}]_{tot}$; see inserts in Fig. 2a, b, respectively) and to construct true binding isotherm of $<X>$ versus $[X]$ or $[X]_{tot}$ (not shown).

The model-independent analysis of multiple titrations of a fluorescent ligand at different concentrations with a nonfluorescent macromolecule ("reverse titrations") can also be performed to obtain the same information, although the analysis is a bit more complex (35, 36).

MBDF analysis of two titrations of *Dr*SSB (at 0.1 and 0.5 μM) with $(dT)_{50}$ shown in the insert to Fig. 2a indicates that the observed fluorescence quenching is directly proportional to the average degree of binding ($<X>$), as expected for a 1:1 binding interaction, and that the binding stoichiometry for this system is 1:1 (as follows from linear extrapolation of the dependence of Q on $<X>$ to the value of $Q_{max} = 0.8$). Another example of an MBDF analysis for a system where the macromolecule binds multiple ligands is shown in Fig. 2b for the binding of *Eco*SSB to $(dT)_{35}$ (14). Two titrations were performed at two protein concentrations, 0.10 and 0.32 μM (buffer T, pH 8.1, 0.2 M NaCl, 25°C). The analysis (see insert in Fig. 2b) shows that two molecules of $(dT)_{35}$ can bind to an *Eco*SSB tetramer at saturation (~90% quenching); however binding of the first and the second molecule is characterized by different fluorescence quenching values of ~50% and ~40%.

3.10. Fitting the Data to a Particular Model to Obtain the Equilibrium Binding Parameters

Model-independent analysis of the titrations in Fig. 2a shows that a 1:1 complex is formed upon interaction of *Dr*SSB with $(dT)_{50}$. The binding isotherm shown in Fig. 2a can be directly fit to the expression for an "*n*" site binding model given in Eqs. 5 and 5a (see Subheadings 1 and 2). A global fit of the data for two titrations performed at *Dr*SSB concentrations of 0.1 and 0.5 μM (see Fig. 2a and see Note 21) is consistent with this model and yields the following binding parameters: $n = 0.94 \pm 0.02$; $Q_{max} = 0.80 \pm 0.02$; and $K = (8.09 \pm 0.73) \times 10^7\ M^{-1}$.

One can also examine binding by monitoring the fluorescence intensity or anisotropy of a DNA molecule if it is labeled with an extrinsic fluorophore. To demonstrate this we show in Fig. 3 the results of a titration of F-$(dT)_{50}$ ($(dT)_{50}$ labeled with fluorescein at its 5′ end) with *Dr*SSB performed at the same solution conditions as for the experiment in which the *Dr*SSB Trp fluorescence was monitored in Fig. 2a (buffer T, pH 8.1, 1.0 M NaCl, 25°C). In this case, binding is accompanied by a quenching of the fluorescein fluorescence (0.2 μM F-$(dT)_{50}$ in the cell, $\lambda_{ex} = 494$ nm, $\lambda_{ex} = 520$ nm) upon binding protein. The data have been corrected as described above for titrations of the SSB with DNA and the resulting binding isotherm shown in Fig. 3a was fitted to the expression for an "*n*" independent and identical sites model given in Eqs. 5 and 5a, where *Dr*SSB is now considered the ligand (X) and F-$(dT)_{50}$ is the macromolecule (M). The fit describes the data well and provides the following binding parameters: $n = 0.98 \pm 0.01$; $Q_{max} = 0.55 \pm 0.01$; and $K = (2.42 \pm 0.35) \times 10^8\ M^{-1}$. The results indicate that the affinity of $(dT)_{50}$ when labeled with fluorescein is increased threefold compared to the binding of an unlabeled $(dT)_{50}$. This is likely due to additional interactions of the fluorescein with the protein. Hence, although the use of extrinsically labeled protein or DNA is convenient, the extrinsic probes will generally influence the energetics of the interaction. In fact, this is common for SSB interactions with DNA labeled with different fluorescent dyes. For example, the affinity of *Eco*SSB binding to $(dT)_{70}$ (2.0 M NaBr) increases three- and tenfold, when the DNA is labeled with fluorescein or Cy3, respectively (22). However, one can always use the equilibrium of the SSB with the labeled DNA in a competition experiment to probe the interaction of SSB with an unlabeled DNA (37, 38).

The data in Fig. 2b (14) can be fit to the equilibrium model presented in Scheme 2 using Eqs. 6 and 6a. Under the conditions of this experiment (buffer T, 0.2 M NaCl) the following parameters were obtained (12): $Q_1 = 0.50 \pm 0.02$; $Q_2 = 0.90 \pm 0.01$; and

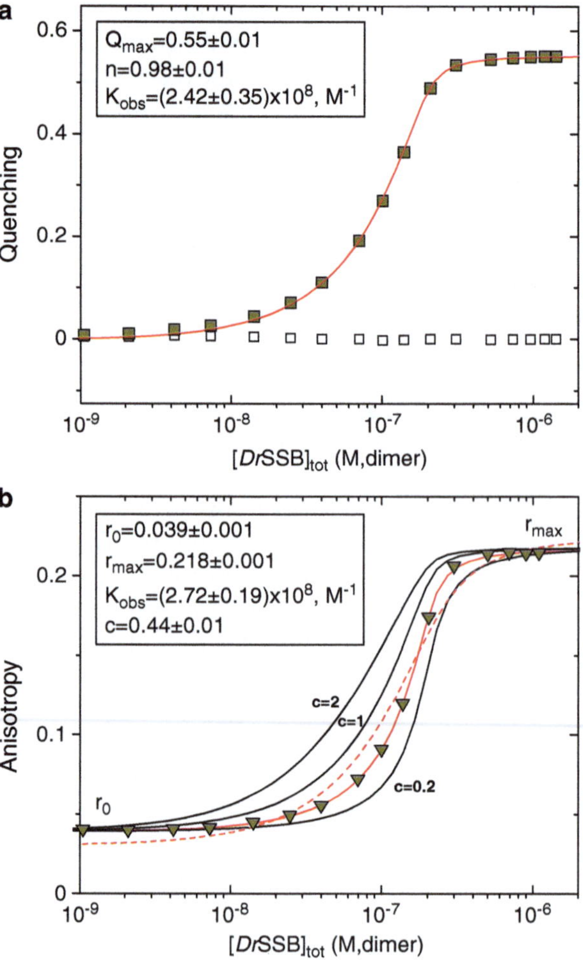

Fig. 3. Titrations of fluorescein-labeled DNA (F-(dT)$_{50}$) with SSB monitoring (**a**) quenching of total fluorescence intensity or (**b**) fluorescein anisotropy. (**a**) Binding isotherm for the titration of 0.2 μM of F-(dT)$_{50}$ with *Dr*SSB (buffer T, pH 8.1, 0.2 M NaCl, 25 °C) plotted as relative fluorescence quenching versus total concentration of *Dr*SSB (dimer). The smooth curve through the data points represents a fit of the data to an n-independent and identical sites model (see Eqs. 5 and 5a and text for details) with the binding parameters indicated. The residuals for the fit are shown as open squares. (**b**) Binding isotherm for the titration of 0.2 μM of F-(dT)$_{50}$ with *Dr*SSB (buffer T, pH 8.1, 0.2 M NaCl, 25 °C) plotted as anisotropy versus total concentration of *Dr*SSB (dimer) (*yellow triangles*). The *dashed red line* shows the best nonlinear least squares fit of the data to a 1:1 binding model (Eq. 9), assuming no fluorescence change between the bound and free states of DNA (i.e., $c = F_{bound}/F_{free}$ was constrained to equal 1) and demonstrates that the fit is not adequate. The *solid red curve* shows the best nonlinear least squares fit of the same data to Eq. 9 when c is allowed to float yielding $c = 0.44 \pm 0.01$ and $K_{obs} = (2.72 \pm 0.19)\times10^8$ M^{-1} (see insert), which are in excellent agreement with the binding parameters obtained for the same experiment when total fluorescence quenching is monitored (Fig. 3a) (see text for more details). The titration curves shown in *solid black lines* are simulations showing the effect of different values of c (different extents of fluorescence quenching/enhancement) on the shape and position of the titration curves. The simulations were performed using Eq. 9 with $K_{obs} = 2.4 \times 10^8$ M^{-1} and different values of c ($c = 1$, no fluorescence change), ($c = 2$, 100 % fluorescence enhancement), and ($c = 0.2$, 80 % fluorescence quenching).

$K_{2,obs} = (3.3 \pm 0.3) \times 10^8\ M^{-1}$. Determination of $K_{1,obs}$ was not possible at this salt concentration due to the fact that the binding affinity is too high ($K_{1,obs} > 10^{10}\ M^{-1}$). These data show that an *Eco*SSB tetramer can bind two $(dT)_{35}$ molecules, but with negative cooperativity. Similar titrations performed over a range of salt concentrations (12) indicate that this negative cooperativity decreases as the salt concentration increases and contributes to the fact that SSB displays salt-dependent binding modes (12).

3.11. Analysis of Titration Data When Monitoring a Change in Fluorescence Anisotropy

If a change in fluorescence anisotropy of one species accompanies binding of a second species, then this can also be used to monitor binding. In fact, this approach has become increasingly popular due to the ease with which DNA can be labeled with an extrinsic fluorescence probe. However, there are some important caveats to be considered when using this approach. These include the likelihood that the binding energetics will be influenced by the addition of the extrinsic fluorophore, as shown above for F-$(dT)_{50}$. In addition, if there is also a change in fluorescence intensity of the probe, this must be considered in the analysis of the data.

As an example of a binding system that is accompanied by a fluorescence anisotropy change we show results for the binding of *Dr*SSB to F-$(dT)_{50}$ (where F is fluorescein). This experiment is identical to that shown in Fig. 3a, in which the change in fluorescein fluorescence intensity of F-$(dT)_{50}$ was monitored. However, in Fig. 3b we plot the change in fluorescein fluorescence anisotropy as a function of added *Dr*SSB (see Note 22). For this system, the fluorescence anisotropy for the free F-$(dT)_{50}$ is $r = 0.039$, whereas $r = 0.218$ for the SSB–DNA complex. (We note that the maximum fluorescence anisotropy for a system is $r = 0.4$; see also Note 3).

We have chosen this system as an example because it allows us to demonstrate the need for caution when using fluorescence anisotropy to analyze a binding equilibrium if there is also a change in fluorescence intensity that accompanies binding. For this system the fluorescence intensity of the bound F-$(dT)_{50}$ is quenched by 55% relative to the free F-$(dT)_{50}$ ($F_0 - F_{min}/F_0 \approx 0.55$) as shown in Fig. 3a. Since the observed anisotropy is a sum of the individual species' anisotropies weighted by the fractional fluorescence contribution (see Eq. 8), this will influence the measured values of the anisotropies (29, 39) and thus the shape of the titration curve, a problem that seems not to be widely appreciated. Indeed, a direct comparison of the titration curves monitoring fluorescence intensity (Fig. 3a) and fluorescence anisotropy (Fig. 3b) shows that the titration curve obtained by monitoring the anisotropy change is steeper and shifted to higher total protein concentrations compared to the titration curve obtained by monitoring the fluorescence intensity change. As a result, an attempt to fit the anisotropy titration curve to a simple 1:1 binding model without accounting for the fluorescence intensity change fails (red dashed curve in Fig. 3b) in the direction that would suggest some positive cooperativity, which is impossible for a simple 1:1 binding system such as this.

To analyze the anisotropy titration curve correctly, one must account for the change in fluorescence intensity of the free and bound DNA. This is taken into account in Eq. 9 by the factor $c = F_{bound}/F_{free} = F_{MX}/F_{M}$ (in this case we consider the DNA as the macromolecule (M) and *Dr*SSB as the ligand (X)).

The experimental data shown in Fig. 3b are well described by a simple 1:1 binding model Eq. 9 as long as the correction factor ($c = 0.44 \pm 0.01$) is included and the resulting best fit equilibrium binding parameters ($r_0 = 0.039 \pm 0.001$; $r_{max} = 0.218 \pm 0.00$; $K_{obs} = (2.72 \pm 0.19) \times 10^8\ M^{-1}$) are in excellent agreement with the parameters obtained by directly fitting the fluorescence intensity titration curve in Fig. 3a ($n = 0.98 \pm 0.01$; $K_{obs} = (2.42 \pm 0.35) \times 10^8\ M^{-1}$; and $Q_{max} = 0.55 \pm 0.01$) (recall that Q_{max} in Fig. 3a is related to c in Fig. 3b by $c = 1 - Q_{max}$).

In order to further illustrate the effect on the observed anisotropy titration curve when a fluorescence quenching/enhancement accompanies binding, we also show several simulations for a 1:1 binding system where $K_{obs} = 2.4 \times 10^8\ M^{-1}$ but with different values of c ($c = 1$, no fluorescence change), ($c = 2$, 100% fluorescence enhancement), and ($c = 0.2$, 80% fluorescence quenching). The solid lines in Fig. 3b show the resulting shifts in apparent binding affinity and shape of the isotherms. As a result, significant errors in binding parameters and even conclusions about cooperativity can result if these corrections are not considered. This also points out that if there is a significant change in fluorescence intensity of the labeled DNA it is simpler to use that signal directly to monitor binding rather than the anisotropy change.

4. Notes

1. Precise values of the extinction coefficients for the protein and ssDNA are required for determination of their concentrations which, in turn, are needed for accurate analysis of the titration data. The methods for the determination of extinction coefficients of proteins and ssDNA are described in detail in Note 5 of the ITC Chapter.
2. The macroscopic binding constants in Eqs. 2 and 6 do not contain the statistical factors and can be converted to microscopic binding constants, k_i, which account for statistical factors as follows: $k_1 = K_1/2$ and $k_2 = 2\ K_2$ (1).
3. For this reason in designing the anisotropy titration experiments for SSB–ssDNA binding we recommend labeling ssDNA with the fluorophore (r values usually range from 0.04 to 0.1) allowing one to observe a detectable increase in

anisotropy upon protein binding. We also note that the theoretical upper limit for r is 0.4 (40); hence apparent experimental values of r that exceed this value likely indicate experimental problems such as aggregation resulting in an increase in light scattering.

4. The G-factor accounts for the differential sensitivities of the detection system for vertically and horizontally polarized light (29) and can be determined experimentally by measuring horizontally and vertically polarized emission with the excitation polarizer in the horizontal position, I_{HH} and I_{HV}, respectively, and then calculated as $G = I_{HH}/I_{HV}$. It is important to note that the G factor is dependent upon emission wavelength and to some extent the bandpass of the monochromator. Therefore, the G factor must be determined for each particular experimental setup for the wavelengths being used at least once by measuring I_{HH} and I_{HV} for the fluorescent species in the cell before the titration is started.

5. The polarization (p) is defined by Eq. 16:

$$p = \frac{I_{VV} - GI_{VH}}{I_{VV} + GI_{VH}} \tag{16}$$

Although the polarization and anisotropy contain the same information, use of the anisotropy for studies of interacting systems is preferred because it is normalized to the total fluorescence intensity $I_T = I_{VV} + 2\ GI_{VH}$. This simplifies its use in defining the average anisotropy of a mixture of fluorescent species $\bar{r} = \sum r_i f_i$ which was used for the derivation of Eq. 9 (see Subheading 3). The corresponding expression using p becomes very cumbersome (29). When data are obtained as polarization, Eq. 17 should be used to convert to anisotropy:

$$r = \frac{2p}{3 - p} \tag{17}$$

6. To obtain information about the assembly state(s) and relative stability of the protein under study (particularly for a new system) the protein should be characterized using both sedimentation velocity and sedimentation equilibrium methods over a range of protein concentrations under the same solution conditions that are to be used for the titrations.

7. We recommend that such titration tables are prepared in the process of designing a titration experiment. This can be easily accomplished using Excel spreadsheet that can also be used for recording the experimentally observed fluorescence intensities corresponding to each titration point.

8. An important test to determine if the binding is stoichiometric under the conditions and for the protein and DNA concentrations used is to perform a second titration at a lower or higher concentration of the species in the cell. If the binding is stoichiometric no change in the shape of the binding isotherm should be observed, whereas for non-stoichiometric binding the isotherm becomes more shallow as seen in Fig. 1c for the titration with $(dT)_{50}$ performed at a fivefold lower *Dr*SSB concentration in 1.0 M NaCl.
9. Preparation of concentrated solutions of the titrant (DNA) as suggested here will certainly leave some extra DNA unused, although any unused DNA can be safely stored at 4°C for long periods of time and used in future studies.
10. *Eco*SSB–ssDNA interactions are characterized by a very large and temperature-dependent enthalpy of binding (41). Hence, even small changes in ambient temperature will generally influence the binding affinity. Moreover, the fluorescence intensity itself is also temperature dependent; hence it is critical to maintain control of the temperature throughout the titration.
11. The use of a stir bar is recommended for thorough mixing of the solution in the cuvette. However, the stirring speed should be maintained as low as possible while maintaining a stable fluorescence signal. We use "Tube" (or cylindrical) type stir bars (http://www.stirbars.com) specifically designed to fit in 10 mm cuvettes. If the cuvette holder does not have a built-in stirring plate, mixing can be achieved by gently redistributing the cell solution back and forth using micropipette; however, this approach generally leads to poorer results.
12. If your instrument has only one cuvette holder, the reference titration should be performed after completing the experimental titration, but without changing any of the instrument settings. Subtraction of the buffer background signal arising from light scattering can be important when the fluorescence intensity of the sample is low.
13. For a new system we recommend that both excitation and emission spectra are run on the fluorescent sample first to determine the optimal excitation and emission wavelengths. This is also important for choosing an excitation wavelength to minimize any inner filter effects (see Subheading 7 and Note 20). For example, when monitoring Trp fluorescence, we typically use a higher excitation wavelength (e.g., 296 nm), rather than the maximum excitation wavelength of 280 nm, to reduce the inner filter effect.
14. The fluorescence signal should be directly proportional to the concentration of the fluorophore. It can deviate from linearity if the photon count reaches the saturation limit of the PMT

(for our instrument this limit is 10^6 s^{-1}). Therefore, before starting the experiment it is necessary to ensure that the signal change during the titration will stay within the appropriate range. This can be adjusted by increasing or decreasing the slit width (preferably for the emission monochromator). For a new system we recommend using an intermediate value at the beginning of the titration.

15. Some proteins are more susceptible to photobleaching than others. Photobleaching occurs when the protein sample is exposed to the excitation light for an extended period of time. Therefore, it is useful to close the excitation shutter and not expose the protein sample to the excitation beam unless a reading is to be taken. Use of a smaller excitation slit width will also reduce photobleaching. In addition, the signal can decrease if protein has a tendency to stick to the cuvette walls. The latter situation is observable at low concentrations of the protein (usually <0.1 μM) (this is typically not observed for DNA labeled with external dye). Both phenomena lead to a decrease in fluorescence intensity. Therefore, before starting an experiment you should determine if the fluorescence signal of your sample is stable or changes with time. A slight decrease in fluorescence intensity due to photobleaching can be corrected (see Subheading 7) by monitoring the fluorescence of a protein solution in a separate cuvette when exposed to the excitation light for the same periods of time as during the titration experiment. Sticking of a nonself-associating protein to the cuvette walls can generally be detected by titrating increasing protein concentrations into buffer while monitoring the fluorescence intensity. If the fluorescence intensity is not directly proportional to protein concentration (e.g., if the molar intensity is lower at lower protein concentrations), then this suggests that protein is sticking to the cuvette walls. If either of these effects is severe it is possible that the system cannot be studied using fluorescence. One alternative is to change the design of the experiment. For example if the protein can be labeled with an external fluorophore that can be excited at a wavelength above 300 nm, then one can add BSA to the solution (usually ~0.1 mg/mL) to prevent sticking of the labeled protein to the cuvette (38). However, any such labeling may affect the DNA binding properties of the protein (20). If this is the case, then one can examine the binding of the unlabeled protein using competition methods (37, 38).
16. For our titrations we use 2, 10, and 20 μl Gilson micropipettes with disposable plastic tips.
17. It is sometimes difficult to determine from the raw titration data when saturation of the signal has been achieved, especially if binding is weak or shows negative cooperativity. In fact, this is

often only determined after all of the corrections to the data have been made and the data plotted. In that case, one knows that when a second titration is performed higher DNA concentrations need to be achieved to approach saturation. One example is the case of a much weaker binding of a second molecule of $(dT)_{35}$ to *Eco*SSB (13).

18. It is important to know the precise values of the concentrations of SSB and DNA used in the experiment for meaningful quantitative analyses of the data. Mistakes can be made when diluting the samples to the experimental concentrations. An accurate determination of the titrant concentration (DNA) and SSB in the cell is of particular importance under stoichiometric binding conditions (see Fig. 1) since it is used for determination of the stoichiometry or occluded site size.

19. If a few different titrant stock concentrations are used during a titration (e.g., as in Fig. 2a, see also Table 2) Eq. 12 can be modified to account for this. It is also recommended that the total volume of the added titrant (V_i) at the end of the titration not exceed ~5–10% of the initial volume (V_0) in the cell.

20. The inner filter correction factor, C_i, can be calculated for absorbances $A_i < 0.3$ (which is usually the case for the experiments described here) using the formula in Eq. 18 (30, 31):

$$C_i = \frac{1 - 10^{-A_i}}{2.303\, A_i} \tag{18}$$

where A_i is the absorbance at titration point i determined using Eq. 19:

$$A_i = \varepsilon_D D_{\mathrm{tot},i} + \varepsilon_P P_{\mathrm{tot},i} \tag{19}$$

where $D_{\mathrm{tot},i}$ and $P_{\mathrm{tot},i}$ are the concentrations of DNA and SSB at particular titration point and ε_D and ε_P are defined in Eqs. 20a and 20b:

$$\varepsilon_D = \varepsilon_{D,\mathrm{ex}} + \varepsilon_{D,\mathrm{em}} \tag{20a}$$

$$\varepsilon_P = \varepsilon_{P,\mathrm{ex}} + \varepsilon_{P,\mathrm{em}} \tag{20b}$$

where $\varepsilon_{D,\mathrm{ex}}$, $\varepsilon_{D,\mathrm{em}}$, $\varepsilon_{P,\mathrm{ex}}$, and $\varepsilon_{P,\mathrm{em}}$ are the extinction coefficients for DNA and protein at excitation and emission wavelengths. These can be easily determined from the corresponding absorbance spectra if the extinction coefficient at one wavelength is known. For higher absorbances more complicated expressions taking into account geometric factors of the fluorescence cell compartment need to be taken into account or the inner filter correction can be determined empirically (30).

21. Nonlinear least squares fitting of the data was performed using the nonlinear regression package in Scientist (MicroMath Scientist Software, St. Louis, MO). This software allows implicit solutions of equations of the type shown in Eqs. 5 or 6 to be used to fit multiple data sets simultaneously.
22. The experiment was performed essentially as described in Subheading 3.6; however F-$(dT)_{50}$ in the cuvette was titrated with *Dr*SSB with the excitation and emission polarizers in place and I_{VV} and I_{VH} were measured for each titration point. The measurements of I_{HH} and I_{HV} were performed before the experiment was started and after completion of the experiment and yielded the same value of $G = 0.66$.

Acknowledgements

This work was supported in part by grants to T.M.L. from NIH (GM030498 and GM045948) and R.G. from NIH (GM098509).

References

1. Wyman J, Gill SJ (1990) Bindng and linkage. University Science Books, Mill Valley, CA
2. Lohman TM, Overman LB (1985) Two binding modes in Escherichia coli single strand binding protein-single stranded DNA complexes. Modulation by NaCl concentration. J Biol Chem 260:3594–3603
3. Bujalowski W, Lohman TM (1986) Escherichia coli single-strand binding protein forms multiple, distinct complexes with single-stranded DNA. Biochemistry 25:7799–7802
4. Bujalowski W, Overman LB, Lohman TM (1988) Binding mode transitions of Escherichia coli single strand binding protein-single-stranded DNA complexes. Cation, anion, pH, and binding density effects. J Biol Chem 263:4629–4640
5. Overman LB, Bujalowski W, Lohman TM (1988) Equilibrium binding of Escherichia coli single-strand binding protein to single-stranded nucleic acids in the (SSB)65 binding mode. Cation and anion effects and polynucleotide specificity. Biochemistry 27:456–471
6. Overman LB, Lohman TM (1994) Linkage of pH, anion and cation effects in protein-nucleic acid equilibria. Escherichia coli SSB protein-single stranded nucleic acid interactions. J Mol Biol 236:165–178
7. Lohman TM, Ferrari ME (1994) Escherichia coli single-stranded DNA-binding protein: multiple DNA-binding modes and cooperativities. Annu Rev Biochem 63:527–570
8. Wei TF, Bujalowski W, Lohman TM (1992) Cooperative binding of polyamines induces the Escherichia coli single-strand binding protein-DNA binding mode transitions. Biochemistry 31:6166–6174
9. Lohman TM, Bujalowski W (1994) Effects of base composition on the negative cooperativity and binding mode transitions of Escherichia coli SSB-single-stranded DNA complexes. Biochemistry 33:6167–6176
10. Bujalowski W, Lohman TM, Anderson CF (1989) On the cooperative binding of large ligands to a one-dimensional homogeneous lattice: the generalized three-state lattice model. Biopolymers 28:1637–1643
11. Bujalowski W, Lohman TM (1987) Limited co-operativity in protein–nucleic acid interactions. A thermodynamic model for the interactions of Escherichia coli single strand binding protein with single-stranded nucleic acids in the "beaded," (SSB)65 mode. J Mol Biol 195:897–907
12. Bujalowski W, Lohman TM (1989) Negative co-operativity in Escherichia coli single strand binding protein–oligonucleotide interactions. II. Salt, temperature and oligonucleotide length effects. J Mol Biol 207:269–288

13. Lohman TM, Bujalowski W (1988) Negative cooperativity within individual tetramers of Escherichia coli single strand binding protein is responsible for the transition between the (SSB)35 and (SSB)56 DNA binding modes. Biochemistry 27:2260–2265
14. Bujalowski W, Lohman TM (1989) Negative co-operativity in Escherichia coli single strand binding protein–oligonucleotide interactions. I. Evidence and a quantitative model. J Mol Biol 207:249–268
15. Ferrari ME, Lohman TM (1994) Apparent heat capacity change accompanying a nonspecific protein–DNA interaction. Escherichia coli SSB tetramer binding to oligodeoxyadenylates. Biochemistry 33:12896–12910
16. Kozlov AG, Lohman TM (1999) Adenine base unstacking dominates the observed enthalpy and heat capacity changes for the Escherichia coli SSB tetramer binding to single-stranded oligoadenylates. Biochemistry 38:7388–7397
17. Ferrari ME, Bujalowski W, Lohman TM (1994) Co-operative binding of Escherichia coli SSB tetramers to single-stranded DNA in the (SSB)35 binding mode. J Mol Biol 236:106–123
18. Kozlov AG, Eggington JM, Cox MM, Lohman TM (2010) Binding of the dimeric Deinococcus radiodurans single-stranded DNA binding protein to single-stranded DNA. Biochemistry 49:8266–8275
19. Kumaran S, Kozlov AG, Lohman TM (2006) Saccharomyces cerevisiae replication protein A binds to single-stranded DNA in multiple salt-dependent modes. Biochemistry 45:11958–11973
20. Dillingham MS, Tibbles KL, Hunter JL, Bell JC, Kowalczykowski SC, Webb MR (2008) Fluorescent single-stranded DNA binding protein as a probe for sensitive, real-time assays of helicase activity. Biophys J 95:3330–3339
21. Kozlov AG, Lohman TM (2002) Stopped-flow studies of the kinetics of single-stranded DNA binding and wrapping around the Escherichia coli SSB tetramer. Biochemistry 41:6032–6044
22. Kozlov AG, Lohman TM (2002) Kinetic mechanism of direct transfer of Escherichia coli SSB tetramers between single-stranded DNA molecules. Biochemistry 41:11611–11627
23. Roy R, Kozlov AG, Lohman TM, Ha T (2007) Dynamic structural rearrangements between DNA binding modes of E. coli SSB protein. J Mol Biol 369:1244–1257
24. Kunzelmann S, Morris C, Chavda AP, Eccleston JF, Webb MR (2010) Mechanism of interaction between single-stranded DNA binding protein and DNA. Biochemistry 49:843–852.
25. Bujalowski W, Lohman TM (1991) Monomer-tetramer equilibrium of the Escherichia coli ssb-1 mutant single strand binding protein. J Biol Chem 266:1616–1626
26. Bujalowski W, Lohman TM (1991) Monomers of the Escherichia coli SSB-1 mutant protein bind single-stranded DNA. J Mol Biol 217:63–74
27. Lohman TM, Green JM, Beyer RS (1986) Large-scale overproduction and rapid purification of the Escherichia coli ssb gene product. Expression of the ssb gene under lambda PL control. Biochemistry 25:21–25
28. Eggington JM, Haruta N, Wood EA, Cox MM (2004) The single-stranded DNA-binding protein of Deinococcus radiodurans. BMC Microbiol 4:2
29. Lakowicz JR (2006) Principles of fluorescence spectroscopy, 3 edn. Springer, New York
30. Birdsall B, King RW, Wheeler MR, Lewis CA Jr, Goode SR, Dunlap RB, Roberts GC (1983) Correction for light absorption in fluorescence studies of protein–ligand interactions. Anal Biochem 132:353–361
31. Lohman TM, Mascotti DP (1992) Nonspecific ligand-DNA equilibrium binding parameters determined by fluorescence methods. Methods Enzymol 212:424–458
32. McGhee JD, von Hippel PH (1974) Theoretical aspects of DNA–protein interactions: co-operative and non-co-operative binding of large ligands to a one-dimensional homogeneous lattice. J Mol Biol 86:469–489
33. Raghunathan S, Kozlov AG, Lohman TM, Waksman G (2000) Structure of the DNA binding domain of E-coli SSB bound to ssDNA. Nat Struct Biol 7:648–652
34. Bujalowski W, Lohman TM (1987) A general method of analysis of ligand-macromolecule equilibria using a spectroscopic signal from the ligand to monitor binding. Application to Escherichia coli single-strand binding protein–nucleic acid interactions. Biochemistry 26:3099–3106
35. Lohman TM, Bujalowski W (1991) Thermodynamic methods for model-independent determination of equilibrium binding isotherms for protein–DNA interactions: spectroscopic approaches to monitor binding. Methods Enzymol 208:258–290
36. Bujalowski W (2006) Thermodynamic and kinetic methods of analyses of protein–nucleic acid interactions. From simpler to more complex systems. Chem Rev 106:556–606
37. Jezewska MJ, Bujalowski W (1996) A general method of analysis of ligand binding to competing macromolecules using the

spectroscopic signal originating from a reference macromolecule. Application to Escherichia coli replicative helicase DnaB protein nucleic acid interactions. Biochemistry 35:2117–2128

38. Wong CJ, Lucius AL, Lohman TM (2005) Energetics of DNA end binding by E.coli RecBC and RecBCD helicases indicate loop formation in the 3′-single-stranded DNA tail. J Mol Biol 352:765–782

39. Eftink MR (1997) Fluorescence methods for studying equilibrium macromolecule–ligand interactions. Methods Enzymol 278:221–257

40. Cantor CR, Schimmel PR (1980) Biophysical chemistry. Freeman W.H., New York

41. Kozlov AG, Lohman TM (2006) Effects of monovalent anions on a temperature-dependent heat capacity change for Escherichia coli SSB tetramer binding to single-stranded DNA. Biochemistry 45:5190–5205

Chapter 5

Single-Molecule Analysis of SSB Dynamics on Single-Stranded DNA

Ruobo Zhou and Taekjip Ha

Abstract

SSB proteins bind to and control the accessibility of single-stranded (ss) DNA generated as a transient intermediate during a variety of cellular processes. For subsequent DNA processing, however, SSB needs to be removed and yield to other proteins while avoiding ssDNA exposure to nucleases. Using single-molecule two- and three-color fluorescence resonance energy transfer (FRET) and fluorescence-force spectroscopy, we recently showed that the SSB/DNA complex is a highly dynamic system and SSB functions as a sliding platform that migrates on ssDNA for recruiting other proteins in DNA repair, replication, and recombination. Here, we present the activity assays in detail for observing the transitions between different SSB binding modes and SSB diffusion on ssDNA in real time by using single-molecule FRET microscopy and for studying how mechanical forces regulate SSB–DNA interactions using fluorescence-force spectroscopy. These single-molecule approaches are generally applicable to many other protein–nucleic acid systems.

Key words: SSB, ssDNA, Translocation, Single molecule, FRET, Optical tweezers

1. Introduction

Recent advances in single-molecule techniques enable us to directly observe the conformational dynamics of proteins and nucleic acids and interactions between them in real time (1–3). In particular, single-molecule Förster (or fluorescence) resonance energy transfer (smFRET) (4) and optical tweezers (5) are two mainstream single-molecule techniques that have been widely used. In contrast to conventional ensemble methods, single-molecule methods can look at a biological system even when it is out of equilibrium and hence allow us to explore the heterogeneity among molecules and detect the transition and intermediate states which are otherwise hidden in ensemble measurements (6, 7).

Single-stranded DNA binding protein (SSB) binds single-stranded DNA (ssDNA) specifically to protect it from degradation and to recruit other proteins necessary for downstream functions.

James L. Keck (ed.), *Single-Stranded DNA Binding Proteins: Methods and Protocols*, Methods in Molecular Biology, vol. 922, DOI 10.1007/978-1-62703-032-8_5, © Springer Science+Business Media, LLC 2012

Using single-molecule two- and three-color FRET, we have shown that SSB/DNA complexes are highly dynamic: (1) The SSB-bound ssDNA undergoes rapid rearrangements between two major binding modes of *E. coli* (*Eco*)SSB in certain conditions; (2) Once *Eco*SSB binds to ssDNA at the initial site of binding, it migrates along the ssDNA with a diffusion coefficient of ~300 nt^2/s (at 37°C), and this SSB activity transiently melts DNA secondary structures and stimulates RecA filament glowth (8). Additionally, using a recently developed hybrid instrument combining smFRET with optical tweezers, we showed that SSB diffuses on ssDNA primarily via a reptation mechanism involving transient DNA loop-bulge formation and propagation with a mean step size of three nucleotides (9).

In this chapter, we describe the detailed protocols for the experimental assays mentioned above, including the materials and the steps for sample preparation, data acquisition, and analysis. We recommend the readers to read the comprehensive guides elsewhere for assembling a home-built smFRET total internal reflection (TIR) setup (10) and a home-built fluorescence-force spectroscopy setup (11).

2. Materials

2.1. DNA

DNA oligonucleotides used for the single- molecule experiments are listed below. "iAmMC6T" represents amine-modified thymine that is used to label the DNA with Cy3.

1. Biotin strand for making 3′ tail constructs: 5′-/Cy5/GCC TCG CTG CCG TCG CCA-/biotin/-3′. Biotin modification is used for the surface immobolization.
2. 5′-TGG CGA CGG CAG CGA GGC $(T)_{70}$-Cy3-T-3′.
3. 5′-TGG CGA CGG CAG CGA GGC-Cy3-$(T)_m$-3′, where $m = 65, 70$.
4. 5′-TGG CGA CGG CAG CGA GGC $(T)_{68}$/iAmMC6T/$(T)_n$-3′, where $n = 4, 8, 12, 18$.
5. 5′-TGG CGA CGG CAG CGA GGC $(T)_{65}$/iAmMC6T/T-3′.
6. 5′-TGG CGA CGG CAG CGA GGC $(T)_{59}$/iAmMC6T/$(T)_4$-3′.
7. 5′-/Cy5.5/GCC TCG CTG CCG TCG CCA-/biotin/-3′.
8. 5′-TGG CGA CGG CAG CGA GGC $(T)_{130}$-Cy5-T-3′.
9. Biotin strand for making 5′ tail constructs: 5′-/biotin/TGG CGA CGG CAG CGA GGC/Cy5/-3′.
10. 5′-/5Phos/GGG CGG CGA CCT T/iAmMC6T/$(T)_{68}$ GCC TCG CTG CCG TCG CCA-3′.

11. 5′-GGG CGG CGA CCT $(T)_{13}$/iAmMC6T/$(T)_{68}$ GCC TCG CTG CCG TCG CCA-3′.
12. Dig oligo: 5′-AGG TCG CCG CCC TTT/digoxigenin/-3′. Digoxigenin modification was used to attach the DNA linker to an anti-digoxigenin-coated bead.

The sequences underlined are the 12 nt cohesive end site of phage lambda DNA that we use to attach lambda DNA to the biomolecules under investigation using fluorescence-force spectroscopy.

2.2. Reagents and Buffers

1. Acetone.
2. Methanol.
3. 1 M KOH in MilliQ water.
4. Amino silane (*N*-(2-Aminoethyl)-3-Aminopropyltrimethoxysilane).
5. Acetic acid.
6. Biotin-PEG–NHS ester (Bio-PEG-SC; MW 5000).
7. PEG-NHS ester (mPEG-SVA; MW 5000).
8. PEGylation buffer: 0.1 M $NaHCO_3$, pH 8.5 (see Note 1).
9. Diamond drill bits (0.75 mm).
10. Permanent double-sided tape.
11. 5 min Epoxy.
12. T50 buffer: 10 mM Tris–HCl, pH 8.0, 50 mM NaCl.
13. Neutravidin.
14. Rectangular glass coverslips (24 × 40 mm).
15. Quartz microscope slides (3″ × 1″, 1 mm thick).
16. Imaging buffer A: 20 mM Tris–HCl, pH 8.0, 0.1 mM EDTA, 0.1 mg/ml BSA, 0.8% (w/v) dextrose, 165 U/ml glucose oxidase, 2,170 U/ml catalase, 2–3 mM Trolox, and indicated amounts of NaCl and SSB (see Note 2).
17. ~500 μg/ml Lambda-phage DNA.
18. Glass microscope slides (3″ × 1″, 1 mm thick).
19. Protein G-carboxylate beads.
20. Anti-Digoxigenin.
21. MES buffer: 100 mM MES-NaOH, pH 6.5 (see Note 3).
22. EDC hydrochloride (3-Dimethylaminopropyl)-*N′*-ethylcarbodiimide hydrochloride).
23. NHS (*N*-Hydroxysuccinimide).
24. Blocking buffer: 10 mM Tris–HCl, pH 8.0, 50 mM NaCl, 1 mg/ml tRNA, 1 mg/ml BSA.

25. Antibody reconstitution buffer: 0.019 M NaH_2PO_4, 0.081 M Na_2HPO_4, 0.14 M NaCl, 2.7 mM KCl.
26. Bead storage buffer: 0.039 M NaH_2PO_4, 0.061 M Na_2HPO_4, 0.14 M NaCl, 2.7 mM KCl, 0.1 mg/ml BSA, 0.1% (v/v) Tween-20 and 0.02% NaN_3.
27. Imaging buffer B: 20 mM Tris–HCl, pH 8.0, 0.1 mM EDTA, 0.5 mg/ml BSA, 0.01 mg/ml anti-digoxigenin, 0.8% (w/v) dextrose 165 U/ml glucose oxidase, 2,170 U/ml catalase, 3 mM Trolox, and 0.1% (v/v) Tween 20 (see Note 2).

2.3. Single-Molecule FRET with TIR Microscopy

An extensive description of assembling a home-built TIR microscopy can be found elsewhere (10). Most of the experiments are performed at room temperature (22 ± 1°C). In some circumstances where the experiment needs to be performed at different temperatures, a water-circulating bath circulator (NESLAB RTE-7 Digital One; Thermo Scientific, Newington, NH; BOM # 271103200000) is used to heat up or cool down the sample stage, objective, and prism-holder all together. The temperature of the outer surface of the chamber channel (see Note 4) is measured by a handhold digital thermometer as the best estimation of the experimental temperature inside the sample channel.

2.4. Single-Molecule Fluorescence-Force Spectroscopy

An extensive description of assembling and calibrating a home-built hybrid setup combining optical tweezers and confocal microscopy can be found elsewhere (9, 11, 12).

3. Methods

3.1. Slide/Coverslip Cleaning and PEG Surface Preparation

A polymer-passivated surface coated with polyethyleneglycol (PEG) is generally used in single-molecule experiments to eliminate nonspecific surface adsorption of proteins and other biological macromolecules. Here we provide the protocol for preparing microscope slides and coverslips coated with PEG.

3.1.1. Precleaning and Surface Activation

1. Drill a pair of inlet/outlet holes in the glass slide for later assembly of a flow chamber.
2. Put slides and coverslips into a clean glass container. Fill the container with acetone and sonicate for 20 min.
3. Pour out acetone and fill the glass container with 1 M KOH. Sonicate for 30 min.
4. Pour out KOH and rinse the coverslips and slides thoroughly with MilliQ H_2O.

5. Burn the coverslips/slides with a propane torch (see Note 5). Then put them in a dry glass container and fill the container with methanol.

3.1.2. Aminosilanization

1. Pour out methanol from the glass container.
2. Pour 100 ml of methanol into a flask. Add 5 ml of acetic acid and 1 ml of amino silane into the flask. Mix well and immediately pour the mixture into the glass container. Incubate for 10 min. Sonicate for 2 min. Incubate for another 10 min.
3. Rinse the coverslips and slides thoroughly with methanol.
4. Rinse the coverslips and slides thoroughly with MilliQ H_2O.
5. Dry the coverslips and slides with gentle nitrogen flow.

3.1.3. PEGylation (Coating the Amino-Modified Surface with PEG-NHS Esters)

1. Measure 2 mg of biotin-PEG-SC and 80 mg of mPEG-SVA in a 1.5 ml centrifuge tube (for five slides/coverslips).
2. Place the slides on a level surface. A clean pipette tip box is recommended for use and the slides can be put on a level surface inside the box.
3. Add 320 μl of PEGylation buffer into the centrifuge tube. Mix well and centrifuge for 1–2 min at ~10,000 × *g*.
4. Apply 70 μl of mixture from the previous step to each slide. Gently place a coverslip onto each slide (see Note 6).
5. Put some water in the box and beneath the slides to prevent drying.
6. Incubate 5–8 h in the dark room at room temperature.
7. Rinse the slides and coverslips with MilliQ H_2O and dry them with gentle nitrogen flow.
8. Put each slide/coverslip pair into a 50 ml plastic tube and store them in −20°C freezer.

3.2. Flow Chamber Assembly

In order to make a flow chamber with a single channel, the double-sided tape (~100 μm thick) is used as spacer between the slide and the coverslip (see Note 7). More detailed guide can be found elsewhere (10). Briefly, put two strips of double-sided tape on the PEG-coated slide bordering a single channel and leaving the two drilled holes near two ends of the channel. Then place the coverslip on top of the slide with the PEG-coated surface facing towards the slide. Reinforce the tape adhesion by pressing the coverslip with a pipette tip. Finally, seal the two openings of the channel with 5-min epoxy. One can make a flow chamber with multiple channels in a similar way.

3.3. Single-Molecule FRET with TIR Microscopy

Here we provide the protocols to use single-molecule FRET to study (1) the dynamics between two different SSB binding modes and (2) SSB migration on DNA. The protocols for studying different SSB activities share a general sample preparation method. Below we present the general sample preparation protocol followed by the description of the activity assays. The different activity assays mainly differ in using different partial duplex DNA constructs (the length of the ssDNA tail, Cy3/Cy5 labeling positions on DNA, etc).

3.3.1. General Sample Preparation Protocol

1. Make a sample chamber with multiple channels as described above.
2. Infuse 20 μl (roughly the channel volume) of 0.25 mg/ml Neutravidin in T50 buffer into the chamber channel and incubate for 5 min.
3. Flush the channel with 50 μl of T50 buffer to remove excess unbound Neutravidin molecules.
4. Infuse 50 μl of 30–50 pM Cy3-Cy5-labeled DNA construct (see Note 8). Incubate for 5 min.
5. Flush the channel with 50 μl of T50 buffer to remove excess unbound DNA molecules.
6. Infuse imaging buffer A with desired concentrations of NaCl and SSB.
7. Image the channel surface (10). Typically, one can observe 300–500 well-separated individual fluorescent spots at a time through the camera. Each individual spot represents a Cy3-Cy5-labeled DNA molecule. One can record the Cy3 and Cy5 emission intensities as a function of time by the camera and calculate the corresponding FRET-time trajectory.

3.3.2. Binding Assays to Monitor SSB Binding Mode Transitions

The SSB binding assay is used to study the conformational dynamics of SSB-bound ssDNA when SSB switches between different binding modes (13). In *E. coli*, the *Eco*SSB tetramer has multiple DNA binding sites and can bind in two major binding modes (14): at low salt concentrations (<10 mM Na^+ or <0.2 mM Mg^{2+}) and high SSB to ssDNA ratios, an SSB tetramer binds with $(SSB)_{35}$ mode where 35 nt of ssDNA interacts with only two subunits and this mode can form long cooperative clusters along ssDNA. At moderately high salt conditions (>200 mM Na^+ or >2 mM Mg^{2+}), a low cooperativity mode ($(SSB)_{65}$) is dominant in which ~65 nt of ssDNA wraps fully around all four subunits so that the two ends of the ssDNA exit the protein in close proximity.

For the binding assay, a partial duplex DNA construct with a poly (dT) tail is used with Cy3 and Cy5 attached near the two ends of the ssDNA tail. For *Eco*SSB, we use a DNA construct containing 70 nt poly (dT) ssDNA ($(dT)_{70}$). 70 nt of ssDNA can

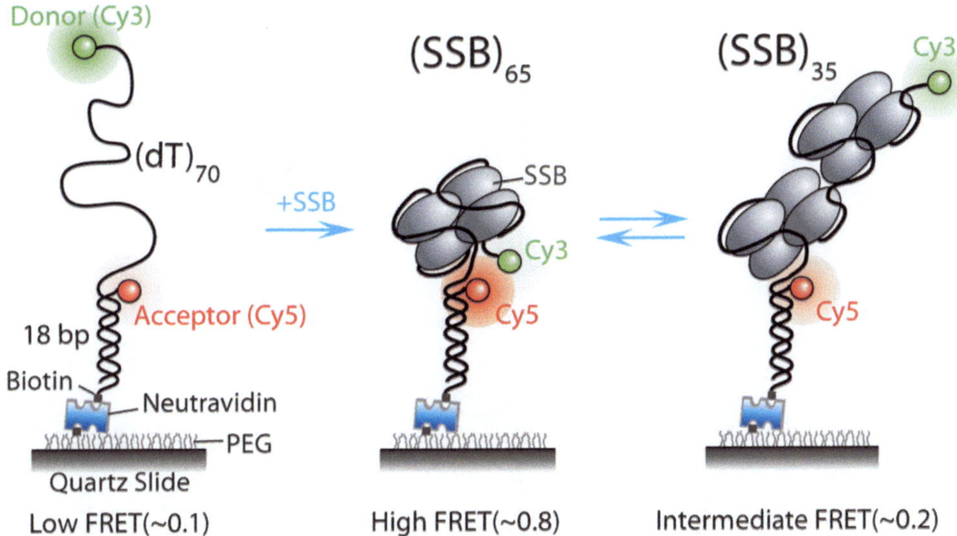

Fig. 1. SSB binding assay to monitor SSB binding mode transitions. A partial duplex DNA construct immobilized on a PEG-coated slide surface has low FRET. Binding of a single *E. coli* SSB tetramer in $(SSB)_{65}$ binding mode results in high FRET. Binding of another SSB (in $(SSB)_{35}$ binding mode) will result in an intermediate FRET state.

accommodate one SSB tetramer in the ($(SSB)_{65}$ mode or two SSB tetramers in the $(SSB)_{35}$ mode, resulting in a high FRET efficiency (~0.8) and an intermediate FRET efficiency (~0.2), respectively (Fig. 1). The bare DNA gives a low FRET efficiency close to zero (~0.1). Therefore, one can monitor the transitions between the three states (bare DNA, $(SSB)_{65}$ and $(SSB)_{35}$) by monitoring the FRET efficiency of individual molecules in real time. In this assay, the imaging buffer should contain some SSB proteins in the solution because for the transition from $(SSB)_{65}$ to $(SSB)_{35}$ mode, a second SSB tetramer needs to bind to $(dT)_{70}$. One can image the same channel and record the FRET-time trajecteries from individual DNA molecules at varying salt/SSB concentrations. For titrating the salt/SSB concentations, a series of imaging buffers with desired salt/SSB concentrations need to be made and infused into the channel for data acquisition.

3.3.3. Probing SSB Diffusion on DNA

The second type of activity assay monitors SSB diffusion on ssDNA. Generally, we have designed three different assays to study a single SSB tetramer diffusing on DNA: (1) helix destabilization assays; (2) diffusion detection assays; (3) step-size determination assays. Because SSB diffusion occurs in the low cooperativity mode, moderately high salt concentration (>200 mM NaCl) is used in all three assays to achieve the $(SSB)_{65}$ mode. After following the general sample preparation protocol and incubating the immobilized DNA in imaging buffer A containing 1 nM SSB tetramer for 1 min, 100 μl of another imaging buffer A containing no SSB is infused to the channel to remove the unbound free SSB in solution,

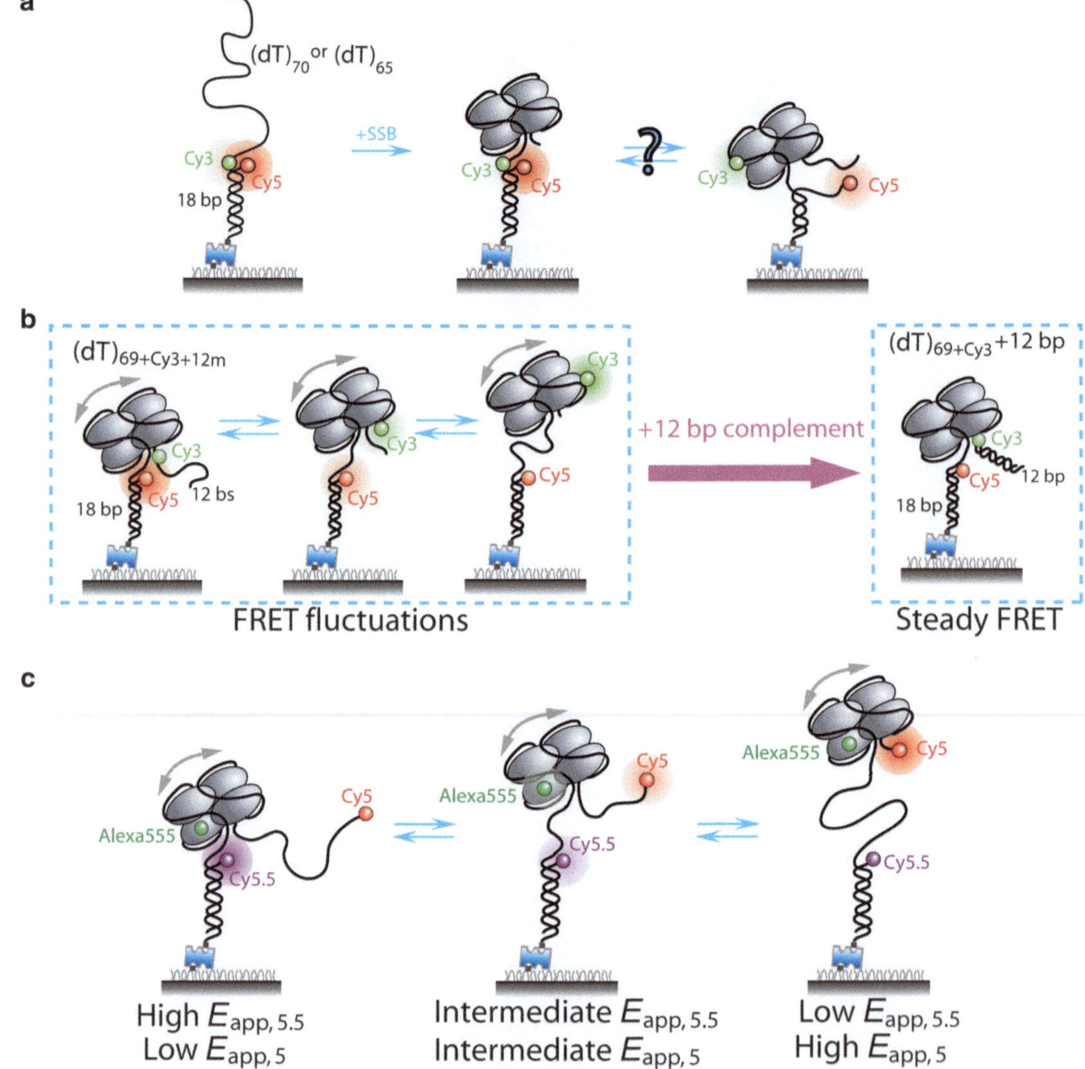

Fig. 2. Single-molecule FRET assays to demonstrate SSB diffusion. (**a**) Helix destabilization assays. (**b**) Diffusion detection assays. (**c**) Step-size determination assays.

excluding binding and dissociation of additional SSB tetramers to the DNA construct as the cause of the dynamics (or fluctuations) observed in the FRET-time trajectories. *Eco*SSB remains bound to $(dT)_{70}$ or longer stretch of ssDNA at least 5 h after flushing (8).

Helix Destabilization Assays

In the helix destabilization assays, a partal duplex DNA construct with a poly (dT) tail is used with Cy3 and Cy5 attached in the vicinity of the duplex region (Fig. 2a). This scheme is used as a negative control to rule out the possibility that the FRET fluctuation observed is due to the transient destabilization of the double-stranded region of the DNA construct. Our 18 bp helix region is commonly used in all the assays presented in this article. We observed no FRET fluctuations using this assay when the ssDNA tail was 65 or 70 nt long (8).

Diffusion Detection Assays

The validation for SSB diffusion needs various designs of DNA constucts, which are usually protein specific. We present two kinds of diffusion detection assays to demostrate *Eco*SSB diffusion in this chapter. One is based on two-color FRET and the other is based on three-color FRET.

In the two-color FRET assay, we use a partial duplex DNA with a 82 nt of ssDNA tail ($(dT)_{70}$ + 12 nt extension) Cy3 and Cy5 are attached near the two ends of the $(dT)_{70}$ region (about the $(SSB)_{65}$ binding site size). The 12 nt mixture sequence beyond $(dT)_{70}$ provides extra ssDNA extension for a single SSB to diffuse. In this DNA construct, one should observe FRET fluctuations if SSB tetramer diffuses along DNA back and forth (Fig. 2b). For a negative control, the DNA construct described above is annealed to a complementary sequence of the 12 nt extension (see Note 9), and a steady high FRET efficiency (~0.8) without any fluctuations beyond measurement noise should be observed after SSB binds to this second DNA construct. The comparison between the FRET trajecteries obtained from the two DNA constructs provides evidence that the FRET fluctuations observed in the first DNA construct are caused by transient excursions of SSB to the 12 nt extension. In addition, one may also vary the extension length (for example, make the extension 4, 8, 12, 18 nt) while keeping the ssDNA between Cy3 and Cy5 at 69 nt. If SSB indeed diffuses on the DNA, larger excursions of SSB away from the high FRET state are expected for longer extensions.

In the three-color FRET assay, we use a partial duplex DNA with a long ssDNA tail (130 nt poly (dT)) with two acceptors (Cy5 and Cy5.5) attached near the two ends of $(dT)_{130}$. The donor (Alexa555) is attached to SSB in this assay (Fig. 2c). The detailed description for three-color FRET can be found elsewhere (15). We record the emission intensities of Alexa555, Cy5, and Cy5.5 as a function of time when Alexa555-labeled SSB binds to this DNA construct and calculate the FRET efficiencies to the two acceptors ($E_{app,5}$ and $E_{app,5.5}$; see Note 10). When a single SSB tetramer diffuses on $(dT)_{130}$, $E_{app,5}$ and $E_{app,5.5}$ should show significant anti-correlated fluctuations as a function of time (i.e., when one FRET efficiency increases, the other FRET efficiency decreases). A similar three-color FRET assay with the donor labeling position moved from SSB to the middle of the $(dT)_{130}$ can also demonstrate EcoSSB diffusion on ssDNA (8).

Step-Size Determination Assays

In order to determine the apparent step size for SSB diffusion on DNA, we use a series of partial duplex DNA constructs containing a 64, 67, or 73 nt of ssDNA tail ($(dT)_{60+4}$, $(dT)_{66+1}$, and $(dT)_{69+4}$).

Cy3 and Cy5 are separated by 60, 66, and 69 nt, respectively. In this assay, we use a hidden Markov model (HMM)-based statistical tool (16) to determine the most likely number of distinct FRET states involved in the FRET-time trajectories obtained from each DNA construct. SSB is considered to take one step when FRET changes from one state to another. Therefore, if we can validate that one more FRET state can be determined every time we increase the ssDNA tail length by *N* nucleotides, *N* would be the apparent step size for SSB to diffuse. HMM analysis yielded 2, 3, and 5 FRET states for $(dT)_{60+4}$, $(dT)_{66+1}$, and $(dT)_{69+4}$, respectively, from which we concluded that the apparent SSB step size is 3 nt (8). In practice, one has to try a sufficient number of DNA constructs with various lengths of ssDNA tail before the apparent step size can be convincingly determined. If the SSB diffusion-induced FRET fluctuation is too fast to be determined by HMM analysis (see Note 11), one may decrease the experimental temperature (for example, down to 6°C) to slow down the FRET fluctuations by using the water-circulating bath (10). A free software (HaMMy) for HMM analysis can be downloaded at http://bio.physics.illinois.edu/HaMMy.html.

3.4. Single-Molecule Fluorescence-Force Spectroscopy

The TIR measurements in the previous section are useful for studying SSB diffusion at zero force. With a tension applied to the SSB-bound DNA, one can study the process of SSB dissociation from DNA and observe how mechanical forces regulate SSB diffusion on DNA. This method has been successfully used to map the reaction energy landscape of a DNA four-way (Holliday) junction structure (12), to calibrate an in vivo force sensor (17), and to study SSB/DNA interactions (9). After calibrating the optical trap and co-aligning the confocal laser beam with the trapping laser beam as described (11), we can synchronize the movement of the confocal spot with that of the piezo stage such that the confocal spot is able to keep track of the fluorescently labeled molecule under investigation. A thorough guide of assembling a fluorescence-force spectroscopy microscope can be found elsewhere (11).

3.4.1. Preparation of the Anti-digoxigenin-Coated Beads

A detailed protocol to cross-link anti-digoxigenin to protein-G beads can be found elsewhere (9, 11). Items 21–26 in Subheading 2 are required for this cross-linking reaction.

3.4.2. General Sample Preparation and Data Acquisition Protocol

For making the DNA templates for fluorescence-force measurements, we first make biotinylated partial duplex DNA with a 5′ ssDNA tail containing the sequence of 5′-GGG CGG CGA CCT which is complementary to the 12 nt *cos* site of λ-phage DNA. Then a previously described protocol (11) is followed to attach the partial duplex DNA to a long λ-DNA (~16 μm contour length) through one end of the λ-DNA, and to attach a digoxigenin-tagged oligonucleotides to it though the other end of the λ-DNA. The final products are the DNA templates we use in the following sample preparation and data acquisition protocol.

1. Infuse 25 μl of 0.25 mg/ml Neutravidin in T50 buffer into chamber channel and incubate for 5 min.
2. Flush the channel with 50 μl of T50 buffer.
3. Infuse 50 μl blocking buffer. Incubate for 1 h.
4. Infuse 10–20 pM of DNA templates containing λ-DNA, biotin, and digoxigenin tags (see Note 12). Incubate for 30 min.
5. Flush the channel with 100–120 μl of T50.
6. Take 1 μl of the anti-digoxigenin-coated beads (original concentration) and 99 μl of T50 buffer in a centrifuge tube. Complete the buffer exchange from bead storage buffer to T50 buffer by resuspending and centrifuging twice. Infuse 25 μl of the 100-times diluted beads into the channel. Incubate for 30 min.
7. Flush the channel with 100–120 μl T50 buffer.
8. Infuse the imaging buffer B containing desired SSB concentrations and seal the inlet/outlet holes with 5-min epoxy.
9. Load the sample chamber to the fluorescence-force spectroscopy microscope. Use bright field to look at the coverslip surface through a camera and trap a surface-tethered bead using optical tweezers.
10. Find the tethered point (see Note 13) and reset the piezo-stage origin in the tethered point.
11. Move the piezo-stage to a starting position to avoid interference between optical trapping and FRET measurements. Typically the separation is 13–14 μm between the tethered point and the optical trap center (see Note 14).
12. The confocal image around the tethered point is taken by scanning the confocal spot in the sample plane (scan area, 3.2 μm × 3.2 μm) using a steering mirror (11).
13. For the fluorescence-force measurements, the piezo-stage is moved back and forth between the starting position (typically 13–14 μm separation between the tethered point and the trap center) to an end position (16.5–16.8 μm separation between the tethered point and the trap center) at a constant stage-moving speed (455 or 910 nm/s) for several stretching and relaxing cycles. In the meanwhile, the photon counts for Cy3 and Cy5 emissions are recorded by two APDs and the force signal is obtained from QPD signals as a function of time.

3.4.3. Probing SSB Dissociation from DNA

Here we describe two assays using fluorescence-force spectroscopy to study different stages of SSB dissociation from ssDNA.

Initial Stage of ssDNA Unraveling

In the first assay, the experimental configuration is sketched in Fig. 3a. A partial duplex DNA with a 5′ 82 nt of ssDNA overhang ($(dT)_{70}$ + 12 nt *cos* site of λ-DNA) is annealed to λ-DNA through one end of the λ-DNA as mentioned above. A Cy3-Cy5 FRET pair is attached near the two ends of the ssDNA region, separated by 68 nt of ssDNA. A bead held in an optical trap is attached to the other end of λ-DNA. The ssDNA fully wrapping around a SSB tetramer in the $(SSB)_{65}$ mode should result in high FRET due to the closed wrapping topology (Fig. 3a). When applied forces increase from 0.3 pN, SSB-bound ssDNA is unraveled from the protein surface. The extremely long λ-DNA linker prevents one from seeing ssDNA unraveling through the force spectroscopy (see Note 15). FRET efficiency is then used to monitor the initial ssDNA unraveling from SSB surface in this assay.

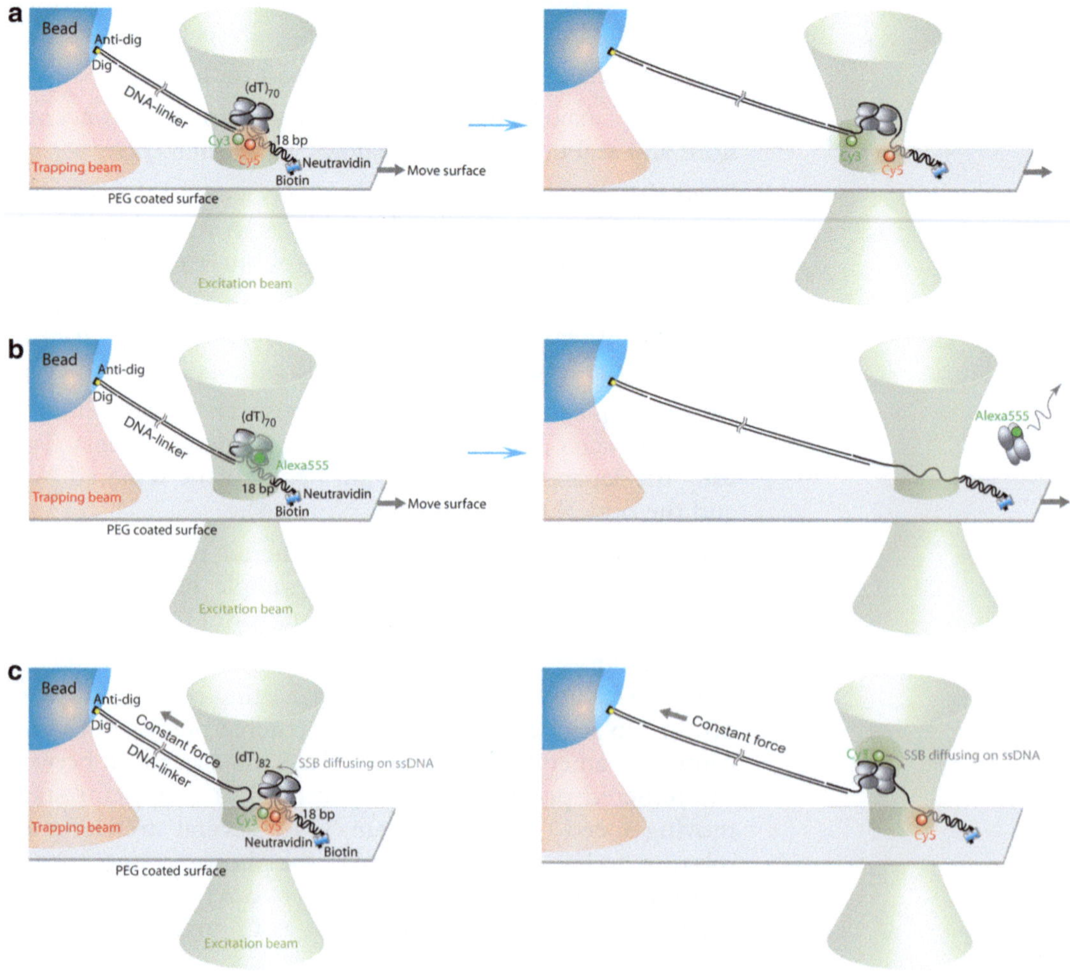

Fig. 3. The assays based on Fluorescence-force spectroscopy. (**a**) The assay to study the initial stage of ssDNA unraveling from a SSB tetramer surface. (**b**) The assay to observe directly the force-induced SSB dissociation. (**c**) The assay to study the mechanical regulation of SSB diffusion on ssDNA.

Direct Observation of Final SSB Dissociation

In the previous assay, FRET efficiency, the only indicator of ssDNA unraveling, drops down to zero if Cy3-Cy5 separation is ~10 nm (this happens at 5–6 pN). However, the final dissociation of SSB occurs at higher forces. Therefore, another assay is required to detect SSB dissociation. In this second assay, the DNA template is the same as in the previous assay but the A122C SSB mutant labeled with ~one Alexa555 fluorophore per SSB tetramer (8) is used instead of the unlabeled wild-type SSB. Cy3 and Cy5 on the DNA template are used for confocol imaging to determine the ssDNA position in the Step 12 of Subheading 3.4.2. Once the ssDNA position is determined, Cy3 and Cy5 are photobleached by the confocol (532 nm) laser. Alexa555 fluorescence becomes observable only upon Alexa555-labeled SSB binding to the surface-tethered ssDNA because proteins free in solution contribute only to the overall background fluorescence. Therefore, the Alexa555 emission recorded by ADP is used as a real-time indicator for SSB binding. The sudden appearance and disappearance of the fluorescence signal represent SSB binding and dissociation events, respectively (Fig. 3b).

3.4.4. Mechanical Regulation of SSB Diffusion

Finally, we present an assay to study how tensions in ssDNA regulate the SSB diffusion rate, which has implications for the SSB diffusion mechanism. The DNA template is the same as in Subheading 3.4.3, step 1 except that a 13 nt poly (dT) extension is inserted beyond Cy3 (Fig. 3c). Different from previous assays, the piezo-stage is kept in several positions for certain period of time for constant force measurements. For example, the piezo-stage can be moved from origin by 13.2, 14.5, 15.0, 15.5, 15.8 μm, respectively and stay in each position for 6 s while recording the Cy3/Cy5 signals as a function of time. Cross-correlation analysis previously described (18) can be used to quantify the SSB diffusion rate under each applied force.

4. Notes

1. The PEGylation buffer needs to be prepared freshly each time before use.
2. Glucose oxidase and catalase has to be added into the imaging buffer immediately before injecting the imaging buffer into the chamber channel in a single-molecule experiment to avoid acidification of solution (19). We make a high concentration of glucose oxidase and catalase mixture (termed "gloxy" solution (10)) in advance.
3. The MES buffer needs to be prepared freshly each time before use.

4. The temperature measured from sample chamber can be different from the temperature set in the water bath circulator.
5. The burning step is to remove the fluorescent organic dirt remaining on the slides or coverslips. For single-molecule prism-based TIR experiments, we image the quartz slide surface so only quarts slides require the burning step. On the other hand, for single-molecule fluorescence-force spectroscopy, we image the coverslip surface so only coverslips require the burning step.
6. This step needs to be done very gently to prevent the bubble generation between the slide and the coverslip.
7. For single-molecule prism-based TIR experiments, we use quartz slides for chamber assembly, whereas for single-molecule fluorescence-force experiments, we use glass slides.
8. Duplex DNA construct is typically annealed by mixing 5 μM of the short (18 nt) biotin strand and 7 μM of the long poly(dT) strand in T50 (sequences are indicated in Subheading 2) and slowly cooling the mixture from 90°C to room temperature for 2–3 h. T50 buffer is used to dilute the annealed DNA product to 30–50 pM.
9. To obtain the DNA construct ($(dT)_{70}$ flanked with duplex DNA) for TIR measurement, one can first immobilize 30–50 pM of biotinylated partial duplex DNA with the 82 nt tail ($(dT)_{70}$ + 12 nt extension) on a sample channel surface as we do in the general sample preparation protocol. After flushing away unbound DNA molecules with T50 buffer, 10 μM of 12 nt oligos (in T50) complementary to the 12 nt extension is infused to the channel. Incubate for 10–20 min at room temperature and flush away the excess unbound 12 nt oligos with 100 μl of T50. Then add imaging buffer A containing 1 nM SSB and incubate for 1 min, followed by a buffer flush with an imaging buffer A containing no SSB.
10. When the separation between the two acceptors is very large compared to their Förster distance, one can ignore any significant FRET between them. This simplifies the calculation of $E_{app,5}$ and $E_{app,5.5}$. However, if the separation between the two acceptors is comparable to their Förster distance, the alternating laser excitation (ALEX) technique should be applied for the three-color FRET measurements (20, 21).
11. The highest time resolution can be achieved is about 30 ms using the electron multiplying CCD camera (iXon, Andor Technology; Model # DV887DCS-BV) for 512 × 512 pixels without any binning. With 2 × 2 binning, we can obtain 8-ms time resolution.

12. Large orifice pipette tips should be used when handling λ-phage DNA to avoid high shearing forces. Additionally, in Steps 4–8, an automated pump (PHD 22/2000 series syringe pump; Harvard Apparatus) and a syringe/tubing system is used for solution injections at a speed of 30 μl/min (10).
13. Move the piezo-stage and obtain the stretching curves in two orthogonal directions in the sample plane. The central position of the stretching curves is the tethering position.
14. The separation of 13–14 μm between the tethered point and the optical trap center typically corresponds to 0.3–1 pN of force applied to the SSB-bound DNA. The reason for such large separation is that the fluorophores (Cy3 and Cy5) are photobleached extremely fast in the presence of both the trapping laser and the 532 nm excitation laser (22).
15. This feature offers a natural constant force clamp for measuring the conformational changes of biomolecules at a constant force without adding a feedback loop system to the instrument.

Acknowledgements

We thank all the members of Ha laboratory for experimental help and discussions. These studies were supported by grants from the National Institutes of Health (RR025341 and GM065367) and the National Science Foundation (0822613 and 0646550). TH is an employee of the Howard Hughes Medical Institute.

References

1. Ha T (2001) Single-molecule fluorescence methods for the study of nucleic acids. Curr Opin Struct Biol 11:287–292
2. Ha T (2001) Single-molecule fluorescence resonance energy transfer. Methods 25:78–86
3. Neuman KC, Nagy A (2008) Single-molecule force spectroscopy: optical tweezers, magnetic tweezers and atomic force microscopy. Nat Methods 5:491–505
4. Ha T, Enderle T, Ogletree DF, Chemla DS, Selvin PR et al (1996) Probing the interaction between two single molecules: fluorescence resonance energy transfer between a single donor and a single acceptor. Proc Natl Acad Sci U S A 93:6264–6268
5. Ashkin A, Dziedzic JM, Bjorkholm JE, Chu S (1986) Observation of a single-beam gradient force optical trap for dielectric particles. Opt Lett 11:288
6. Joo C, Balci H, Ishitsuka Y, Buranachai C, Ha T (2008) Advances in single-molecule fluorescence methods for molecular biology. Annu Rev Biochem 77:51–76
7. Roy R, Hohng S, Ha T (2008) A practical guide to single-molecule FRET. Nat Methods 5:507–516
8. Roy R, Kozlov AG, Lohman TM, Ha T (2009) SSB protein diffusion on single-stranded DNA stimulates RecA filament formation. Nature 461:1092–1097
9. Zhou RB, Kozlov AG, Roy R, Zhang JC, Korolev S et al (2011) SSB functions as a sliding platform that migrates on DNA via reptation. Cell 146:222–232

10. Joo C, Ha T (2008) Single molecule FRET with total internal reflection microscopy. In: Selvin PR, Ha T (eds) Single molecule techniques: a laboratory manual. Cold Spring Harbor Laboratory Press, New York, p 507
11. Zhou R, Schlierf M, Ha T (2010) Force-fluorescence spectroscopy at the single-molecule level. Methods Enzymol 475:405–426
12. Hohng S, Zhou R, Nahas MK, Yu J, Schulten K et al (2007) Fluorescence-force spectroscopy maps two-dimensional reaction landscape of the holliday junction. Science 318:279–283
13. Roy R, Kozlov AG, Lohman TM, Ha T (2007) Dynamic structural rearrangements between DNA binding modes of E. coli SSB protein. J Mol Biol 369:1244–1257
14. Lohman TM, Ferrari ME (1994) Escherichia coli single-stranded DNA-binding protein: multiple DNA-binding modes and cooperativities. Annu Rev Biochem 63:527–570
15. Hohng S, Joo C, Ha T (2004) Single-molecule three-color FRET. Biophys J 87:1328–1337
16. McKinney SA, Joo C, Ha T (2006) Analysis of single-molecule FRET trajectories using hidden Markov modeling. Biophys J 91:1941–1951
17. Grashoff C, Hoffman BD, Brenner MD, Zhou RB, Parsons M et al (2010) Measuring mechanical tension across vinculin reveals regulation of focal adhesion dynamics. Nature 466:263–U143
18. Kim HD, Nienhaus GU, Ha T, Orr JW, Williamson JR et al (2002) Mg2 + -dependent conformational change of RNA studied by fluorescence correlation and FRET on immobilized single molecules. Proc Natl Acad Sci U S A 99:4284–4289
19. Shi X, Lim J, Ha T (2010) Acidification of the oxygen scavenging system in single-molecule fluorescence studies: in situ sensing with a ratiometric dual-emission probe. Anal Chem 82:6132–6138.
20. Lee S, Lee J, Hohng S (2010) Single-molecule three-color FRET with both negligible spectral overlap and long observation time. Plos One 5:e12270
21. Lee NK, Kapanidis AN, Koh HR, Korlann Y, Ho SO et al (2007) Three-color alternating-laser excitation of single molecules: monitoring multiple interactions and distances. Biophys J 92:303–312
22. van Dijk MA, Kapitein LC, van Mameren J, Schmidt CF, Peterman EJG (2004) Combining optical trapping and single-molecule fluorescence spectroscopy: enhanced photobleaching of fluorophores. J Phys Chem B 108:6479–6484

Chapter 6

Sample Preparation Methods to Analyze DNA-Induced Structural Changes in Replication Protein A

Chris A. Brosey, Susan E. Tsutakawa, and Walter J. Chazin

Abstract

Propagation and maintenance of the cellular genome are among the most fundamental cellular processes, encompassing pathways associated with DNA replication, damage response, and repair. Replication Protein A (RPA), the primary single-stranded DNA-binding protein (SSB) in eukaryotes, serves to protect ssDNA generated during these events and to recruit and organize other DNA-processing factors requiring access to ssDNA substrates. RPA engages ssDNA in distinct, progressive binding modes, which are thought to correspond to different functional states of the protein during the course of DNA processing. Structural characterization of these unique complexes has remained challenging, however, as RPA is a multi-domain protein characterized by a flexible, modular organization. Biophysical approaches that are well suited to probing time-varying architectures, such as NMR and small-angle X-ray and neutron scattering (SAXS/SANS), when integrated with computational methods, can provide critical insights into the architectural changes associated with RPA's different DNA-binding modes. The success of these methods, however, is highly contingent upon the purity, homogeneity, and stability of the sample under study. Here we describe a basic protocol for characterizing and optimizing sample conditions for RPA/ssDNA complexes prior to study by SAXS and/or SANS.

Key words: Replication protein A, Single-stranded DNA-binding protein, Oligonucleotide-/oligosaccharide-binding fold, DNA processing, Protein modularity, Solubility screening, Small-angle X-ray scattering, Small-angle neutron scattering, Size-exclusion chromatography, Multi-angle light scattering

1. Introduction

Every organism requires a ssDNA-binding activity to protect ssDNA from chemical and enzymatic assault, while resolving nucleic acid secondary structure that can derail DNA-processing machinery (1–3). In addition to this critical role, the primary eukaryotic single-stranded DNA-binding protein (SSB), Replication Protein A (RPA), also serves as a central scaffold for the regulated assembly, exchange and disassembly of multi-protein

James L. Keck (ed.), *Single-Stranded DNA Binding Proteins: Methods and Protocols*, Methods in Molecular Biology, vol. 922, DOI 10.1007/978-1-62703-032-8_6, © Springer Science+Business Media, LLC 2012

DNA-processing complexes at ssDNA substrates. A heterotrimer, RPA consists of three polypeptide subunits (RPA70, RPA32, RPA14), which contain a total of seven globular domains connected by flexible linkers and a single disordered domain. Three of these domains (70C, 32D, 14) interface through a hydrophobic three-helix bundle to form the trimeric core of RPA from which emanate the N-terminal domains of RPA70 (70B, 70A, 70N), the C-terminal domain of RPA32, and the disordered N-terminus of RPA32 (32N) (4). As with the majority of SSBs, all domains are oligonucleotide-/oligosaccharide-binding (OB) folds with the exception of RPA32C (a winged helix domain). All RPA domains have been characterized individually or in tandem at high structural resolution (5–11). The high-affinity ssDNA binding (K_d 10^{-9} M) of RPA is mediated by a "DNA-binding core" (DBC) containing the four central OB-fold domains (70A, 70B, 70C, and 32D) with an occluded site size of 30 nucleotides and a 5′ to 3′ binding polarity from 70A to 32D (1, 2). Protein recruitment is mediated primarily by domains 70N and 32C and to some extent by the principal DNA-binding domains 70A and 70B (3).

RPA is believed to proceed through three different interaction modes upon binding ssDNA: an initial 8–10 nucleotide binding mode that includes domains 70A and 70B, a 12–23 nucleotide mode that proceeds to engage 70A–70C, and a final 28–30 nucleotide mode that encompasses all four DNA-binding domains (3). Collectively, these discrete interaction modes form a "DNA-binding trajectory" that RPA traverses as it progressively engages ssDNA and presumably modulate interaction with other DNA-processing factors. The intrinsic inter-domain flexibility within RPA-DBC (12, 13), however, has made structural characterization of these protein–DNA complexes particularly challenging. Integration of low-resolution spatial information from small-angle X-ray and neutron scattering (SAXS/SANS) with computational simulation and existing high-resolution structures (14–19) is a powerful approach for obtaining information on both the global architecture and disposition of individual domains in modular, multi-domain proteins (14). Successful application of this integrated approach to any system, though, depends heavily upon the ability to acquire high-quality scattering data, which necessitates a pure, homogeneous, and stable sample (16, 20). DNA-binding proteins such as RPA present a number of challenges for such sample production, due to their sensitivity to ionic strength and reducing environment, their potential to aggregate at high concentrations (in excess of 2–3 mg/mL), and their capacity to bind DNA in multiple states during attempts to form stoichiometric complexes.

Here we outline a general strategy for preparing protein–DNA systems for study by small-angle scattering, using RPA-DBC as a specific example. This strategy is organized into three stages: (1)

screening stable buffer conditions, (2) assessing sample aggregation and stoichiometric complex formation by SEC-MALS, and (3) preparative purification of specific protein–DNA complexes for both SAXS and SANS.

2. Materials

2.1. Optimal Buffer Solubility Screening

2.1.1. Primary pH Solubility Screening

1. Screening buffers (100 mM buffer, 50 mL each, refer to Fig. 1 for complete list).
2. 60-mL syringes and 0.45-μm syringe filters.
3. 50 mL sterile conical tubes.
4. 200 μL RPA-DBC protein prepared at 1 mg/mL in 10 mM HEPES (pH 7.5), 100 mM NaCl, and 10 mM β-mercaptoethanol (βME) (21).
5. 10.0 mM dC_{30} oligonucleotide stock in sterile water.
6. 96-well sitting drop crystallography plates.
7. Transparent packing tape.
8. Tape applicator.
9. Dissection microscope.

		15 min.		1 hour		3 hours		24 hours	
Buffer (100 mM)	pH	1 mg/mL	2 mg/mL	1 mg/mL	2 mg/mL	1 mg/mL	2 mg/mL	1 mg/mL	2 mg/mL
Na/K phosphate	5.0								
Na citrate	5.5								
Na/K phosphate	6.0								
MES	6.2								
Cacodylate	6.5								
MOPS	7.0								
Na/K phosphate	7.0								
HEPES (control)	7.5								
Tris	7.5								
Imidazole	8.0								
Tris	8.5								
CHES	9.0								

Date:
Sample:
Concentration:
Initial Buffer:
Temperature:

Score
0 No precipitation
1 Faint precipitation
2 Partial precipitation
3 Complete precipitation throughout drop

Fig. 1. Sample scoring sheet for primary pH solubility screen.

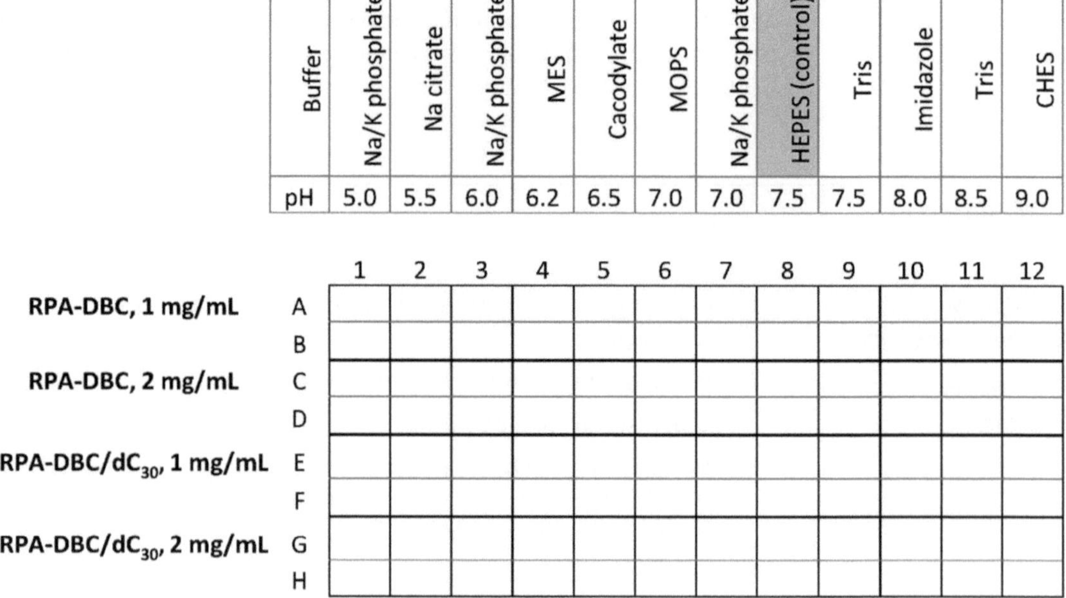

Fig. 2. Organization of 96-well plate for primary pH solubility screen.

2.1.2. Primary Ionic and Additive Solubility Screening

1. Screening buffers (15 mL each, refer to Fig. 2 for complete list).
2. 30-mL syringes and 0.45-μm syringe filters.
3. 15 mL sterile conical tubes.
4. RPA-DBC protein (Subheading 2.1.1).
5. 10.0 mM dC_{30} oligonucleotide (Subheading 2.1.1).
6. 96-well sitting drop crystallography plates.
7. Transparent packing tape.
8. Tape applicator.
9. Dissection microscope.

2.1.3. Secondary Solubility Screening

1. Screening buffer (Subheading 3.3, refer to Figs. 1 and 2).
2. 40 mL RPA-DBC protein prepared at 0.5 mg/mL (Subheading 2.1.1).
3. 10.0 mM dC_{30} oligonucleotide (Subheading 2.1.1).
4. Amicon Ultra centrifugal concentrators (15 mL, 30 kDa MWCO, Millipore).
5. 2× SDS Loading Buffer: 100 mM Tris–HCl, pH 6.8, 4% (w/v) SDS (electrophoresis grade), 0.2% (w/v) bromophenol blue, 20% (v/v) glycerol, 200 mM βME (added fresh).

2.2. Assessing Sample Monodispersity by SEC-MALS

2.2.1. Preparing, Equilibrating, and Calibrating the SEC-MALS System

1. In-line system for gel filtration chromatography and measurement of UV absorbance, static light scattering, and differential refractive index (see Note 18).
2. Superdex 200 PC 3.2/30 column (GE Healthcare).
3. Screening buffer (Subheading 3.3), supplemented with 0.05% sodium azide.
4. Bovine albumin, monomer, lyophilized (Sigma).
5. 0.22-µm centrifugal spin filters.
6. 96-well round-bottom plate (0.5 mL) (Agilent Technologies).

2.2.2. Sample Preparation and SEC-MALS Acquisition for Concentration Series

1. Amicon Ultra centrifugal concentrators (15 mL, 30 kDa MWCO, Millipore).
2. 20 mL RPA-DBC protein prepared at 0.5 mg/mL (Subheading 2.1.1).
3. 10.0 mM dC_{30} oligonucleotide (Subheading 2.1.1).

2.2.3. Sample Preparation and SEC-MALS Acquisition for Time Series

1. Amicon Ultra centrifugal concentrators (15 mL, 30 kDa MWCO, Millipore).
2. 30 mL RPA-DBC protein prepared at 0.5 mg/mL (Subheading 2.1.1).
3. 10.0 mM dC_{30} oligonucleotide (Subheading 2.1.1).

2.3. Preparative Sample Purification for SAXS and SANS

2.3.1. Preparative Purification for SAXS

1. FPLC purification system and accessories.
2. Gel filtration buffer (Subheading 3.3).
3. Superdex 200 HR 10/30 column (GE Healthcare).
4. Amicon Ultra centrifugal concentrators (15 mL, 30 kDa MWCO, Millipore).
5. Amicon Ultra centrifugal concentrators (4 mL, 10 kDa MWCO, Millipore).
6. 0.22-µm centrifugal spin filters.
7. 15 mL RPA-DBC protein prepared at 0.5 mg/mL (Subheading 2.1.1).
8. 10.0 mM dC_{30} oligonucleotide (Subheading 2.1.1).

2.3.2. Preparative Purification for SANS

1. FPLC purification system and accessories.
2. Hydrogenated gel filtration buffer (Subheading 3.3).
3. Deuterated gel filtration buffer (Subheading 3.3).
4. Superdex 200 HR 10/30 column (GE Healthcare).
5. Amicon Ultra centrifugal concentrators (15 mL, 30 kDa MWCO, Millipore).
6. Amicon Ultra centrifugal concentrators (4 mL, 10 kDa MWCO, Millipore).

7. 0.22-μm centrifugal spin filters.
8. 15 mL RPA-DBC protein prepared at 0.5 mg/mL (Subheading 2.1.1).
9. 10.0 mM dC_{30} oligonucleotide (Subheading 2.1.1).

3. Methods

This section describes three aspects of optimizing sample homogeneity and monodispersity in preparing a system for study by SAXS or SANS. The first part outlines a screening procedure for selecting and testing buffer conditions to ensure maximum sample solubility and stability under the acquisition environments of each scattering technique. The second part details the application of size-exclusion chromatography in-line with multi-angle light scattering (SEC-MALS) to provide further refinement of buffer conditions by assessing the potential for sample aggregation. The final section describes a scheme for preparative purification of protein–DNA complexes under refined buffer conditions prior to SAXS or SANS data acquisition.

3.1. Optimal Buffer Solubility Screening

The primary buffer screening assay is based upon that described by Howe (22) and implements a 96-well crystal screening format using a minimal amount of protein. Primary screening is performed in two separate, progressive rounds to assay first for optimal pH, then ionic strength and common additives. Successful buffer conditions revealed by primary screening are then tested on a larger scale at concentrations, temperatures, and timescales relevant to SAXS and SANS, respectively. Screening should be performed in parallel on both the protein alone and in complex with DNA, as we have found that binding of ssDNA substrates improves the stability of RPA. Consecutive rounds of primary and secondary screening will each require 2–3 days to complete.

3.1.1. Primary pH Solubility Screening

1. Prepare 50 mL each of 100 mM buffer stocks (Fig. 1), adjust to target pH as needed, and filter at 0.45 μm into sterile conical tubes (see Notes 1 and 2).
2. Prepare 200 μL stocks of *fresh* RPA-DBC and RPA-DBC/dC_{30} complexes at concentrations of 1 and 2 mg/mL in a buffer containing 10 mM HEPES (pH 7.5), 100 mM NaCl, and 10 mM βME. RPA-DBC/dC_{30} stock can be prepared by mixing RPA-DBC with 1.1-fold excess dC_{30} oligonucleotide (see Notes 3–7).
3. Collect two fresh 96-well plates and clear all dust and fibers with pressurized air. Mark and divide the plates into four sections by

rows (i.e., 2 rows per section, Fig. 2). Fill each reservoir of the first row with 100 μL of each test buffer solution and repeat for the remaining rows. For the reservoir designated as "Control," provide the starting buffer for RPA-DBC (see step 2).

4. Distribute 1 μL RPA-DBC stock at 1 mg/mL into the wells of the first two rows (A, B) and repeat for RPA-DBC at 2 mg/mL for rows C and D (Fig. 2). Repeat for RPA-DBC/dC_{30} stock at 1 and 2 mg/mL for the rows that remain. This should provide replicate tests of each buffer at two sample concentrations.
5. Draw up 1 μL buffer from the reservoir, add to the protein drops, and mix *carefully* (see Note 8).
6. Apply transparent packing tape over the surface of each plate and seal thoroughly to prevent evaporation.
7. Incubate plates at 4°C (control) and 25°C and visually assess each drop for precipitation with a dissection microscope (according to the metric provided in Fig. 1) for 15 min, 1 h, 3 h, and overnight incubation (see Note 9).
8. Once the assay has been completed, a target pH should be selected based upon conditions at 25°C that reflect scores in the range of 0–1 (faint or no precipitation) over the course of 3 h (see Note 10) and show an improvement upon the starting buffer (control reservoir). If the screen offers no improvement relative to the starting buffer control, screening may proceed to the next stage with the original buffer and pH (here, HEPES at pH 7.5). Should the screening fail to return any satisfactory conditions, the screen may need to be repeated utilizing a different starting buffer for the protein (i.e., higher ionic strength, etc.).

3.1.2. Primary Ionic and Additive Solubility Screening

1. Prepare 15 mL of a buffer containing 100 mM of the buffer stock selected from the pH screen combined with 100, 200, or 500 mM NaCl or KCl (Fig. 3). Prepare a parallel set of identical buffers containing 100 mM buffer, 100 mM NaCl, and 2, 5, or 10% glycerol. Adjust to target pH as needed and filter at 0.45 μm into sterile conical tubes. This should provide a total of twelve buffers (Fig. 3, see Notes 2 and 11).
2. Collect two fresh 96-well plates and clear all dust and fibers with pressurized air. Mark and divide the plates into four sections by rows (i.e., 2 rows per section, Fig. 3). Fill each reservoir of the first row with 100 μL of each test buffer solution and repeat for the remaining rows. For the reservoir designated as "Control," provide the starting buffer for RPA-DBC (see Subheading 3.1.1).
3. Aliquot 1 μL protein into each well as described in Subheading 3.1.1, carefully mix each well with 1 μL reservoir buffer, and seal the plates with transparent packing tape.

		15 min.		1 hour		3 hours		24 hours	
Buffer (100 mM)	**pH**	**1 mg/mL**	**2 mg/mL**	**1 mg/mL**	**2 mg/mL**	**1 mg/mL**	**2 mg/mL**	**1 mg/mL**	**2 mg/mL**
0 mM NaCl									
100 mM NaCl									
200 mM NaCl									
500 mM NaCl									
0 mM KCl									
100 mM KCl									
200 mM KCl									
500 mM KCl									
(with 100 mM NaCl)									
0% glycerol (control)									
2% glycerol									
5% glycerol									
10% glycerol									

Date:
Sample:
Concentration:
Initial Buffer:
Temperature:

Score
0 No precipitation
1 Faint precipitation
2 Partial precipitation
3 Complete precipitation throughout drop

Fig. 3. Sample scoring sheet for primary ionic and additive solubility screen.

4. Incubate plates at 4°C (control) and 25°C and visually assess each drop for precipitation with a dissection microscope (according to the metric provided in Fig. 1) for 15 min, 1 h, 3 h, and overnight incubation (see Note 9).
5. Once the assay has been completed, a target salt and glycerol concentration should be selected based upon conditions at 25°C that reflect scores in the range of 0–1 (faint or no precipitation) over the course of 3 h (see Note 10) and show an improvement upon the starting buffer (control reservoir). If the screen offers no improvement relative to the starting buffer control, screening may proceed to the next stage with the original buffer conditions. Should the screening fail to return any satisfactory conditions, the screen may need to be refined and repeated with a different set of additives (Table 1, see Note 11).

3.1.3. Secondary Solubility Screening

Once a candidate buffer has been chosen from the primary screens, the buffer must be tested under concentrations and timescales relevant to SAXS (1–10 mg/mL, 15–60 min) and SANS (1–5 mg/mL, 1–12 h). For SAXS, data acquisition is rapid (seconds), and the sample lifetime need only encompass the time required to transport the freshly prepared sample from an on-site preparative laboratory to initiating the experiment at the beamline. For SANS, where the neutron beam flux is much lower and thus acquisition times much longer, the sample lifetime must extend through the hours needed for data collection. In addition to this,

Table 1
Additional additives and counterions for primary screening

$MgCl_2$	$(NH_4)SCN$	NaOAcetate
$MgSO_4$	NaSCN	(NH_4)OAcetate
$CaCl_2$	KSCN	NaCitrate
$MnCl_2$	$NaSO_4$	NaTartrate
LiCl	$(NH_4)SO_4$	NaFormate
Imidazole	Arginine	Glutamate

collection of a SANS contrast variation series (see Note 12) also requires exchanging the sample into a deuterated version of the buffer, which may exhibit pronounced differences in solubility. Below is a procedure for testing sample lifetime under these different conditions, where the sample is first transferred into the target buffer, then monitored for 24 h across a concentration series (1, 2, 5, 10 mg/mL) at room temperature and 4°C.

1. Prepare 4 L of the target buffer, adjust to the target pH, and prechill to 4°C (see Note 13).
2. Dialyze 40 mL of *freshly* prepared RPA-DBC at 0.5 mg/mL into the target buffer overnight at 4°C with gentle stirring.
3. Prerinse a centrifugal concentrator unit (15 mL, 30 kDa MWCO, Millipore) with Milli-Q water (filtered at 18.2 Ω) for 10 min at 3,700 × *g* in a refrigerated (4°C) tabletop centrifuge. Concentrate the dialyzed protein to ~1.6 mL (or restore the concentrate to this volume with flow-through buffer if the volume is less) and measure the concentration by absorbance at 280 nm with a UV–Vis spectrophotometer (see Note 14).
4. Divide the concentrate in half and add 1.1-fold molar excess dC_{30} oligonucleotide to one half of the protein stock to prepare complexes of RPA-DBC and ssDNA. Mix well and incubate for 15–20 min (see Note 15).
5. Prepare two 100-μL aliquots for each sample (RPA-DBC and RPA-DBC/dC_{30}) at the following concentrations: 1, 2, 5, and 10 mg/mL. This should result in 4 concentration series (16 aliquots total, Fig. 4).
6. Collect an initial concentration reading from each aliquot using a Nanodrop (2 μL material required per reading, see Note 16). Incubate one concentration series for each sample (RPA-DBC and RPA-DBC/dC_{30}) at 4°C and the remaining series at room temperature.

		0 min.			30 min.			1 hour		
Temperature	Conc.	Ppt.	Pellet	Conc.	Ppt.	Pellet	Conc.	Ppt.	Pellet	Conc.
4°C	1 mg/mL									
	2 mg/mL									
	5 mg/mL									
	10 mg/mL									
25°C	1 mg/mL									
	2 mg/mL									
	5 mg/mL									
	10 mg/mL									

		3 hours			6 hours			24 hours		
Temperature	Conc.	Ppt.	Pellet	Conc.	Ppt.	Pellet	Conc.	Ppt.	Pellet	Conc.
4°C	1 mg/mL									
	2 mg/mL									
	5 mg/mL									
	10 mg/mL									
25°C	1 mg/mL									
	2 mg/mL									
	5 mg/mL									
	10 mg/mL									

Date:
Sample:
Initial Buffer:

Score
0 No precipitation
1 Faint precipitation
2 Partial precipitation
3 Complete precipitation throughout tube

Fig. 4. Sample scoring sheet for secondary solubility screen.

7. Observe the samples after 15 min for any visible clouding of the solutions. At 30 min, 1 h, 3 h, 6 h, and 24 h (overnight) (Fig. 4) examine the samples for any visible precipitation. Spin the samples 5 min at 16,000 × *g* in a refrigerated Eppendorf tabletop centrifuge. Note if any precipitated material has accumulated at the bottom of each tube, then collect concentration measurements by Nanodrop. Collect 10 μL of each sample and combine with 10 μL 2× SDS-PAGE loading buffer for evaluation by SDS-PAGE.
8. At the completion of the assay, run an SDS-PAGE gel of all samples to ensure an absence of degradation during the incubation periods. Assess fluctuations in sample concentration to select a target concentration and time frame over which sample stability can be assured. As mentioned previously, a minimum sample concentration of 1–2 mg/mL that can maintain stability for 15–60 min (SAXS) or 1–12 h (SANS) is preferred. If the sample fails to do well under scale-up conditions, another round of primary screening can be employed to determine additional additives that may improve stability.

3.2. Assessing Sample Monodispersity by SEC-MALS

Once buffer conditions have been established via the protocol outlined in the previous section, it is critical to confirm that the sample is not invisibly aggregating under these conditions. Size-exclusion chromatography and multi-angle light scattering (SEC-MALS) are arguably the most effective means to confirm sample homogeneity and monodispersity prior to investigation by SAXS or SANS. As in the previous section, monodispersity is assayed for select sample concentrations and incubation periods, dependent upon the results of the secondary screening (Subheading 3.1.3). In the interest of completeness, we describe assessment of the full concentration series (1, 2, 5, 10 mg/mL) and time course (30 min, 1 h, 3 h, 6 h, 24 h) (see Note 17).

3.2.1. Preparing, Equilibrating, and Calibrating the SEC-MALS System

1. Exchange the pump and FPLC system into a solution of 0.05% sodium azide in Milli-Q water filtered fresh at 0.2 μm (see Notes 18, 19). With the flow rate set to 0.01 mL/min, make a drop-to-drop connection to the inlet of the 2.4 mL Superdex 200 PC 3.2/30 column and immediately connect the outlet to the UV detector. Gradually increase the flow rate to 0.04 mL/min in 0.01 mL/min increments (allow 4–5 min equilibration for each new flow rate). Allow the column and system to equilibrate for 1.5 column volumes (CV), approximately 90 min, with purging of the reference cell for the refractive index (RI) unit.
2. Disconnect the column under flow and close off the ends. Exchange the pump and FPLC system into the target test buffer, freshly filtered at 0.2 μm. As before, make a drop-to-drop connection to the inlet of the 2.4 mL Superdex 200 PC 3.2/30 column at 0.01 mL/min and immediately connect the outlet to the UV detector. Gradually increase the flow rate to 0.04 mL/min in 0.01 mL/min increments. Allow the column and system to equilibrate at least 3–4 CV, preferably overnight, to ensure a stable baseline for the RI detector. Do not stop the flow of buffer until all experiments are complete, as the equilibration period will have to be repeated to restabilize the RI baseline.
3. Once the RI baseline has stabilized, prepare a fresh 500-μL solution of monomeric BSA at 10 mg/mL in the target buffer. Filter 50 μL of the stock BSA solution using a 0.22-μm centrifugal spin filter in a refrigerated Eppendorf tabletop centrifuge (see Note 20). Load 45 μL of the filtered BSA solution into a round-bottom 96-well plate and insert into the auto-injection module of the system. Take care not to disturb the column or lines connecting the RI module.
4. Prepare a ChemStation method to inject a 40 μL sample and then run for 98 min at the target flow rate and pressure limit for the Superdex 200 PC 3.2/30 column (here,

0.04 mL/min and 15 bar). This includes 3 min for completing the auto-injection, 90 min of elution time, and 5 min for a COMET cleaning of the light scattering flow cell (see Note 21). Prepare an equivalent recording session in Astra V to record for 90 min with the 5-min COMET cleaning enabled (see Note 22).

5. Implement the Astra recording session, which should pause and await sample auto-injection from the ChemStation software (see Note 22). Initiate the ChemStation method, ensure that there are no issues with the sample auto-injection, and confirm that Astra begins recording data.
6. Once the run has finished, the system should continue to flow buffer through the column according to the rate and pressure limit set prior to starting the run. Save the Astra session (the data is not automatically saved by the program). Apply baseline corrections and define a peak region about the primary monomeric BSA peak, taking care to avoid any contributions from stray higher-order species. Apply peak normalization, alignment, and band broadening against the UV, light scattering, and refractive index traces as described in the Astra V user's guide. Confirm that the average molecular weight (M_r) of the monomeric peak is ~66 kDa and that the polydispersity is 1.000 (see Note 23).
7. Save the calibrated session as a template and start all new experiments from the BSA template.

3.2.2. Sample Preparation and SEC-MALS Acquisition for Concentration Series

1. Prepare 20 mL of *freshly* purified RPA-DBC at 0.5 mg/mL in the target buffer and concentrate to ~1 mL in a prerinsed centrifugal concentrator unit (15 mL, 30 kDa MWCO, Millipore) (see Note 14). Assess protein concentration by absorbance at 280 nm with a UV–Vis spectrophotometer. Divide the RPA-DBC stock in half and add 1.1-fold molar excess dC_{30} oligonucleotide to one half; mix well and incubate for 20–30 min (see Note 15).
2. Prepare 100-μL aliquots of each sample at 1, 2, 5, and 10 mg/mL and store on ice at 4°C. Filter 50 μL of the first sample solution (RPA-DBC, 1 mg/mL) using a 0.22-μm centrifugal spin filter in a refrigerated Eppendorf tabletop centrifuge. Load 45 μL of the filtered sample solution into a round-bottom 96-well plate and insert into the auto-injection module of the system. Take care not to disturb the column or lines connecting the RI module.
3. Initiate the Chemstation and Astra sessions as before (Subheading 3.2.1) to inject the sample and record the subsequent elution profile. Once the run has finished, save the Astra session, apply baseline corrections and define peak regions for all peaks present on the chromatogram. The normalization,

alignment, and band broadening procedures need only be applied to the calibration run and should already be encoded in the experiment template used for the subsequent runs on RPA-DBC.

4. In the results section of the Astra session, examine the average M_r and polydispersity (Is the M_r close to the expected value? Is the polydispersity close to 1.000 with a relatively low percent error?). In the EASI graph section of the program, inspect the light scattering chromatogram versus that for UV absorbance. (Is the primary peak symmetric? Does the peak tilt or are there any shoulders present? Does the sample elute in the void volume?) Using the EASI graph feature, plot the molar mass over the light scattering chromatogram and determine the consistency of the molar mass across the span of the eluted peak. (Is the molar mass estimation flat across the width of the peak or is it sloped?) Compare the molar mass distribution with that seen for the BSA calibration run. Aggregation can be manifest as an upturn in molar mass toward the early eluting edge of the peak or the presence of a peak in the corresponding void volume of the column (see Note 24).
5. Repeat the run procedure for the remainder of the concentration series and overlay the resulting light scattering chromatograms to determine if there is a concentration-dependent appearance of aggregation at the void volume and/or upturn in the molar mass.

3.2.3. Sample Preparation and SEC-MALS Acquisition for Time Series

1. Prepare 30 mL of *freshly* purified RPA-DBC at 0.5 mg/mL in the target buffer and concentrate to ~1.5 mL in a prerinsed centrifugal concentrator unit (15 mL, 30 kDa MWCO, Millipore). Assess protein concentration by absorbance at 280 nm with a UV–Vis spectrophotometer. Divide the RPA-DBC stock in half and add 1.1-fold molar excess dC_{30} oligonucleotide to one half; mix well and incubate for 20–30 min (see Note 15).
2. Prepare 6 100-μL aliquots at the target concentration determined previously (Subheading 3.2.2). These will correspond to incubation periods of 30 min, 1 h, 3 h, 6 h, and 24 h at room temperature. Because each SEC-MALS run requires 1.5–2 h, the incubation periods must be staggered accordingly.
3. Sample filtration and loading are performed as described (Subheading 3.2.2), as well as data analysis, once each run has been saved. Once all runs have been completed, an overlay of light scattering chromatograms should reveal if there is a time-dependent appearance of aggregation at the void volume and/or upturn in the molar mass.

3.3. Preparative Sample Purification for SAXS and SANS

As a final precaution against aggregation and to promote homogeneous formation of protein–DNA complexes, all scattering samples should be passed over a final gel filtration column prior to data acquisition. Because of the reduced sample requirements for SAXS, the elution profile can be finely parsed and sampled to select for the most homogeneous portion of the eluting peak (see Note 25). The increased sample needs for SANS, though, may require combining a broader section of the eluting peak; thus minimizing all forms of aggregation during buffer optimization is particularly critical (see Note 26). In this last section, we describe basic protocols for preparative gel filtration of samples for SAXS and SANS, as well as preparation of the resulting gel filtration fractions for data acquisition.

3.3.1. Preparative Purification for SAXS

Sample loading and elution from the gel filtration column are implemented using an automated FPLC method. This protocol is designed to generate samples across the base of the main eluting peak, with concentration series based upon each fraction (1×, 2×, and 4×) (see Note 27).

1. Equilibrate the Superdex 200 HR 10/30 column with 1.5 CV filtered Milli-Q water (36 mL) and at least 3 CV filtered sample buffer (72 mL). The column should be connected to the system under flow (0.1 mL/min), with gradual increase of the flow rate every 5 min by 0.1 mL/min until the target flow rate of 0.3 mL/min is reached (see Note 28).
2. Prerinse one centrifugal concentrator unit (15 mL, 30 kDa MWCO, Millipore) with Milli-Q water for 10 min at 3,700 × *g* in a refrigerated (4°C) tabletop centrifuge. Concentrate 15 mL *freshly* prepared RPA-DBC protein at 0.5 mg/mL to ~600 μL (see Note 14).
3. Assess protein concentration by absorbance at 280 nm with a UV–Vis spectrophotometer. Divide the RPA-DBC stock in half and add 1.1-fold molar excess dC_{30} oligonucleotide to one half; mix well and incubate for 20–30 min (see Note 15).
4. Filter the first DNA-free sample (~300 μL, ~12 mg/mL) using a 0.22-μm centrifugal spin filter in a refrigerated Eppendorf tabletop centrifuge and load onto the Superdex 200 HR 10/30 column at 0.3 mL/min, collecting 290 μL fractions in Eppendorf tubes for a 1.5 CV (36 mL) elution period (see Note 29). Store the fractions at 4°C as they come off the column.
5. Repeat the purification with the RPA/dC_{30} sample.
6. Using a Nanodrop, assay absorbance for each fraction at 260 and 280 nm to determine protein concentration and to assess DNA-binding for RPA/dC_{30} according to the A_{260}/A_{280} ratio (see Note 30).

7. Select fractions that are 1 mg/mL or greater and reserve 25 μL each for data acquisition (see Note 31). Select an additional 3 fractions from the void volume (typically 7–8 mL into the elution, which should exhibit a flat UV baseline) to serve as samples for buffer background subtraction.
8. Prerinse one centrifugal concentrator unit per fraction (4 mL, 10 kDa MWCO, Millipore) with Milli-Q water for 10 min at 3,700 × *g* in a refrigerated (4°C) tabletop centrifuge. Rinse the concentrators two additional times with gel filtration buffer. Concentrate the remainder of each fraction (~265 μL) to ~130 μL (2×), remove 25 μL for data acquisition, and continue concentration to ~60 μL (4×) (see Note 32).
9. The original fractions (1×) and concentrated fractions (2× and 4×) should be kept at 4°C until transfer to a 96-well plate for SAXS data acquisition (no less than 24 h after gel filtration).

3.3.2. Preparative Purification for SANS

Sample loading and elution from the gel filtration column are implemented using an automated FPLC method. Sample preparation proceeds in three stages: (1) gel filtration exchange of RPA-DBC/dC_{30} into H_2O- and D_2O-based buffers, (2) preparation of concentrated stocks of RPA-DBC/dC_{30} from each purification, and (3) mixing samples for the contrast variation series.

1. Equilibrate the Superdex 200 HR 10/30 column with 1.5 CV filtered Milli-Q water (36 mL) and at least 3 CV filtered hydrogenated sample buffer (72 mL). The column should be connected to the system under flow (0.1 mL/min), with gradual increase of the flow rate every 5 min by 0.1 mL/min until the target flow rate of 0.3 mL/min is reached (see Note 28).
2. Prerinse one centrifugal concentrator unit (15 mL, 30 kDa MWCO, Millipore) with Milli-Q water for 10 min at 3,700 × *g* in a refrigerated (4°C) tabletop centrifuge. Concentrate 15 mL *freshly* prepared RPA-DBC protein at 0.5 mg/mL to ~600 μL (see Note 14).
3. Assess protein concentration by absorbance at 280 nm with a UV–Vis spectrophotometer. Add 1.1-fold molar excess dC_{30} oligonucleotide to the entire stock, mix well, and incubate for 20–30 min. Divide the RPA-DBC/dC_{30} mixture in half.
4. Filter the first RPA-DBC/dC_{30} sample (~300 μL, ~12 mg/mL) using a 0.22-μm centrifugal spin filter in a refrigerated Eppendorf tabletop centrifuge and load onto the Superdex 200 HR 10/30 column at 0.3 mL/min, collecting 500 μL fractions in Eppendorf tubes for a 1.5 CV (36 mL) elution period (see Note 29). Store the fractions at 4°C as they come off the column.

5. Re-equilibrate the Superdex 200 HR 10/30 column with at least 3 CV filtered deuterated sample buffer (72 mL) and proceed with the purification of the remaining RPA/dC_{30} sample as in step 4.
6. Using a Nanodrop, assay absorbance for each fraction at 260 and 280 nm to determine protein concentration and to assess DNA-binding for RPA/dC_{30} according to the A_{260}/A_{280} ratio.
7. A five-point contrast variation series (0, 10, 20, 80, and 100% D_2O buffers) with 325-μL samples will require ~940 μL of H_2O sample stock and ~680 μL D_2O sample stock to generate all five samples. Starting from the fraction containing the peak maximum of each H_2O and D_2O elution profile, determine which fractions from each purification would be required to generate H_2O and D_2O at the target sample concentration (at least 1–2 mg/mL) at the given volumes. Select an additional 3 fractions from the void volume (typically 6–7 mL into the elution, which should exhibit a flat UV baseline) to prepare samples for buffer background subtraction.
8. Prerinse two centrifugal concentrators (4 mL, 10 kDa MWCO, Millipore) with Milli-Q water for 10 min at 3,700 × *g* in a refrigerated (4°C) tabletop centrifuge. Rinse the concentrators an additional time with gel filtration buffer. Pool and concentrate the selected fractions from each purification to the target volumes for each stock (see Note 32).
9. Transfer the concentrates to Eppendorf tubes and spin down any precipitation in a refrigerated Eppendorf tabletop centrifuge for 10 min at 16,000 × *g*. Assess concentration by absorbance at 260 and 280 nm.
10. Prepare 10, 20, and 80% D_2O mixtures of RPA-DBC/dC_{30} and equivalent buffer blanks by volume (325-μL per sample). Measure A_{260} and A_{280} a final time and store samples and the remaining 0 and 100% D_2O stocks on ice at 4°C until acquisition. Transfer samples to clean quartz cells after spinning a final time for 10 min at 16,000 × *g* and allow samples to equilibrate at room temperature for 20–40 min to allow outgassing (see Note 33).

4. Notes

1. Selected buffers for pH testing are based upon those described by Jancarik and colleagues (23) and encompass a range of pH 5–9.
2. Buffer stocks which are not to be used immediately may be stored long term at −80°C.

3. In the absence of ssDNA substrate, RPA-DBC aggregates at concentrations in excess of 2.5 mg/mL. While this critical concentration is likely to vary from system to system, a starting concentration of at least 1–2 mg/mL is preferred, as this is the minimum concentration range recommended for scattering studies. Follow-up screening at higher concentrations can be pursued once stable conditions have been established for the minimum concentration.
4. The concentration of HEPES (10 mM) within the protein buffer is kept low relative to that described in the original purification procedure to ensure that the primary buffering capacity arises from the screening buffer (i.e., 50 mM test buffer versus 5 mM HEPES).
5. Components of the starting buffer will vary and may be subject to optimization as well, depending upon the system of interest. For various RPA constructs, starting buffers have included 50–100 mM NaCl and 2–10 mM BME to provide ionic stabilization to the protein's basic OB-folds and full reduction of the cysteine resides coordinating the metal center of the zinc ribbon of the 70C domain.
6. βME should be added fresh to the buffer prior to use.
7. Protein stocks should be maintained on ice at 4°C.
8. Care is needed in mixing, as this can introduce air bubbles to the protein drops.
9. Precipitation may manifest in a number of ways, whether a fine clouding of the drop or the dramatic appearance of brown/black conglomerates. The metric provided in Fig. 1 reflects the assumption that the severity of the precipitation is reflected by the density of the precipitation within the drop (a few grains versus complete occupation of the drop). For more details, refer to Howe (22).
10. In our experience, the majority of precipitation occurs within the first 15–20 min after mixing; if a drop remains clear beyond this, it is likely to remain relatively clear through the remainder of the incubation period.
11. The choice of the salts NaCl and KCl and the additive glycerol for the second round of screening is based on past successes with various RPA constructs. A list of ionic agents and additives that may be tested in addition to these is provided in Table 1.
12. A SANS contrast variation series encompasses five to six samples of the target protein–DNA complex prepared in buffers containing different mixtures of H_2O and D_2O as solvent (typically 0, 10, 20, 30, 80, and 100% D_2O). Differences in scattering contrast for protein and DNA under these different ratios of H_2O and D_2O allow their respective scattering

contributions to be parsed from the global scattering of the complex (24). Because D_2O has very different physical properties compared to H_2O (including differential hydrogen bonding and acid/base behavior), protein solubility may be negatively affected when substituting D_2O as the primary buffer solvent.

13. Due to the expense of D_2O, we recommend running all secondary solubility testing on hydrogenated buffers first, before proceeding with testing of deuterated buffers. Typically, SANS is contemplated for a protein–DNA complex once successful data acquisition and analysis have been achieved by SAXS; as such, deuterated buffers may not be required for some time. When this stage of the analysis is reached, we recommend embarking on a small-scale version of the secondary testing described here, employing a 100–300 mL dialysis buffer with protein volume scaled accordingly.
14. Concentrators are usually spun in 10–15 min increments with careful mixing upon each addition of protein to prevent build-up and aggregation of protein at the base of the concentrator.
15. In the interest of preserving protein stability prior to exposing a sample to long-term incubation at room temperature, we have historically prepared protein–DNA mixtures on ice. If protein stability is assured, incubation at room temperature is preferred in order to ensure complete equilibration of protein–DNA binding.
16. In our experience, absorbance measurements by UV–Vis spectrophotometer return more accurate assessments of concentration on more dilute protein solutions compared to Nanodrop. Absorbance measurement by Nanodrop, however, requires the expenditure of less protein. In light of this, we describe measuring the initial concentration of the original stock solutions by UV–Vis spectrophotometery, then following subsequent changes in concentration for each series by Nanodrop.
17. As mentioned before for the secondary solubility screening assay, testing of deuterated buffers should be considered only after hydrogenated buffers have been thoroughly optimized and at least one round of study by SAXS has been successfully completed. A 36-h SEC-MALS run using a 2.4 mL Superdex 200 PC 3.2/30 column would be expected to consume ~120 mL of buffer.
18. Our SEC-MALS system includes in-line detectors for ultraviolet absorbance at 280 nm (Agilent 1100 series, Agilent Technologies), static light scattering (DAWN HELEOS 8+, Wyatt Technology), and differential refractive index (Agilent 1200 series, Agilent Technologies). The system is equipped with an automated injection system (Agilent 1100 series, Agilent

Technologies). Operation of the FPLC system is managed by Agilent ChemStation software, while data recording and analysis are carried out by Astra V (version 5.3.4.18).

19. It is critical that all solutions introduced to the FPLC system are freshly filtered and supplemented with 0.05% sodium azide to prevent any particulate matter from potentially clogging the light scattering flow cell and to inhibit microbial growth and contamination of the system.
20. Allow approximately 15 and 30 min, respectively, for the light scattering laser and UV lamp to warm up prior to final preparation of the first sample.
21. Since all modifications to ChemStation methods are implemented in real time, we recommend preparing and uploading the ChemStation method prior to connecting the column to the instrument for equilibration to prevent unexpected changes in flow rate or maximum pressure allowances.
22. Our system has been setup to initiate recording automatically in Astra once sample auto-injection is complete.
23. If there is a discrepancy in the measured and predicted molecular weights, confirm that the definitions of the peak boundaries exclude any trace aggregation that may elute as a shoulder to the main BSA peak. (We observe slight dimer formation, even with the monomeric standard). If the discrepancy still remains, Astra's default value for the buffer refractive index may not be accurate for the test buffer (the default parameters are based upon phosphate buffered saline). This value can be measured with a refractometer and adjusted under the solvent settings of the experiment.
24. In the case of protein–DNA complexes, signs of aggregation by SEC-MALS may in fact highlight the presence of non-stoichiometric binding (two or more DNA molecules per protein molecule or vice versa). The addition of 1.1-fold molar excess oligonucleotide for RPA is intended to ensure full saturation of all binding sites without providing opportunity for multiple DNA molecules to associate with a single RPA molecule. For example, we find that the addition of threefold molar excess of DNA produces an early eluting shoulder in addition to the primary peak of the complex. If a non-stoichiometric interaction is suspected, we recommend varying the molar ratio of DNA to protein when forming the complex or adjusting the salt concentration of the buffer.
25. Each SAXS sample requires ~20 μL at the target concentration (minimum of 1 mg/mL).
26. A five-point SANS contrast variation series will require five 325-μL samples (~1.6 mL) at a minimum of 1–2 mg/mL. Sample volume assumes the use of 1 mm quartz banjo cells.

27. Collecting SAXS data on a concentration series of the sample allows for assessment of inter-particle interference while maximizing signal sensitivity (17). In our case, a concentration series is preferred to a dilution series, as irreversible aggregation that could be induced at higher concentrations is not propagated to more dilute samples.
28. All gel filtration columns will shed a certain amount of their matrix upon first application of liquid flow pressure and then will stabilize. In order to minimize introduction of this matrix material into the eluting sample, it is recommended that flow be maintained after column equilibration and between individual runs. If the column must be left for an extended period of time, simply reduce the flow rate to 0.05 mL/min and ensure that there is sufficient buffer in the pump reservoir.
29. The concentration of the sample will be diluted after passage through the gel filtration column. If the dilution factor of the column is known for a particular sample volume (a series of calibration runs can establish this), then the loading concentration can be adjusted to guarantee a final concentration of 1–2 mg/mL at the maximum point of the eluting peak.
30. We have also found that confirming concentration by Bradford assay is particularly helpful for protein–DNA complexes, as defining an accurate extinction coefficient for the mixed species can be difficult. If sample material is limited prior to data acquisition, the assay can be performed once the experiment is completed.
31. In our case, the peak for RPA-DBC without ssDNA is contained within three 290-μL fractions, each greater than 1.5 mg/mL.
32. The speed of concentration will depend upon the protein. We recommend spinning in 5 min increments with mixing in between to reduce the risk of overshooting the target concentrations and potentially inducing aggregation.
33. As a cold aqueous sample comes to room temperature, excess dissolved gas will come out of solution and form bubbles on the face of the quartz cells that can interfere with sample scattering.

Acknowledgements

The authors would like to thank Dr. Robert Rambo at Lawrence Berkeley National Laboratory for sharing his extensive expertise in the use of SEC-MALS, as well as Dr. Kevin Weiss and Dr. William

Heller at the Center for Molecular and Structural Biology at Oak Ridge National Laboratory for helpful discussions regarding SANS sample preparation strategies. This work was supported by the National Institutes of Health operating grants R01 GM65484 and P01 CA092584.

References

1. Wold MS (1997) Replication protein A: a heterotrimeric, single-stranded DNA–binding protein required for eukaryotic DNA metabolism. Annu Rev Biochem 66:61–92
2. Iftode C, Daniely Y, Borowiec JA (1999) Replication protein A (RPA): the eukaryotic SSB. Crit Rev Biochem Mol Biol 34:141–180
3. Fanning E, Klimovich V, Nager AR (2006) A dynamic model for replication protein A (RPA) function in DNA processing pathway. Nucleic Acids Res 34:4126–37
4. Mer G, Bochkarev A, Chazin WJ, Edwards AM (2000) Three-dimensional structure and function of replication protein A. Cold Spring Harbor Symposia 65:193–200
5. Bochkarev A, Pfuetzner RA, Edwards AM et al (1997) Structure of the single-stranded-DNA-binding domain of replication protein A bound to DNA. Nature 385:176–81
6. Bochkarev A, Bochkareva E, Frappier L et al (1999) The crystal structure of the complex of replication protein A subunits RPA32 and RPA14 reveals a mechanism for single-stranded DNA binding. EMBO J 18:4498–4504
7. Bochkareva E, Belegu V, Korolev S et al (2001) Structure of the major single-stranded DNA-binding domain of replication protein A suggests a dynamic mechanism for DNA binding. EMBO J 20:612–8
8. Bochkareva E, Korolev S, Lees-Miller SP et al (2002) Structure of the RPA trimerization core and its role in the multistep DNA-binding mechanism of RPA. EMBO J 21:1855–63
9. Jacobs DM, Lipton AS, Isern NG et al (1999) Human replication protein A: global fold of the N-terminal RPA-70 domain reveals a basic cleft and flexible C-terninal linker. J Biomol NMR 14:321–331
10. Bochkareva E, Kaustov L, Ayed A et al (2005) Single-stranded DNA mimicry in the p53 transactivation domain interaction with replication protein A. Proc Natl Acad Sci USA 102:15412–7
11. Mer G, Bochkarev A, Gupta R et al (2000) Structural basis for the recognition of DNA repair proteins UNG2, XPA, and RAD52 by replication factor A. Cell 103:449–456
12. Brosey CA, Chagot M-E, Ehrhardt M et al (2009) NMR analysis of the architecture and functional remodeling of a modular multidomain protein, RPA. J Am Chem Soc 131:6346–7
13. Pretto DI, Tsutakawa S, Brosey CA et al (2010) Structural dynamics and single-stranded DNA binding activity of the three N-terminal domains of the large subunit of replication protein A from small angle X-ray scattering. Biochemistry 49:2880–9
14. Tsutakawa SE, Hura GL, Frankel KA et al (2006) Structural analysis of flexible proteins in solution by small angle X-ray scattering combined with crystallography. J Struct Biol 158:214–221
15. Bernadó P (2010) Effect of interdomain dynamics on the structure determination of modular proteins by small-angle scattering. Eur Biophys J 39:769–80
16. Rambo RP, Tainer JA (2010) Bridging the solution divide: comprehensive structural analyses of dynamic RNA, DNA, and protein assemblies by small-angle X-ray scattering. Curr Opin Struct Biol 20:128–137
17. Putnam CD, Hammel M, Hura GL et al (2007) X-ray solution scattering (SAXS) combined with crystallography and computation: defining accurate macromolecular structures, conformations, and assemblies in solution. Q Rev Biophys 40:191–285
18. Pelikan M, Hura GL, Hammel M (2009) Structure and flexibility within proteins as identified through small angle X-ray scattering. Gen Physiol Biophys 28:174–189
19. Bernadó P, Mylonas E, Petoukhov MV et al (2007) Structural characterization of flexible proteins using small-angle X-ray scattering. J Am Chem Soc 129:5656–64
20. Jacques DA, Trewhella J (2010) Small-angle scattering for structural biology-expanding the frontier while avoiding the pitfalls. Protein Sci 19:642–57
21. Brosey CA, Chagot M-E, Chazin WJ (2012) Preparation of the modular multi-domain protein RPA for study by NMR spectroscopy. Meth Mol Biol 831:181–95

22. Howe PWA (2004) A straight-forward method of optimising protein solubility for NMR. J Biomol NMR 30:283–6
23. Jancarik J, Pufan R, Hong C et al (2004) Optimum solubility (OS) screening: an efficient method to optimize buffer conditions for homogeneity and crystallization of proteins. Acta Crystal Sect D 60:1670–1673
24. Whitten AE, Trewhella J (2009) Small-angle scattering and neutron contrast variation for studying bio-molecular complexes. Meth Mol Biol 544:307–23

Chapter 7

Structural Studies of SSB Interaction with RecO

Mikhail Ryzhikov and Sergey Korolev

Abstract

Interaction of recombination protein RecO with single-stranded (ss) DNA-binding protein (SSB) is essential for DNA damage repair and restart of stalled replication (Cox, Crit Rev Biochem Mol Biol 42 (1):41–63, 2007). To understand mechanism of this interaction and its role in DNA repair, we deciphered a high-resolution structure of RecO complex with C-terminal tail of SSB (SSB-Ct). The structure revealed a key role of hydrophobic interactions between two proteins and suggests the mechanism of RecO recruitment to DNA during homologous recombination and strand annealing.

Key words: Protein crystallization, Atomic resolution structure, Peptide binding, DNA recombination and repair, Selenomethionine protein derivative, Fluorescence polarization

1. Introduction

RecO initiates binding of RecA recombinase to single-stranded (ss) DNA in response to DNA damage or stalled replication (1). Alternatively, RecO promotes annealing of DNA strands. Both reactions are inhibited by the ssDNA-binding protein (SSB). Displacement of SSB from ssDNA is thought to be a main function of RecO. However, SSB physically interacts with RecO (2). An SSB mutant lacking its C-terminal tail (SSB-Ct) inhibits RecO function in recombination initiation, suggesting that RecO binds the SSB-Ct (3). To understand the mechanism of RecO interaction with SSB and its role in DNA repair we determined an atomic resolution structure of *Escherichia coli* RecO (see Note 1) bound to SSB-Ct. Since only part of SSB-Ct peptide was visualized in the structure and since there was an additional site with somewhat similar structural features, where a detergent molecule was localized, extensive mutagenesis studies were conducted to verify localization of the binding site and to study the contribution of hydrophobic and electrostatic interactions for complex formation.

James L. Keck (ed.), *Single-Stranded DNA Binding Proteins: Methods and Protocols*, Methods in Molecular Biology, vol. 922, DOI 10.1007/978-1-62703-032-8_7, © Springer Science+Business Media, LLC 2012

2. Materials

2.1. Buffers

1. Purification *buffer A*: 50 mM HEPES (4-(2-hydroxyethyl)-1-piperazineethanesulfonic acid), pH 8.0, 10% Glycerol, 1 M NaCl, 1 mM TCEP.
2. Storage *buffer B*: 50 mM HEPES, pH 8.0, 40% Glycerol, 0.5 M NaCl, 1 mM TCEP.
3. Protein concentration/crystallization *buffer C*: 0.2 M NaCl, 10 mM Bis-Tris Propane, pH 6.5, 5% dimethyl sulfoxide (DMSO), 5% Sucrose, 0.2 mM CHAPS (zwitterionic detergent, 3-[(3-cholamidopropyl)dimethylammonio]-2-hydroxy-1-propanesulfonate), 0.5 mM TCEP.
4. Peptide binding assay *buffers D*: 25% glycerol, 50 mM NaCl, 20 mM Bis-Tris Propane, pH 8.0, 1 mM TCEP.
5. Peptide binding assay *buffers E*: 25% glycerol, 200 mM NaCl, 20 mM Bis-Tris Propane, pH 8.0, 1 mM TCEP.

2.2. Peptides

1. SSB-Ct peptide had an additional N-terminal tryptophan residue for quantification at 280 nm: WMDFDDDIPF.
2. SSB-Ct-FAM: SSB-Ct with the attached fluorescein at N-terminus.
3. Peptides were solubilized with dimethylformamide (DMF) before experiment.

2.3. DNA

1. pMCSG7 plasmid.
2. Oligonucleotide primers for Ligase-independent cloning.
3. Oligonucleotides for site-directed mutagenesis designed accordingly to QuikChange protocol (Strategene).

2.4. Cells

1. *E. coli* BL21(DE3)pLys cells.

2.5. Purification

1. NiNTA agarose resin.
2. 5 ml HiTrap Heparin agarose column.
3. ACTA FPLC purification system.
4. Tobacco etch virus (TEV) protease (4).
5. SDS-PAGE system.

2.6. Crystallization Screens/Plates

1. 96-Well Corning plates, crystallization screens for initial screening, crystal mounting pins, loops, and vials.

2.7. Peptide Binding

1. 96-Well Greiner Sensoplate.
2. Multimode Microplate Reader.

3. Methods

3.1. Cloning

RecO was cloned from genomic DNA into pMCSG7 plasmid as described (5). Plasmid was transformed into *E. coli* BL21(DE3) pLys cells. Site-directed mutagenesis was performed similarly to QuikChange protocol.

3.2. Purification

E. coli RecO wild-type and mutant proteins were purified as previously described for *D. radiodurans* RecO with an additional purification step on a heparin agarose column (6). Selenomethionine (SeMeth) protein derivatives were obtained according to a previously described protocol (7, 8) using inhibition of methionine pathway synthesis method and was purified similarly to the native protein.

1. Grow cell cultures (4 L) at 37°C in LB medium containing ampicillin and chloramphenicol to an *OD*600 of 0.6, induce by addition of IPTG to 0.5 mM for 15 h at 15°C. Harvest cells by centrifugation at 4,000 × *g* for 20 min at 4°C. Resuspend cell paste in 50 ml of Buffer A with an addition of 10 mM imidazole. Flash freeze in liquid nitrogen and store at −80°C.
2. Thaw frozen cells under cold water and adjust protein solution to 0.1% Triton X-100 and 0.5 mg/ml of lysozyme. Incubate for 15 min at 15°C and for 30 min at 4°C, sonicate, and centrifuge at 30,000 × *g* for 40 min.
3. Incubate supernatant with 5 ml Ni-NTA agarose in 50 ml tube for 60 min at 4°C and load onto a 10 ml gravity column. Wash resin using peristaltic pump with 100 ml of Buffer A plus 10 mM imidazole, 50 ml of Buffer A plus 20 mM imidazole. Elute protein in 20 ml of Buffer A plus 200 mM imidazole.
4. Add 1 mg of TEV protease to eluted protein and dialyze mixture against 500 ml of Buffer A overnight at 4°C. Purify dialyzed protein on 5 ml of Ni-NTA agarose column equilibrated in buffer A plus 10 mM imidazole. Collect flow-through fraction.
5. Dilute protein solution four times by Buffer A without NaCl and load onto a 5 ml HiTrap Heparin agarose column mounted on AKTA FPLC system. Elute protein with salt gradient from 0.1 M to 1.0 M NaCl. Pool together peak fractions. Analyze samples from all the above-described steps on SDS-PAGE.
6. Dialyze protein solution overnight against storage buffer at 4°C. Adjust protein concentration to 1 mg/ml, aliquot, flash freeze in liquid nitrogen, and store at −80°C.

3.3. Crystallization

1. Thaw 600 μl of protein stock on ice and concentrate by centrifugation to 200 μl using Ultrafree-0.5 centrifugal filter device.
2. Add 500 μl of crystallization buffer C and concentrate to 200 μl. Repeat this step 2 times to exchange the buffer.
3. Concentrate solution to final volume of 100 μl with protein concentration of 5 mg/ml (see Note 2). Add 8 μl of SSB-Ct peptide solution (4 mM). Centrifuge the mixture at 16,000 × *g* for 10 min.
4. Set up crystallization trays using vapor diffusion sitting drops method in 96-well Corning plates. Dispense 50 μl buffer solutions from 96-well high-throughput crystallization screens (Hampton Research) into bottom reservoirs of each well using 100 μl multichannel pipetman. Dispense 1 μl of protein in the protein wells with multichannel pipetman and immediately mix with 1 μl of crystallization buffer. Repeat for all wells and seal the plate. Store crystallization plates in incubator at 20°C and periodically examine (see Note 3).
5. Design optimization screens based on conditions where microcrystals appeared by systematically varying all parameters of solution using either 24 or 96 format screens.
6. Set up optimization screens using 96-well Corning plates similarly to step 4. If less than 96 conditions were tried, use single-channel pipetman.
7. Perform "streak-seeding" procedure with a subset of conditions to speed up optimization process. Touch the old crystallization drop with microcrystals by clean cat whisker and immediately dip it in fresh crystallization drops. Minimize time of evaporation from crystallization drop (see Note 4).
8. Transfer single crystals with dimensions larger than 20 μm to cryoprotectant solution and flash freeze in crystal mounting loop in liquid nitrogen (see Note 5).
9. Save crystals producing reasonable diffraction pattern in liquid nitrogen. Freeze additional similar crystals and store in liquid nitrogen without testing for collecting data at synchrotron beamline facilities (see Note 6).

3.4. Structure Determination

1. Test crystals on in-house X-ray generator (see Note 7). Collect final data sets at a synchrotron beamline.
2. Collect single wavelength data set from native protein crystals. Use microbeam (20–50 μm) to collect data from different parts of crystal (particularly for rod-shaped crystals) to obtain maximum resolution in case of significant radiation damage.
3. Collect multiwavelength anomalous data sets from SeMeth protein crystals, unless significant radiation damage occurs.

4. Process data (9). Use data from native crystals for molecular replacement (e.g., Phaser within CCP4 program suit (10)) structure solution using known structure of *D. radiodurans* RecO (see Note 6).
5. Use SeMeth protein crystal data for identification of heavy atom positions and calculation of initial phases (see Note 8) using a program suite such as Phenix (11). Enter both SeMeth data and highest resolution native data sets to perform density extension/modification and automated model building to verify correctness of solution.
6. Subject obtained phases for automated model building (e.g., with Arp/Warp program (12)). Complete model building and refinement using programs such as Coot (13) and Refmac (14). Calculate composite omit maps during model building using a program such as CNS (15) (see Note 9).
7. Inspect electron density maps for the presence of additional high-electron-density areas corresponding to potential surface-bound molecules. Model corresponding molecules and refine structures. Verify correctness of the model using annealing omit-map procedure as implemented in CNS (see Note 10).
8. Deposit coordinates into Protein Data Bank (http://www.pdb.org) (16).

3.5. Peptide Binding

Equilibrium binding studies of RecO interaction with SSB-Ct were performed to verify localization of the SSB-Ct binding site (see Note 11) and to reveal the role of hydrophobic and electrostatic interactions.

1. Dialyze 500 μl of 29 μM (0.8 mg/ml) RecO protein in a 2,000 MWCO membrane overnight at 4°C against assay buffer D or E with 50 mM Arg-HCl and 50 mM NaGlu (see Note 12).
2. Post dialysis, aggregates were removed by centrifuging the protein for 30 min at 16,000 × *g*, 4°C (see Note 13).
3. Protein concentration post dialysis was 24 μM (0.7 mg/ml) as measured by the absorbance at 280 nm with extinction coefficient 24,595 M^{-1} cm^{-1} for RecO.
4. Solubilize lyophilized SSB-Ct-FAM with DMF.
5. Dilute the solubilized peptide into assay buffer (see Note 14).
6. Measure the absorbance of the diluted SSB-Ct-FAM at 280 nm and 493 nm. Determine the FAM concentration with 493 nm and excitation coefficient of 70,000 M^{-1} cm^{-1}. Determine the SSB-Ct concentration by subtracting the background FAM fluorescence with the following equation:

$$[\text{SSB-Ct}] = \frac{(\text{SSB} - \text{Ct}_{280} - (C^{*}\text{FAM}_{493}))}{5,500\ \text{M}^{-1}\text{cm}^{-1}}$$

where variable C is the correction factor supplied by manufacturer and 5,500 M^{-1} cm^{-1} is the extinction coefficient of Trp. The SSB-Ct_{280} and FAM_{493} are the absorbencies of SSB-Ct-FAM at 280 nm and 493 nm, respectively.

7. Pipette 153 μl of assay buffer D or E into wells of a 96-well Greiner Sensoplate.
8. Make an 8.8 μM RecO high point by pipetting 112 μl of RecO (24 μM) into 194 μl of assay buffer in one well and mix thoroughly. Make serial dilutions by pipetting 153 μl of the solution from first well to the second well, from second to third, etc. Make alternative dilutions for sufficient coverage of protein concentration range.
9. Pipette 17 μl of 200 nM SSB-Ct-FAM into each well with a final volume of 170 μl per well and final concentration of 20 nM. Mix thoroughly and slowly to avoid air bubbles and aggregation.
10. Incubate the plate for 30 min in the dark at room temperature.
11. Measure the fluorescent polarization (FP) with a plate reader (Synergy 2, BioTek Inc.) using excitation at 485 nm and emission at 528 nm. Convert fluorescent polarization to anisotropy (A) using equation:

$$A = \frac{2^{*}\text{FP}}{3\text{-FP}}$$

Normalize anisotropy using the following equation:

$$A = \frac{A_i \mid A_0}{A_0}$$

where A_0 and A_i are the fluorescence anisotropy values for free substrate and for each titration point correspondingly. Plot the titration curve using SYSTAT SigmaPlot with X-axis set to RecO concentration (μM) and Y-axis set to normalized anisotropy. Determine the apparent binding constant (K_D) with BioKin, Ltd. Dynafit using a single-site binding equilibrium:

$$[\text{RecO}] + [\text{SSB-Ct}] \Leftrightarrow [\text{RecO}][\text{SSB-Ct}]$$

4. Notes

1. Previously, we have solved the structure of RecO homolog from *Deinococcus radiodurans* (6). While this homolog is more soluble and suitable for crystallization experiments, sequence homology is low and interaction between *D. radiodurans* RecO and SSB has not been documented. Therefore, *E. coli* RecO was selected for structural studies of interaction with SSB.

2. *E. coli* RecO is poorly soluble at low salt. Quick buffer exchange was more efficient than overnight dialysis due to a protein precipitation problem. Overconcentration during buffer exchange can cause precipitation and loss of protein. Due to partial loss of protein during buffer exchange and concentration final volume varied in the range of 80–120 μl with protein concentration of 4–5 mg/ml. Extended storage of the protein at −80°C resulted in decrease of ability to form single crystals. The problem was more significant for selenomethionine derivative. Crystallization trays with SeMeth protein were set up in a few days after each protein preparation.
3. Identical trays were also stored in incubators at 15°C and 4°C. Needlelike microcrystals appeared in several conditions after 2 weeks of incubation at 20°C.
4. Initial crystals were obtained with buffer containing 20–25% PEG 4 K and Bis-Tris Propane, pH 5.5. Optimization yielded rod-shaped and rhomboidal crystals with maximum dimensions of 40 × 40 × 100 μm and diffracted to 3.5 Å resolution on X-ray generator.
5. Crystals from conditions described in Note 4 were transferred in solution of 10% Glycerol, 10% Ethylene glycol, 30% PEG 4 K, 0.1 M BTP, pH 5.5 for 2 min before freezing. Alternatively, freezing in liquid nitrogen gas stream was also tried, as well as collecting limited data from crystals at room temperature in quartz capillaries.
6. 3.0 Å resolution data were collected at APS, GM/CA 23ID-D beamline (see below). Initial attempts to solve the structure with the molecular replacement method using the structure of *D. radiodurans* RecO were not successful. Therefore, subsequent crystallization experiments were performed with selenomethionine protein as well. In addition to the above-described crystallization conditions, a second set of crystallization conditions were later identified with 1.9–2.2 M DL-Malic acid, pH 7.0 (Hampton Research). Both native protein and selenomethionine derivative proteins were subjected to an additional optimization using both crystallization conditions. Crystals obtained in sodium malonate conditions were cryoprotected in solution of 3.1 M DL-Malic acid, pH 7.0. Although more difficult to reproduce, these crystals diffracted significantly better and were used for final data collection and to solve the structure of the complex.
7. Due to the low size of crystals, only low-resolution data were collected on in-house generator.
8. Initial phases were obtained using 2.8 Å resolution single wavelength data set (SAD) collected from SeMeth derivative crystal (Na-malonate crystallization conditions). Although three data

sets from SeMeth crystal were collected at selenium absorption peak and remote wavelengths, MAD phasing did not yield solution likely due to radiation damage of the crystal indicated by the partial loss of diffraction power. Density modification with resolution extension was performed using 2.3 Å native data set.

9. Arp/warp calculations resulted in a model comprising ~80% of the structure. There are two molecules of RecO per asymmetric unit. Since high-resolution data provided sufficient amount of experimental data, the non-crystallographic averaging was not implemented during refinement.

10. The strongest density was located between N-terminal OB fold and α-helical domains. While structure of few C-terminal residues of the peptide could be built into this density, the fitting was relatively poor. In contrast, this density ideally fitted the structure of a CHAPS molecule, a detergent presented in crystallization solution. Interestingly, this density was found at the surface of only one molecule in asymmetric unit. Additional areas of electron density were identified in both molecules at the pocket of an α-helical domain near a C-terminus of the protein. The densities were similar in both molecules and last three amino acids of the peptide (Ile-Pro-Phe) where built into each density. Isoleucine was poorly ordered and there was no density for the rest of the peptide.

11. The CHAPS-binding site is characterized by the similar features to that of theSSB-Ct binding site: hydrophobic pocket surrounded by positively charged surface area. Two experiments were conducted to verify that this is not an SSB-Ct binding site. First, it was shown that CHAPS does not inhibit SSB-Ct binding even at ten times higher concentration than used in crystallization. Second, residues involved in CHAPS binding were mutated and tested for the peptide binding.

12. Addition of 50 mM Arg-HCl and 50 mM NaGlu minimizes protein aggregation during dialysis from storage buffer into low-salt assay buffer. Dialysis into low-salt buffer generally results in protein aggregation. This can also be mediated by diluting the protein before dialysis if the concentration is too high. The presence of Arg-HCl and NaGlu at low concentrations after dilution of protein for binding assays did not affect peptide binding.

13. Protein aggregates must be removed prior to initiating next steps in order to insure accuracy of the binding assay. Avoid disturbing the bottom of the Eppendorf tube post spin to prevent resuspension of any pelleted aggregates.

14. The peptide must be diluted in an aprotic solvent such as DMF prior to further dilution into an aqueous buffer. Use of DMSO

resulted in gel formation at room temperature. Storage of DMF-solubilized peptide in aqueous buffers containing DMSO also resulted in apparent degradation. Peptide solubility is dependent on the amino acid composition and may require alternative solvents for different peptides.

Acknowledgement

This work was supported by NIH grant GM073837. The coordinates of RecO–SSB-Ct complex and structure factors were deposited to PBD with ID 3Q8D. We are thankful to Olga Koroleva, who performed all cloning and participated in optimization of protein purification and crystallization procedures, to the staff of GM/CA beamline at APS, and, particularly, to Dr. R. Sanishvili for help in optimization of data collection from small crystals.

References

1. Cox MM (2007) Regulation of bacterial RecA protein function. Crit Rev Biochem Mol Biol 42(1):41–63
2. Umezu K, Chi NW, Kolodner RD (1993) Biochemical interaction of the Escherichia coli RecF, RecO, and RecR proteins with RecA protein and single-stranded DNA binding protein. Proc Natl Acad Sci USA 90 (9):3875–3879
3. Sakai A, Cox MM (2009) RecFOR and RecOR as distinct RecA loading pathways. J Biol Chem 284(5):3264–3272
4. Kapust RB, Waugh DS (2000) Controlled intracellular processing of fusion proteins by TEV protease. Protein Expr Purif 19(2):312–318
5. Stols L et al (2002) A new vector for high-throughput, ligation-independent cloning encoding a tobacco etch virus protease cleavage site. Protein Expr Purif 25:8–15
6. Makharashvili N, Koroleva O, Bera S, Grandgenett DP, Korolev S (2004) A novel structure of DNA repair protein RecO from Deinococcus radiodurans. Structure 12 (10):1881–1889
7. Walsh MA, Dementieva I, Evans G, Sanishvili R, Joachimiak A (1999) Taking MAD to the extreme: ultrafast protein structure determination. Acta Crystallogr D Biol Crystallogr 55 (Pt 6):1168–1173
8. Doublie S (1997) Preparation of selenomethioninyl proteins for phase determination. Methods Enzymol 276:523–530
9. Otwinowski Z, Minor W (1997) Processing of X-ray diffraction data collected in oscillation mode. Methods Enzymol 276:307–326
10. McCoy AJ et al (2007) Phaser crystallographic software. J Appl Crystallogr 40(Pt 4):658–674
11. Adams PD et al (2010) PHENIX: a comprehensive Python-based system for macromolecular structure solution. Acta Crystallogr D Biol Crystallogr 66(Pt 2):213–221
12. Langer G, Cohen SX, Lamzin VS, Perrakis A (2008) Automated macromolecular model building for X-ray crystallography using ARP/wARP version 7. Nat Protoc 3(7):1171–1179
13. Emsley P, Lohkamp B, Scott WG, Cowtan K (2010) Features and development of Coot. Acta Crystallogr D Biol Crystallogr 66(Pt 4):486–501
14. Winn MD, Murshudov GN, Papiz MZ (2003) Macromolecular TLS refinement in REFMAC at moderate resolutions. Methods Enzymol 374:300–321
15. Brunger AT (2007) Version 1.2 of the Crystallography and NMR system. Nat Protoc 2 (11):2728–2733
16. Berman HM et al (2000) The Protein Data Bank. Nucleic Acids Res 28(1):235–242

Chapter 8

Investigation of Protein–Protein Interactions of Single-Stranded DNA-Binding Proteins by Analytical Ultracentrifugation

Natalie Naue and Ute Curth

Abstract

Bacterial single-stranded DNA-binding (SSB) proteins are essential for DNA metabolism, since they protect stretches of single-stranded DNA and are required for numerous crucial protein–protein interactions in DNA replication, recombination, and repair. At the lagging strand of the DNA replication fork of *Escherichia coli*, for example, SSB contacts not only DnaG primase but also the χ subunit of DNA polymerase III, thereby facilitating the switch between primase and polymerase activity. Here, we describe a powerful method that allows the study of interactions between SSB and its binding partners by sedimentation velocity experiments in an analytical ultracentrifuge. Whenever two molecules interact, a complex of a higher mass forms that can usually be distinguished from free binding partners by its different sedimentation behavior. As an example, we show how sedimentation velocity experiments of purified proteins can be employed to determine the binding parameters of the interaction of SSB and the χ subunit of DNA polymerase III from *E. coli*.

Key words: Single-stranded DNA-binding protein, Analytical ultracentrifugation, Sedimentation velocity experiments, Protein–protein interaction, DNA polymerase III, DNA replication, Binding isotherm

1. Introduction

1.1. Protein–Protein Interactions of EcoSSB

The tetrameric single-stranded DNA-binding (SSB) protein of *Escherichia coli* (*Eco*SSB) is essential for the survival of the *E. coli* cell, since it is involved in processes such as DNA replication, recombination, and repair (1). By binding to single-stranded DNA (ssDNA), *Eco*SSB protects the nucleic acid from nucleolytic digestion and configures it for the action of enzymes involved in DNA metabolism (2). *Eco*SSB is comprised of three different regions: An N-terminal single-stranded DNA-binding domain, a highly

James L. Keck (ed.), *Single-Stranded DNA Binding Proteins: Methods and Protocols*, Methods in Molecular Biology, vol. 922, DOI 10.1007/978-1-62703-032-8_8, © Springer Science+Business Media, LLC 2012

flexible glycine-rich linker, and a highly conserved C-terminal region, ending in the sequence Asp-Phe-Asp-Asp-Asp-Ile-Pro-Phe. Not only the negatively charged amino acids are highly conserved, but also the hydrophobic tripeptide at the very C-terminus with its last two amino acids being nearly invariant within the bacterial SSB proteins (3). This region has been identified to be essential for the interaction of *Eco*SSB with several proteins involved in DNA replication, repair, and recombination, such as the DNA repair enzymes exonuclease I (4, 5) and uracil DNA glycosylase (6), DNA polymerase V, which is involved in translesion synthesis on damaged DNA (7), the replication restart protein PriA, (8) and the recombination proteins RecQ (9) and RecO (10). Furthermore, the highly conserved C-terminus of *Eco*SSB has been shown to interact with the χ subunit of the main bacterial replication enzyme, DNA polymerase III (11–13). This interaction is of particular importance for the switch between primase and polymerase activity at the lagging strand of the DNA replication fork, since DnaG primase competes with the DNA polymerase for binding to *Eco*SSB (14). The χ subunit of DNA polymerase III is a component of the clamp loader complex, which loads the processivity conferring β-sliding clamp onto the primer-template. Whereas χ is not involved in the actual clamp-loading mechanism, it serves as a linker between the DNA polymerase III holoenzyme and the lagging strand template decorated by *Eco*SSB (15).

1.2. Studying Protein–Protein Interactions by Analytical Ultracentrifugation

A method well suited for the characterization of protein–protein interactions is analytical ultracentrifugation (AUC). It allows for the investigation of reactions in solution, without the necessity of immobilizing one of the interaction partners, which otherwise could result in the occlusion of binding sites. Furthermore, even weak interactions can be detected by AUC, since the complex always sediments in the presence of excess reactants and can therefore dynamically reform during the experiment. This is in contrast to analytical size exclusion chromatography (SEC) where complex and reactants normally get separated, which, if the interaction is weak, will result in a dissociation of the complex. Therefore, for instance, the interaction of *Eco*SSB and the χ subunit of DNA polymerase III in the absence of ssDNA could not be detected by analytical SEC (12), but can be examined using AUC and surface plasmon resonance (12, 13).

Whereas there are two fundamentally different techniques that can be used in AUC, namely sedimentation equilibrium and sedimentation velocity experiments, we will focus on the latter one for the characterization of protein–protein interactions. In this method, an initially uniform solution containing the proteins of interest is subjected to a high gravitational force and the sedimentation velocity of the proteins and complexes is analyzed. There are three optical systems commercially available for AUC at the moment: Absorbance optics, Rayleigh interference optics, and a fluorescence

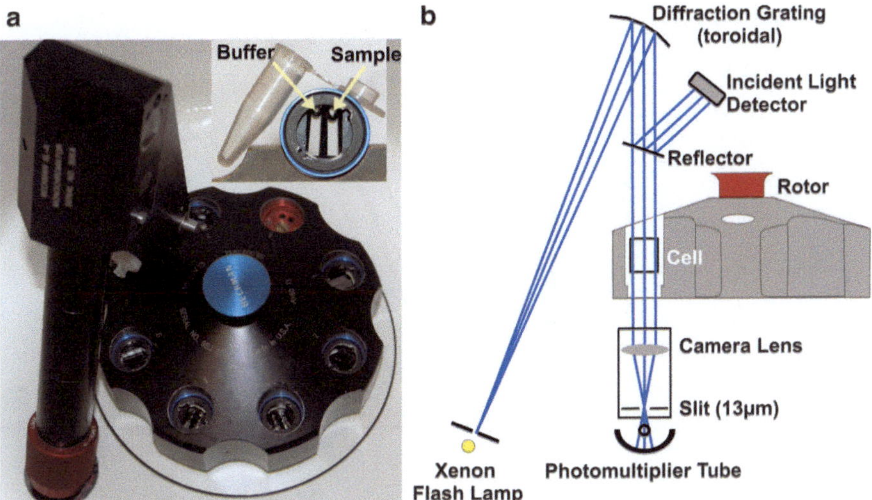

Fig. 1. Analytical ultracentrifugation: Equipment and optics. (**a**) Interior view of the vacuum chamber of a Beckman Optima XL-A containing an 8-hole An-50 Ti rotor and a mounted monochromator. The *inset* shows the top view of a filled analytical ultracentrifugation cell with a double-sector centerpiece. (**b**) Schematic representation of the absorbance optics of the Beckman XL-A/XL-I instruments. During the run, light of a selectable wavelength passes either through the sample or the buffer sector of each cell and the remaining light intensity is measured as a function of radial distance allowing the calculation of the absorbance with a radial resolution of up to 10 μm.

detection system, which allows for excitation at 488 nm (16). Figure 1 shows the absorbance system of a Beckman Optima XL-A. The centrifugation cells contain two sector-shaped compartments, which allow for the measurement of the sample absorbance relative to that of the respective buffer (Fig. 1a). Below the rotor, a movable slit on top of a photomultiplier tube provides a radial resolution of the concentration profile of up to 10 μm (Fig. 1b). Therefore, the movement of the boundary between pure solvent and the protein-containing solution generated by the gravitational force can be easily observed in the analytical ultracentrifuge (Fig. 2a) and can be used to determine the sedimentation coefficient and concentration of the respective species.

Whenever two proteins interact, the molar mass of the complex is higher than that of the reactants, and therefore the complex usually sediments faster than each of the proteins alone. If the equilibration is fast compared to the time scale of sedimentation [dissociation rate constant $>10^{-3}\ s^{-1}$ (17) or kinetic relaxation time <20 s (18)], as observed for the interaction of *Eco*SSB and χ, the law of mass action holds during the whole centrifugation experiment (19). In such a case, the titration of an excess of the slower sedimenting species (referred to as ligand in the following), high enough to detect unbound ligand, to a constant concentration of the faster sedimenting species can be used to evaluate the binding parameters (19). In this scenario, the faster sedimenting species sediments in presence of a constant concentration of ligand and only two sedimenting

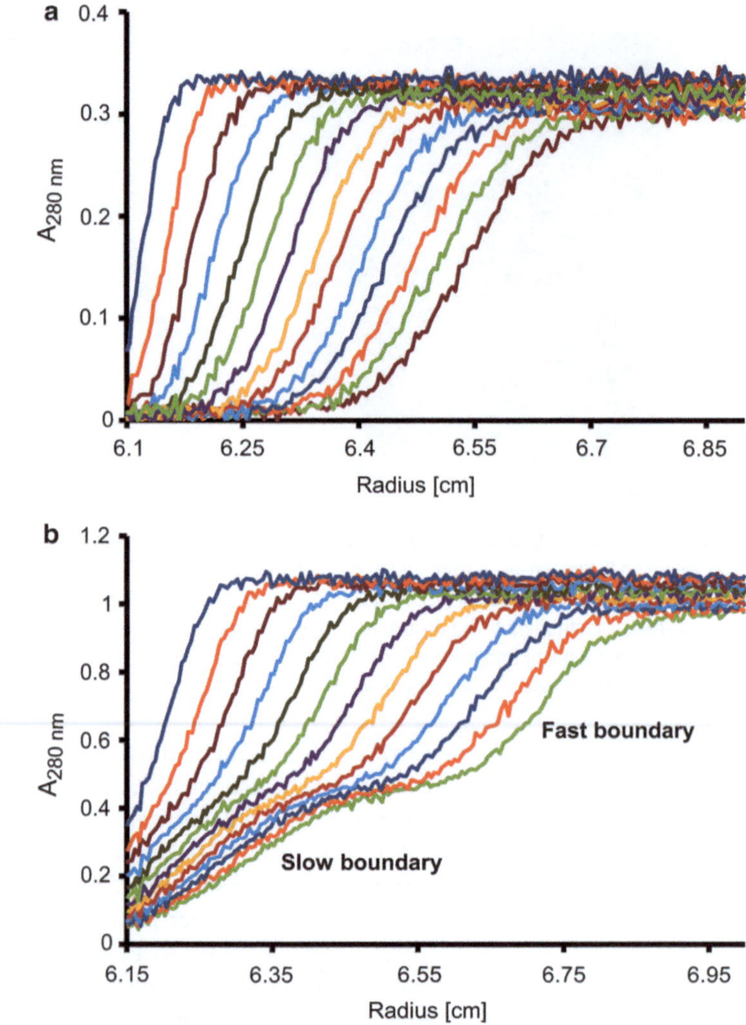

Fig. 2. Sedimentation profiles of free *Eco*SSB and a mixture of *Eco*SSB and *E. coli* χ measured with the absorbance optics of a Beckman Coulter ProteomeLab XL-I. The proteins were sedimented in an An-50 Ti rotor at 50,000 rpm and 20 °C under high salt conditions (20 mM potassium phosphate pH 7.4, 0.3 M NaCl and 0.5 mM DTT). Scans were taken every 8.5 min at $\lambda = 280$ nm. (**a**) When *Eco*SSB (2.5 μM) was investigated, a single sedimenting boundary formed. (**b**) Addition of *E. coli* χ protein (22.5 μM), however, resulted in the formation of two sedimenting boundaries, a slow one, containing free χ protein and a fast one (reaction boundary), representing complexes of *Eco*SSB and χ and free *Eco*SSB.

boundaries can be observed [constant bath approximation (20), Fig. 2b]: A slower one, representing free ligand, and a faster one, the so-called reaction boundary, in which all complexes and the free faster sedimenting component will sediment with a single intermediate sedimentation coefficient (19). The magnitude of this sedimentation coefficient reflects the fractional time that the faster sedimenting species spends in the complex state (17). Therefore, with increasing concentration of free ligand, the s-value of the reaction boundary rises from that of the pure faster sedimenting species to the s-value of the completely saturated complex. From the amplitude of the slower sedimenting boundary, however, the concentration of free ligand can be estimated and since the total ligand concentration is known, the concentration of bound ligand can be calculated (19).

1.3. Determination of Binding Parameters of SSB/Protein Interactions by Sedimentation Velocity Experiments

A very powerful method to evaluate sedimentation velocity data obtained in AUC is the calculation of a diffusion-corrected differential sedimentation coefficient distribution [$c(s)$ distribution] by the program SEDFIT (21). In the $c(s)$ method, the experimentally derived sedimentation data are matched with the best possible combination of sedimentation patterns from differently sized species, taking the diffusion broadening of the sedimenting boundary into account (17). Therefore, even species that differ only slightly in sedimentation coefficient can be separated. Examples of $c(s)$ distributions obtained from the analysis of sedimentation velocity data of *Eco*SSB in the presence of several concentrations of *E. coli* χ are given in Fig. 3a. It can be seen that the peaks of free χ and free *Eco*SSB are well separated and that in the presence of χ the sedimentation coefficient of *Eco*SSB shifts to higher values, indicative of a fast reaction, resulting in complex formation (see above). Furthermore, the area under the faster sedimenting peak increases with increasing χ concentration, since bound χ and *Eco*SSB co-sediment in this reaction boundary. In contrast, the same experiment performed with an *Eco*SSB mutant impaired in χ binding (22), shows only a marginal shift in the sedimentation coefficient of mutant *Eco*SSB and, correspondingly, the area under the faster sedimenting peak remains nearly constant (Fig. 3b).

Since the area under a peak in a $c(s)$ distribution gives the total signal of the material sedimenting within this peak (17), a $c(s)$ distribution can be used to determine the concentrations of the species that are separated by the centrifugation process. When calculating the concentration from the total absorbance signal of a single peak from a $c(s)$ distribution, one has to take into account that the photometer of an XL-A/XL-I instrument only yields relative absorbances that are, however, linear with concentration in the cell (23). Therefore, values obtained from integration of the peaks of a $c(s)$ distribution have to be corrected by determination of the total absorbance of the samples in a standard spectrophotometer before the run. If $A_{\text{slow}}^{\text{AUC}}$ denotes the absorbance of free ligand determined by integration of the peak of the slower sedimenting boundary and $A_{\text{total}}^{\text{AUC}}$ denotes the sum of the absorbances of slow and fast boundary determined from the $c(s)$ distribution, the corrected absorbance of free ligand A_L can be calculated

$$A_L = \frac{A_{\text{slow}}^{\text{AUC}}}{A_{\text{total}}^{\text{AUC}}} \cdot A_{\text{total}} \quad (1)$$

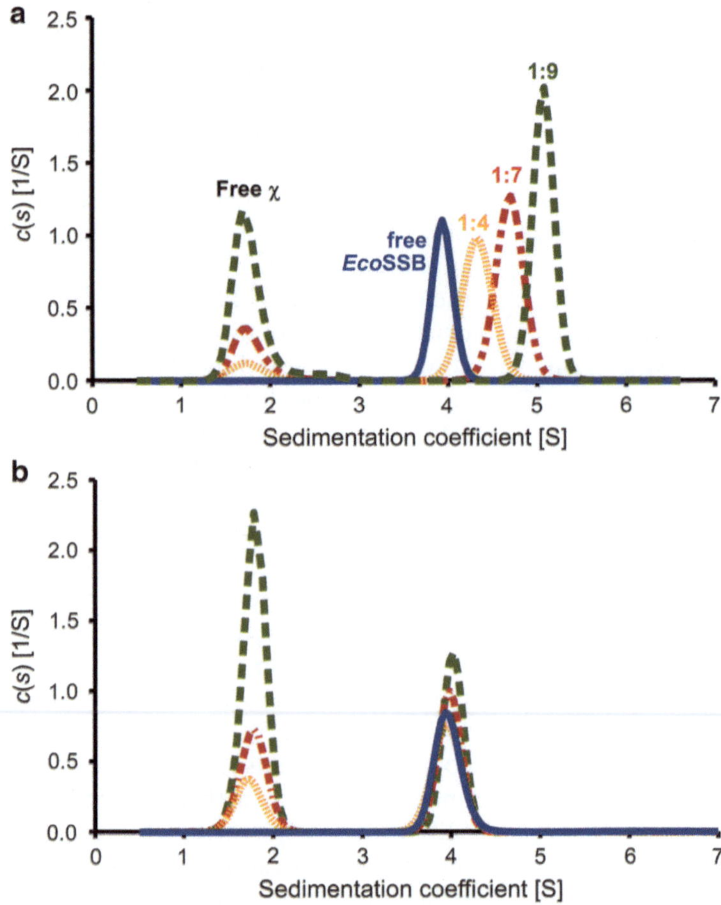

Fig. 3. $c(s)$ distributions for the interaction of *Eco*SSB and *Eco*SSB+Gly with the χ subunit of DNA polymerase III as obtained by SEDFIT analysis. 2.5 μM wild-type *Eco*SSB (**a**) or *Eco*SSB+Gly (**b**) were titrated with different concentrations of *E. coli* χ protein [10 μM: *fine dashed line* (SSB/χ = 1:4), 17.5 μM: *dashed and dotted line* (SSB/χ=1:7) and 22.5 μM: *dashed line* (SSB/χ=1:9)] and analyzed by sedimentation velocity experiments essentially as described in Fig. 2. Free wild-type *Eco*SSB and free *Eco*SSB+Gly sediment with about 3.8 S (*solid lines*). Whereas the peak of wild-type *Eco*SSB shifts significantly upon addition of χ protein (*discontinuous lines* in panel **a**), indicating complex formation, the sedimentation coefficient of the *Eco*SSB+Gly mutant remains nearly unchanged when *E. coli* χ is added.

whereupon A_{total} denotes the total absorbance of the solution measured spectrophotometrically before the run or, if only stock solutions of the interaction partners were measured, the sum of the absorbances of total ligand and total SSB after accounting for dilution. This information together with the molar extinction coefficient of the ligand ε_L and the total concentration of ligand $[L_{\text{total}}]$ can be used to calculate the concentrations of free ($[L_{\text{free}}]$) and bound ($[L_{\text{bound}}]$) ligand

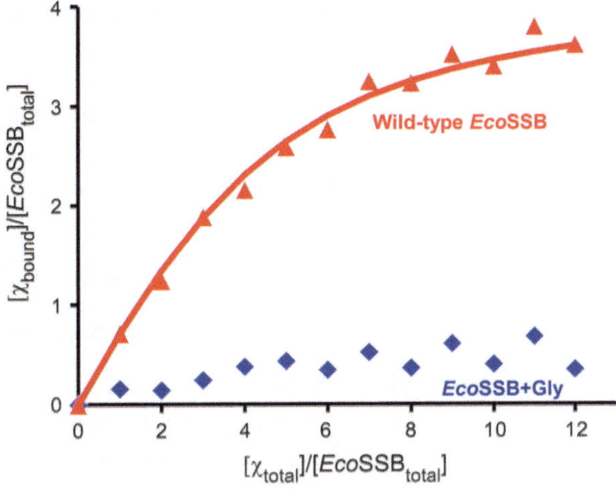

Fig. 4. Binding isotherms for the interaction of *E. coli* χ with *Eco*SSB or *Eco*SSB +Gly under high salt conditions. 2.5 μM of wild-type *Eco*SSB (*triangles*) or *Eco*SSB+Gly (*diamonds*) were mixed with different amounts of *E. coli* χ and analyzed in sedimentation velocity experiments essentially as described in Fig. 2. Subsequent *c*(*s*) analysis was used to determine the concentrations of free and bound χ (for details see text). The *solid line* represents a theoretical binding isotherm calculated for a simple interaction model of *n* molecules of χ with one *Eco*SSB tetramer using the following parameters: $n = 4.2$, $K = 2.9 \times 10^5\ \mathrm{M}^{-1}$. Extension of *Eco*SSB by a C-terminal glycine residue (*Eco*SSB+Gly) diminished the interaction with χ, so that no binding isotherm could be fitted to the data.

$$[L_{\text{free}}] = \frac{A_L}{\varepsilon_L} \quad \text{and} \quad [L_{\text{bound}}] = [L_{\text{total}}] - [L_{\text{free}}] \tag{2}$$

If this analysis is performed at a constant total concentration of SSB [$\text{SSB}_{\text{total}}$] and at several total concentrations of the slower sedimenting ligand, a binding isotherm can be constructed by, e.g., plotting $[L_{\text{bound}}]/[\text{SSB}_{\text{total}}]$ versus $[L_{\text{total}}]/[\text{SSB}_{\text{total}}]$. Examples of such binding isotherms are shown in Fig. 4, where wild-type *Eco*SSB and the C-terminal extension mutant *Eco*SSB+Gly, which carries a single additional glycine residue at the C-terminus of *Eco*SSB (22), were titrated with increasing amounts of *E. coli* χ. In the case of wild-type *Eco*SSB, the concentration of bound χ rises and nearly reaches saturation at the highest χ concentration used. In the case of *Eco*SSB+Gly, however, the concentration of bound χ increases only marginally and saturation cannot be reached. Since the last two C-terminal amino acid residues of *Eco*SSB bind to a well-defined pocket on the surface of χ (22), the addition of a single glycine residue to the C-terminus of *Eco*SSB dramatically decreases the binding affinity.

In order to determine the binding parameters for the interaction of SSB and its interaction partner, a theoretical binding isotherm can be fitted to the experimental data. Assuming that all binding sites on SSB ($SSB * BS_{free}$) are identical and binding to these sites occurs in a noncooperative manner, the microscopic association constant K_A for each binding site on SSB is given by

$$K_A = \frac{[L_{bound}]}{[SSB * BS_{free}] \cdot [L_{free}]} \quad \text{with} \qquad (3)$$
$$SSB * BS_{free} + L_{free} \rightleftharpoons L_{bound}$$

where $[SSB^*BS_{free}]$ denotes the concentration of free binding sites on SSB. If n is the total number of binding sites on an SSB tetramer and taking mass conservation into account ($c_{total} = c_{bound} + c_{free}$), the association constant can be given as

$$K_A = \frac{[L_{bound}]}{(n \cdot [SSB_{total}] - [L_{bound}]) \cdot ([L_{total}] - [L_{bound}])} \qquad (4)$$

For calculation of the theoretical concentration of bound ligand $[L^{calc}_{bound}]$, a simple quadratic equation has to be solved, yielding

$$[L^{calc}_{bound}] = \frac{n \cdot [SSB_{total}] + [L_{total}] + \frac{1}{K_A}}{2} - \sqrt{\left(\frac{(n \cdot [SSB_{total}] + [L_{total}] + 1/K_A)}{2}\right)^2 - n \cdot [SSB]_{total} \cdot [L_{total}]} \qquad (5)$$

If the total concentrations of SSB and ligand are known, this equation can be used to calculate a theoretical binding isotherm for given values of K_A and n. Therefore, assuming reasonable start values for K_A and n, a nonlinear least-squares fit can be performed in order to obtain the values of n and K_A which fit the data obtained from the sedimentation experiment best. The result of such a fit can be seen in Fig. 4 (solid line) for the interaction of wild-type *Eco*SSB and *E. coli* χ. About 4 χ molecules can bind to one *Eco*SSB tetramer, one to each C-terminus, with a binding affinity of $3 \times 10^5 M^{-1}$. Since in the case of the interaction of *Eco*SSB+Gly and χ the binding is so weak that the binding parameters cannot be determined (Fig. 4), no binding isotherm could be fitted to the data.

2. Materials

2.1. Sample Preparation

1. *Eco*SSB and the χ subunit of DNA polymerase III from *E. coli* were over-expressed in *E. coli* and purified to homogeneity as described previously (22).
2. The stock solutions were diluted to concentrations between 20 and 60 μM and extensively dialyzed against a buffer containing 20 mM potassium phosphate pH 7.4, 0.3 M NaCl, 0.5 mM DTT (high salt buffer).
3. Protein concentrations were determined by measuring the absorbance of the protein solutions at 280 nm in a Jasco V-560 UV/VIS Spectrophotometer equipped with a double monochromator.

2.2. Performing Analytical Ultracentrifugation Experiments

1. Standard 12 mm double sector centerpieces, filled with 400 μl of the respective sample, were used. In order to obtain a reliable binding isotherm, 14 different samples (including free *Eco*SSB and free χ) were examined in two runs using an 8-hole An-50 Ti rotor.
2. AUC experiments were performed in a Beckman Coulter ProteomeLab XL-I at 50,000 rpm and 20°C. The g-force depends on the radial position. Therefore, in preparative ultracentrifugation typically g-force at maximum distance or an average value of the g-force is given. In analytical ultracentrifugation, however, the concentration of a sedimenting protein is measured as a function of the radial position (see Fig. 2). Therefore, the protein experiences at each radial distance a different g-force and no single value for the g-force can be given. Since the rotor speed is indicated together with the radial positions (see Fig. 2), no further information is required.
3. The concentration profiles were measured with the absorbance optics of the centrifuge at a wavelength of 280 nm using the ProteomeLab™ software delivered with the centrifuge.

2.3. Data Analysis

1. Data evaluation was performed with the program SEDFIT (21), which provides a model for diffusion-corrected differential sedimentation coefficient distributions [$c(s)$ distributions].
2. Integration of the peaks of the $c(s)$ distributions in order to determine the concentration of the species represented by the respective peak was carried out with the program Microsoft Excel.
3. Nonlinear least-squares fit of the binding isotherms to the experimental data was performed with the solver function of Microsoft Excel.

3. Methods

3.1. Sample Preparation

1. Centrifuge protein solutions after dialysis against high salt buffer (see Notes 1 and 2) for 15 min at 15,500 × *g* and 4°C in order to remove any aggregates which would otherwise interfere with the correct determination of protein concentration.
2. Measure absorbance spectra of your samples in the range of 320–220 nm and make sure that there are no contributions due to light scattering (see Note 3) or to absorbance of contaminating impurities such as nucleic acids or nucleotides (see Note 4). Use the absorbance measured at 280 nm to determine the concentration of your proteins. The molar extinction coefficient of *Eco*SSB at this wavelength is 113,000 M^{-1} cm^{-1} (24). The molar extinction coefficient of the interaction partner at 280 nm can be estimated from amino acid composition using, e.g., the algorithm of Pace et al. (25), implemented in the program SEDNTERP (26), which can be downloaded free of charge from http://jphilo.mailway.com/download.htm. For the χ subunit of DNA polymerase III of *E. coli* this yields a value of 29,450 M^{-1} cm^{-1}. In order to determine the correct stoichiometry and affinity of the interaction, it is crucial to know the exact protein concentrations.
3. To perform the analysis in the way described here, it is essential to titrate increasing amounts of the slower sedimenting component to a constant amount of the faster sedimenting species (see Note 5). In the case of the interaction of *Eco*SSB and *E. coli* χ, it is *Eco*SSB which sediments faster and therefore a constant concentration of *Eco*SSB, e.g., 2.5 μM, can be titrated with 0–30 μM χ protein. In order to obtain a reliable binding isotherm, at least 14 samples should be analyzed (two runs in an An-50 Ti rotor). For each of the samples add the respective volume of both protein solutions to high salt buffer to yield a final volume of 400 μl.

3.2. Performing Analytical Ultracentrifugation Experiments

1. Assemble seven analytical centrifugation cells using standard 12 mm double-sector centerpieces (see Note 6) according to rotor manual LXL/A-TB-003 of Beckman Coulter delivered with the centrifuge.
2. Fill sample sectors of the centrifugation cells with 400 μl of your sample each and apply 405 μl high salt buffer to the reference sectors. Balance the counterbalance to the cell that will be placed into hole 4 and use cells and the counterbalance to load an An-50 Ti rotor according to rotor manual LXL/A-TB-003 of Beckman Coulter.
3. Install the rotor in the vacuum chamber of the centrifuge, mount the monochromator and start a run with the speed set

to 0 rpm (see Note 7) and wait until a temperature of 20°C (see Note 8) is reached for at least 1 h. Use the ProteomeLab™ software of the XL-A/XL-I instrument to start a single absorbance scan at 280 nm (see Note 9) and 3,000 rpm to check whether your cells are properly filled and whether all settings are correct. Make sure that "Radial calibration before first scan" is checked in order to calibrate the radial scanning system before starting the scan. Do not forget to check off this option before increasing the speed, since rotor stretching at higher speeds will result in incorrect radius determination.

4. Set the speed to 50,000 rpm (see Note 10) and start a velocity run comprising about 60–80 scans using default parameters (continuous mode, one replicate, radial resolution of 0.003 cm) and scan without delay. It will take about 9 h until free χ is completely sedimented to the bottom of the cell.
5. Redissolve the sedimented proteins by shaking the ultracentrifugation cells and transfer the sample with a Hamilton syringe to a reaction tube. Apply an aliquot of each sample corresponding to 5–10 μg protein on an SDS-PAGE and check whether any protein degradation occurred during the experiment.

3.3. Data Analysis

1. Load the data obtained for a single sample into the program SEDFIT (21), which can be downloaded free of charge from http://www.analyticalultracentrifugation.com. In addition, a detailed step-by-step tutorial is available at this Web site, which allows even newcomers to use this powerful program. Choose "continuous $c(s)$ distribution" as a model and set minimal and maximal s-value and resolution, check the parameters which should be fitted (typically frictional ratio, time-independent noise, and meniscus position for absorbance data). First press "run" to check whether the parameters are correctly chosen for your actual data. If you obtain proper results, press "fit" to optimize the selected parameters. Check whether the residuals are evenly distributed and whether your data are well represented by the calculated curves. If the reaction is fast (see Note 11) compared to the time scale of the centrifugation experiment, only two peaks should appear in the $c(s)$ distribution (see Fig. 3a). For further analysis, it is important that the two peaks are well separated in order to determine the concentration of the free slower sedimenting species. Repeat data evaluation for each of your samples.
2. Copy the $c(s)$ distribution tables obtained for the different ratios of SSB and ligand into Microsoft Excel or the spreadsheet program of your choice. Overlay them in a diagram and inspect whether the sedimentation coefficient of SSB increases in presence of increasing amounts of ligand (compare Fig. 3a and 3b).

An increase in sedimentation coefficient of the faster moving boundary is a clear indication for an interaction of both proteins; furthermore, it denotes that the interaction is fast compared to the time scale of the sedimentation experiment (17).

3. Estimate the areas under the peaks of free ligand (slower sedimenting boundary) and the reaction boundary for each $c(s)$ distribution by numerical integration in Microsoft Excel (see Note 12). The peak area of the free ligand depicts the absorbance $A_{\mathrm{slow}}^{\mathrm{AUC}}$ and the sum of the areas of both peaks $A_{\mathrm{total}}^{\mathrm{AUC}}$ (see Eq. 1). Together with the total absorbance A_{total} calculated from your photometric measurement by considering the dilution of both protein solutions, this information can be used to calculate free (see Note 13) and bound ligand concentrations ($[L_{\mathrm{free}}]$ and $[L_{\mathrm{bound}}]$, see Eqs. 1 and 2).

4. When plotting $[L_{\mathrm{bound}}]/[\mathrm{SSB}_{\mathrm{total}}]$ versus $[L_{\mathrm{total}}]/[\mathrm{SSB}_{\mathrm{total}}]$, you will receive a binding isotherm as depicted in Fig. 4. A theoretical binding isotherm can be obtained by calculating $[L_{\mathrm{bound}}^{\mathrm{calc}}]$ using Eq. 5, dividing it by $[\mathrm{SSB}_{\mathrm{total}}]$ and plotting it against $[\mathrm{L}_{\mathrm{total}}]/[\mathrm{SSB}_{\mathrm{total}}]$. Assume reasonable start values for the number of binding sites n and the association constant K_A. To obtain the best fit of the theoretical curve to your experimental data, calculate the sum of square errors $\sum_{i=1}^{k}\left(\frac{[L_{\mathrm{bound},i}]}{[\mathrm{SSB}_{\mathrm{total}}]}-\frac{[L_{\mathrm{bound},i}^{\mathrm{calc}}]}{[\mathrm{SSB}_{\mathrm{total}}]}\right)^2$ and minimize the latter, using the solver function of Microsoft Excel by changing the variables n and $\log_{10}(K_A)$ (see Note 14). In order to receive a reliable value for n, the experimentally derived data should almost reach saturation at the highest excess of ligand (see Fig. 4, triangles). If this is not the case, the experiment has to be repeated at higher protein concentrations (see Note 15). If the protein concentration is too high, however, this will result in a stoichiometric binding curve which no longer allows for the determination of K_A. Therefore, the protein concentration at which such a titration has to be performed depends strongly on the affinity of the interaction and it might be necessary to try several concentrations in order to determine the binding parameters.

4. Notes

1. Since the highly conserved C-terminus of SSB, which is involved in most of the SSB/protein interactions, is negatively charged, it should be checked whether the protein–protein interaction of interest depends on the ionic strength of the

buffer. Some interactions might be markedly reduced under high salt conditions.

2. When using the absorbance optics of the analytical ultracentrifuge, you have to make sure that your buffer does not contain any components that strongly absorb light at the detection wavelength. At 280 nm, higher concentrations of, e.g., DTT, benzamidine and nucleotides are problematic. If it is necessary to use such buffer conditions, the XL-I Rayleigh interference optical system of the centrifuge should be used. In this case, you have to make sure that the concentration of your buffer substances in sample and reference sector are identical by applying methods such as dialysis or SEC.
3. If your sample contains protein aggregates, these will induce light scattering, which will result in an incorrect absorbance reading. This can be easily checked by measuring the absorbance spectrum of the sample relative to a buffer reference. Light scattering will result in an off-set of the spectrum at 320 nm where protein normally does not absorb unless it contains an extrinsic chromophore. Aggregates can be removed by high speed centrifugation. If this does not solve the problem, the samples can be passed through a sterile filter (PVDF membrane, 0.22 μm pore size).
4. Due to their high extinction coefficient at 260 nm, even small contaminations of nucleic acids or nucleotides will interfere strongly with photometric determination of protein concentration. Especially when purifying proteins involved in DNA metabolism, care should be taken that no nucleic acids are co-purified. In absorbance spectra, this kind of contamination will result in a shift of the absorbance maximum from 280 nm to 260 nm. In many cases, nucleic acid impurities can be removed by $(NH_4)_2SO_4$ precipitation or anion exchange chromatography.
5. Before investigating the protein–protein interaction it is recommended to check both protein preparations in a preliminary sedimentation velocity experiment. Thus, sedimentation coefficients of both proteins, the homogeneity of the preparations and any concentration dependence of the oligomerization state of the proteins can be examined. If the sedimentation coefficients are already known, at least free SSB and free ligand at the highest concentration used should be included in the 14 samples of the analysis.
6. The absorbance of each reaction mixture should be below 1.5, otherwise linearity of the photometer is no longer given (17). Especially if the interaction is weak, it might be desirable to measure at high concentrations, which result in absorbance readings above this value. In this case, centerpieces with a pathlength of 3 mm instead of 12 mm can be used, which will

reduce the absorbance by a factor of four. Additionally, the use of 3 mm centerpieces will reduce the required volume to 100 μl, which will save material at high concentrations compared to detection at a wavelength where the extinction coefficient is lower (see Note 9).

7. For temperature equilibration it is advantageous to start a run at 0 rpm instead of just pressing the vacuum button. In the latter case, only the mechanical vacuum pump but not the diffusion pump is switched on. Since the actual rotor temperature can only be measured if the chamber pressure is below 100 μm, this might delay temperature equilibration. Above 100 μm and if the vacuum system is activated the chamber temperature is detected.

8. Temperatures between 0 and 40°C can be used in the XL-A/XL-I instruments. If there are no other requirements, a temperature of 20°C is most convenient, since the use of much higher or especially lower temperatures requires longer thermal equilibration times.

9. Besides changing the optical pathlength of the centerpiece used (see Note 6), there is also the possibility to change the wavelength at which absorbance is measured in order to get access to other concentration ranges. If absorbance at 280 nm is too high, a wavelength of 250 nm may be used in order to stay within the linear range of the photometer. If the interaction is strong, so that measurements at lower concentrations are necessary in order to determine the binding constant, the absorbance of the peptide bond at 230 nm can be used to detect the sedimentation of the protein. Depending on the content of aromatic amino acids of the proteins used, the absorbance at this wavelength is usually about six times higher than at 280 nm. However, you have to make sure that no buffer components are used which strongly absorb at 230 nm, such as EDTA, DTT, benzamidine or HEPES.

10. This rotor speed is optimal for the *Eco*SSB/χ interaction. In case different SSB proteins or interaction partners are used which have markedly increased sedimentation coefficients, it might be necessary to lower the rotor speed in order to measure an adequate number of scans before the sample is sedimented to the bottom of the cell.

11. If the reaction is slow compared to the time scale of the centrifugation experiment (dissociation rate constant $< 10^{-3}$ to $10^{-4}\ s^{-1}$), three peaks should appear in the $c(s)$ distribution, which correspond to the sedimentation coefficients of the free proteins and the complex (17). In this case, the peak areas can be directly used to determine the concentrations of reactants and complex and a binding isotherm can be calculated (17).

However, one has to make sure that the samples reached chemical equilibrium before starting the experiment. This can, e.g., be achieved by over night incubation of the reaction partners.

12. Alternatively, the program SEDFIT can be used to determine the integrals of the respective peaks.
13. When using this method to determine binding parameters, it is important to have an excess of free ligand in the reaction mixture. This assures that the concentration of free ligand, as measured from the slower moving boundary, does not increase significantly in the reaction boundary (19, 27). At the lowest χ concentration used here, concentrations of free and bound χ were similar and possible deviations from the theoretical binding isotherm were within the limits of experimental error (see Fig. 4).
14. It is preferable to vary the logarithm of K_A instead of K_A directly, in order to keep both variables within the same order of magnitude.
15. The concentration range in which this kind of interaction experiments in the analytical ultracentrifuge can be performed is determined by the detectable signal the interaction partners produce. This can be influenced by several parameters such as optical pathlength of the centerpiece (see Note 6), detection wavelength (see Note 9), and the detection system used. To date there are two further detection systems commercially available for XL-A/XL-I centrifuges apart from the absorbance optics: the XL-I Rayleigh interference optical system and a fluorescence optics. Whereas the interference optics is less sensitive than absorbance measurement at 230 nm, where the peptide bond absorbs, it can be employed at much higher protein concentrations than the absorbance system due to its linearity over a wide concentration range. For lower protein concentrations in the nanomolar range, however, fluorescence optics can be used. The only system commercially available at the moment (AU-FDS, Aviv-Biomedical) allows for an excitation at exclusively 488 nm (16), but yields an excellent signal to noise ratio even in the lower nanomolar range. Although proteins normally do not absorb at this wavelength, they can be specifically detected after labeling with fluorophores such as Alexa Fluor 488 or fluorescein or as GFP fusion proteins. However, it has to be checked by, e.g., performing competition experiments with the unlabeled protein, whether protein labeling results in an undesired change in binding parameters. Before fluorescence data can be used to construct a binding isotherm, it has to be investigated whether the quantum yield of the fluorophore stays constant upon the protein–protein

interaction. This can be analyzed, e.g., in a static fluorimeter. Therefore, AUC allows for the investigation of protein concentrations from the nanomolar to the millimolar range and hence for the determination of a wide range of affinities of protein–protein interactions.

Acknowledgements

We gratefully thank Lidia Litz for excellent technical assistance, Dr. Claus Urbanke for valuable discussions and for reading the manuscript, and Dr. Dietmar J. Manstein for scientific and financial support.

References

1. Porter R, Black S, Pannuri S, Carlson A (1990) Use of the *Escherichia coli ssb* gene to prevent bioreactor takeover by plasmidless cells. Bio-Technology 8:47–51
2. Lohman TM, Ferrari ME (1994) *Escherichia coli* single-stranded DNA-binding protein: multiple DNA-binding modes and cooperativities. Annu Rev Biochem 63:527–570
3. Shereda RD, Kozlov AG, Lohman TM, Cox MM, Keck JL (2008) SSB as an organizer/mobilizer of genome maintenance complexes. Crit Rev Biochem Mol Biol 43:289–318
4. Genschel J, Curth U, Urbanke C (2000) Interaction of *E. coli* single-stranded DNA binding protein (SSB) with exonuclease I. The carboxy-terminus of SSB is the recognition site for the nuclease. Biol Chem 381:183–192
5. Lu D, Keck JL (2008) Structural basis of *Escherichia coli* single-stranded DNA-binding protein stimulation of exonuclease I. Proc Natl Acad Sci USA 105:9169–9174
6. Handa P, Acharya N, Varshney U (2001) Chimeras between single-stranded DNA-binding proteins from *Escherichia coli* and *Mycobacterium tuberculosis* reveal that their C-terminal domains interact with uracil DNA glycosylases. J Biol Chem 276:16992–16997
7. Arad G, Hendel A, Urbanke C, Curth U, Livneh Z (2008) Single-stranded DNA-binding protein recruits DNA polymerase V to primer termini on RecA-coated DNA. J Biol Chem 283:8274–8282
8. Cadman CJ, McGlynn P (2004) PriA helicase and SSB interact physically and functionally. Nucleic Acids Res 32:6378–6387
9. Shereda RD, Reiter NJ, Butcher SE, Keck JL (2009) Identification of the SSB binding site on *E. coli* RecQ reveals a conserved surface for binding SSB's C terminus. J Mol Biol 386:612–625
10. Hobbs MD, Sakai A, Cox MM (2007) SSB protein limits RecOR binding onto single-stranded DNA. J Biol Chem 282:11058–11067
11. Glover BP, McHenry CS (1998) The chi psi subunits of DNA polymerase III holoenzyme bind to single-stranded DNA-binding protein (SSB) and facilitate replication of an SSB-coated template. J Biol Chem 273:23476–23484
12. Kelman Z, Yuzhakov A, Andjelkovic J, O'Donnell M (1998) Devoted to the lagging strand-the χ subunit of DNA polymerase III holoenzyme contacts SSB to promote processive elongation and sliding clamp assembly. EMBO J 17:2436–2449
13. Witte G, Urbanke C, Curth U (2003) DNA polymerase III chi subunit ties single-stranded DNA binding protein to the bacterial replication machinery. Nucleic Acids Res 31:4434–4440
14. Yuzhakov A, Kelman Z, O'Donnell M (1999) Trading places on DNA – a three-point switch underlies primer handoff from primase to the replicative DNA polymerase. Cell 96:153–163
15. Johnson A, O'Donnell M (2005) Cellular DNA replicases: components and dynamics at the replication fork. Annu Rev Biochem 74:283–315
16. MacGregor IK, Anderson AL, Laue TM (2004) Fluorescence detection for the XLI

analytical ultracentrifuge. Biophys Chem 108:165–185

17. Brown PH, Balbo A, Schuck P (2008) Characterizing protein-protein interactions by sedimentation velocity analytical ultracentrifugation. Curr Protoc Immunol Chapter 18, Unit 18.15
18. Urbanke C, Ziegler B, Stieglitz K (1980) Complete evaluation of sedimentation velocity experiments in the analytical ultracentrifuge. Fres Z Anal Chem 301:139–140
19. Urbanke C, Witte G, Curth U (2005) Sedimentation velocity method in the analytical ultracentrifuge for the study of protein-protein interactions. Methods Mol Biol 305:101–114
20. Krauss G, Pingoud A, Boehme D, Riesner D, Peters F, Maass G (1975) Equivalent and non-equivalent binding sites for tRNA on aminoacyl-tRNA synthetases. Eur J Biochem 55:517–529
21. Schuck P (2000) Size-distribution analysis of macromolecules by sedimentation velocity ultracentrifugation and lamm equation modeling. Biophys J 78:1606–1619
22. Naue N, Fedorov R, Pich A, Manstein DJ, Curth U (2011) Site-directed mutagenesis of the χ subunit of DNA polymerase III and single-stranded DNA-binding protein of *E. coli* reveals key residues for their interaction. Nucleic Acids Res 39:1398–1407
23. ProteomeLab XL-A/XL-I protein characterization system (2007) Beckman Coulter, Inc., Miami, FL
24. Lohman TM, Overman LB (1985) Two binding modes in *Escherichia coli* single strand binding protein-single stranded DNA complexes. Modulation by NaCl concentration. J Biol Chem 260:3594–3603
25. Pace CN, Vajdos F, Fee L, Grimsley G, Gray T (1995) How to measure and predict the molar absorption coefficient of a protein. Protein Sci 4:2411–2423
26. Laue MT, Shah BD, Rigdeway TM, Pelletier SL (1992) Computer-aided interpretation of analytical sedimentation data for proteins. In: Analytical ultracentrifugation in biochemistry and polymer science. Cambridge, UK, pp 90–125
27. Dam J, Schuck P (2005) Sedimentation velocity analysis of heterogeneous protein-protein interactions: sedimentation coefficient distributions c(s) and asymptotic boundary profiles from Gilbert-Jenkins theory. Biophys J 89:651–666

Chapter 9

Ammonium Sulfate Co-precipitation of SSB and Interacting Proteins

Aimee H. Marceau

Abstract

Bacterial single-strand DNA-binding protein (SSB) interacts with many proteins involved in the diverse process of genome maintenance. The interactions are mediated by the essential and conserved amphipathic C-terminus (SSB-Ct). SSB plays a critical role in localizing and stimulating the activity of a wide variety of DNA-processing proteins. The interaction partners have been identified and studied using a variety of methods, one of which, ammonium sulfate co-precipitation, is described here.

Key words: SSB, SSB-Ct, Interaction, Co-precipitation, Ammonium sulfate

1. Introduction

The process of genome maintenance in the cell is complex; SSB functions as an organizational hub for the many proteins involved in DNA replication, recombination, and repair (1). SSB binds to the single-strand DNA (ssDNA) exposed during replication and in the process of repair (2). While the N-terminal domain of SSB binds to ssDNA protecting it from damage and preventing secondary structure formation, the SSB-Ct interacts with the proteins involved in processing the ssDNA (1). Proteins that interact with SSB were discovered through tandem affinity purification, His-tag pulldowns, and co-purification (1, 3, 4). These interactions were further validated by several quantitative and qualitative methods. One of the simplest qualitative methods takes advantage of SSB precipitation in a low concentration of ammonium sulfate (5–7). If an interaction partner of SSB is added to the solution when SSB precipitates it will pull the protein into the pellet. Ammonium sulfate co-precipitation provides a quick qualitative method to access the interaction of SSB with other proteins.

James L. Keck (ed.), *Single-Stranded DNA Binding Proteins: Methods and Protocols*, Methods in Molecular Biology, vol. 922, DOI 10.1007/978-1-62703-032-8_9, © Springer Science+Business Media, LLC 2012

2. Materials

1. 450 g/L Ammonium sulfate.
2. 1 M Tris, pH 8.0.
3. Glycerol.
4. 5 M NaCl.
5. Centrifuge (1.5–0.6 mL tubes) at 4°C.
6. Purified *E. coli* SSB (see Notes 1 and 4).
7. Purified protein(s) of interest.
8. SDS running buffer: 0.1% SDS.
9. 10–15% gradient SDS-PAGE gel.
10. Coomassie Brilliant Blue.

Buffers (all buffers at 4°C).

Co-precipitation buffer: 10 mM Tris–HCl, pH 7.0–8.0, 10% glyercol, 150 mM NaCl.

450 g/L ammonium sulfate.

3. Methods

SSB Co-precipitation:

1. Mix *E. coli* SSB (20 μM/monomer) with 20 μM of its binding partner in co-precipitation buffer. Prepare two additional tubes containing SSB at 20 μM or protein partner at 20 μM as controls. (seeNotes 2–4).
2. Incubate together for 15 min on ice.
3. Add ammonium sulfate from a 450 g/L stock to a final concentration of 150 g/L for a final volume of 20 μL, and mix the solution well.
4. Incubate the proteins for 15 min on ice and collect the precipitate by centrifugation in a microcentrifuge.
5. Wash the pellet three times with co-precipitation buffer plus 150 g/L ammonium sulfate.
6. Resuspend the pellet in SDS running buffer and resolve the proteins by SDS-PAGE.

4. Notes

1. SSB co-precipitation experiments have only been published using *E. coli* SSB and the binding partners (5–7). Attempts have been made using *B. subtilis* SSB; these have thus far been unsuccessful.
2. It is important to establish that the binding partner does not precipitate at the concentration of ammonium sulfate used in the assay. SSB will precipitate at somewhat lower ammonium sulfate concentrations as well, so this may be reduced.
3. If a different buffer is used it is helpful to run a panel of SSB precipitation experiments from 100 to 200 g/L of ammonium sulfate. This panel allows you to establish the best concentration to use under your conditions. This same ammonium sulfate panel can be run with the binding partner to establish if it precipitates under similar conditions without SSB.
4. Purified SSB lacking the SSB-Ct or with mutations in the C-terminus can also be used in the assay to verify it is the site of the interaction.

References

1. Shereda RD et al (2008) SSB as an organizer/mobilizer of genome maintenance complexes. Crit Rev Biochem Mol Biol 43(5):289–318
2. Lohman TM, Overman LB (1985) Two binding modes in Escherichia coli single strand binding protein-single stranded DNA complexes. Modulation by NaCl concentration. J Biol Chem 260 (6):3594–603
3. Butland G et al (2005) Interaction network containing conserved and essential protein complexes in Escherichia coli. Nature 433(7025):531–7
4. Purnapatre K et al (1999) Differential effects of single-stranded DNA binding proteins (SSBs) on uracil DNA glycosylases (UDGs) from Escherichia coli and mycobacteria. Nucleic Acids Res 27(17):3487–92
5. Genschel J, Curth U, Urbanke C (2000) Interaction of E. coli single-stranded DNA binding protein (SSB) with exonuclease I. The carboxy-terminus of SSB is the recognition site for the nuclease. Biol Chem 381(3):183–92
6. Shereda RD, Bernstein DA, Keck JL (2007) A central role for SSB in Escherichia coli RecQ DNA helicase function. J Biol Chem 282 (26):19247–58
7. Marceau AH et al (2011) Structure of the SSB-DNA polymerase III interface and its role in DNA replication. EMBO J 30(20):4236–47

Chapter 10

Analyzing Interactions Between SSB and Proteins by the Use of Fluorescence Anisotropy

Duo Lu

Abstract

Fluorescence anisotropy is a useful tool to detect SSB interaction with SSB binding partners. Titrating an SSB protein partner into a solution containing fluorescently labeled SSB C-terminal tail (SSB-Ct) peptide results in formation of the protein complex with molecular weights greatly higher than SSB peptide alone. Thus the tumbling of the complexes in solution is slower, and the fluorescence anisotropy (FA) signal would be increased, indicating the binding. Based on the FA binding curve, the apparent dissociation constant ($K_{d,app}$) of the binding reaction can also be calculated.

Key words: Fluorescence anisotropy, SSB, SSB C-terminal peptide, SSB protein binding partners, Exonuclease I

1. Introduction

The C-terminal region of the SSB protein is a structurally flexible polypeptide comprising ~60 residues. The extreme C-terminal residues, referred to as the SSB-Ct in this chapter, form an evolutionarily conserved amphipathic sequence that includes an acidic-rich sequence followed by a hydrophobic sequence (Met-Asp-Phe-Asp-Asp-Asp-Ile-Pro-Phe in *Escherichia coli* SSB). Studies from numerous labs have shown that the SSB-Ct functions as a hub at which heterologous proteins involved in genome maintenance associate with SSB/ssDNA substrates. At least a dozen different heterologous proteins have been shown to physically associate with SSB, including Exonuclease I (ExoI), RecQ DNA helicase, and PriA DNA helicase (1–4).

The small size of the SSB-Ct makes it ideal for using fluorescence anisotropy to detect its association with other proteins. Since flurorescently labeled SSB-Ct (F-SSB-Ct) alone tumbles fast in solution, polarized light is greatly depolarized after passing through

James L. Keck (ed.), *Single-Stranded DNA Binding Proteins: Methods and Protocols*, Methods in Molecular Biology, vol. 922, DOI 10.1007/978-1-62703-032-8_10, © Springer Science+Business Media, LLC 2012

the solution with F-SSB-Ct alone, giving a low FA signal. However, F-SSB-Ct is bound to another protein; the total molecular weight of the complex would be greatly increased and tumbling of fluor slowed. Under this condition, the polarized light hitting a protein/F-SSB-Ct complex remains polarized, leading to a relatively high FA signal.

Using fluorescence anisotropy, I have been able to successfully detect SSB-Ct interaction with wild-type (WT) and variant *E. coli* ExoI proteins, and to determine apparent dissociation constants ($K_{d,app}$) for these interactions (4). Fluorescence anisotropy is also very useful in examining binding inhibition by adding components (non-fluorescently labeled SSB-Ct variants or small molecule compounds) into solutions containing fluorescently labeled SSB-Ct and its binding partners (5, 6). Here, I describe the use of FA in measuring ExoI binding to the F-SSB-Ct peptide from *E. coli* as an example.

2. Materials

All solutions are prepared with MilliQ water.

1. Synthesize and purify peptides corresponding to the C-terminal tail of SSB or variants thereof using a commercial vendor. The “F-SSB-Ct” peptide (Trp-Met-Asp-Phe-Asp-Asp-Asp-Ile-Pro-Phe) combines an N-terminal Trp residue for quantification with the 9°C-terminal-most residues from *E. coli* SSB, and also it adds an N-terminal fluorescein moiety. The SSB peptide powder is stored in a dark tube or in an aluminum foil-covered tube at −20°C (see Note 1).
2. Purified ExoI is expressed from an over-expression vector encoding His-tagged *E. coli* ExoI and the recombinant protein is purified as described (7). Protein concentrations are determined by measuring their A_{280nm} in 6.0 M guanidine–HCl (8).
3. FA buffer: 20 mM Tris–HCl, pH 8.0, 100 mM NaCl, 1 mM $MgCl_2$, 4% (v/v) glycerol, 1 mM 2-mercaptoethanol, and 0.1 g/L bovine serum albumin. Store in 4°C, and warm it to room temperature before using.
4. Glass tubes for FA.
5. A fluorimeter or dedicated fluorescence anisotropy system. The experiments described in this chapter use a Panvera Beacon 2000 fluorescence anisotropy system.

3. Methods

1. Prepare a 1 μM F-SSB-Ct solution. Weigh a proper amount of F-SSB-Ct powder and dissolve in MilliQ water in dark tubes or aluminum-covered Eppendorf tubes. Mix well to make sure that all the powder is dissolved (see Note 2).
2. Dilute ExoI (or a variant) within 0.1–10,000 nM range in FA buffer. For the initial test, tenfold dilution could be used. For a more accurate FA binding curve, finer dilutions are requested to get data at more concentration points (see Note 3).
3. Prepare a sample for blank test: Add 0.1 mL FA buffer in a glass tube.
4. Prepare a sample for background test: Add 99 μL of FA buffer in a glass tube, and add 1 μL 1 μM F-SSB-Ct solution in the same tube, and mix well. The final concentration for F-SSB-Ct is 10 nM.
5. Prepare sample tubes: Add 99 μL diluted ExoI protein solutions in different glass tubes, and add 1 μL 1 μM F-SSB-Ct solution in each tube, and mix well. The final concentration for F-SSB-Ct is 10 nM.
6. Use aluminum foil to cover all the glass tubes, and incubate them at room temperature for 10 min.
7. Turn on the Panvera Beacon 2000 fluorescence anisotropy system to let it warm up.
8. Set the excitation wavelength at 490 nm and emission wavelength at 535 nm (appropriate for fluorescein). Set the chamber temperature at 25°C.
9. Put the blank test sample into the chamber, push it all the way down to the bottom, and close the lid. Wait until the system notifies ready for test. Then take the tube out and discard properly.
10. Put the background test sample inside, close the lid, and wait for the system to measure the fluorescence anisotropy (FA) signal. Then take it out and discard.
11. Repeat until all the samples are tested.
12. If necessary, triplicate the tests by using new samples at each time to get average FA values and more accurate binding curves.
13. Data analysis: Export FA values from the system into Excel (or similar) file. The average FA value is plotted with one standard deviation of the mean shown as error. FA values are converted to the fraction of peptide bound and apparent dissociation constants are the [ExoI] needed for 50% binding saturation (Fig. 1) (see Note 4).

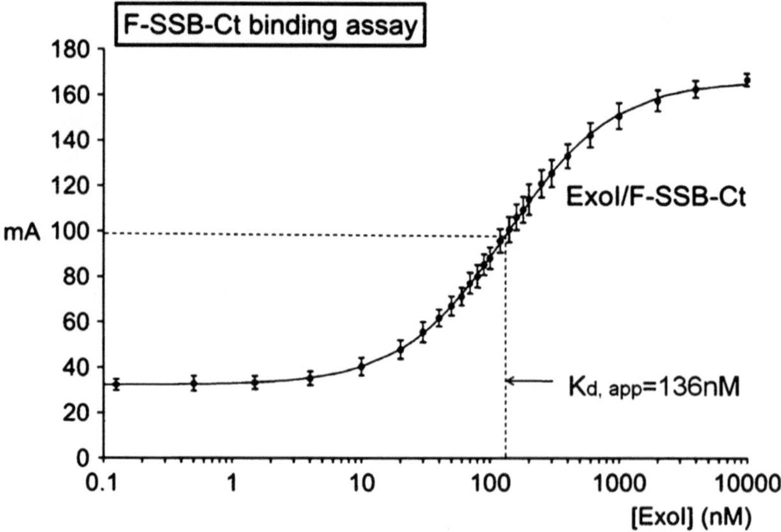

Fig. 1. Equilibrium binding isotherm of F-SSB-Ct association with ExoI as monitored by fluorescence anisotropy.

4. Notes

1. Because fluorescein moiety is light sensitive all the fluorescent-labeled materials need to be stored in either dark containers or containers with aluminum foil covered.
2. The F-SSB-Ct solution is better if dissolved fresh. Extra can be stored in 4°C for further use within days.
3. Due to the varied association affinities of SSB and protein partners, different proteins may need to use different ranges of concentration to get the best results.
4. If the binding saturation cannot be reached with the highest protein concentration, the less accurate estimated value or range of apparent dissociation constant may still be able to be measured by the shape of the current binding curve.

References

1. Kowalczykowski SC et al (1994) Biochemistry of homologous recombination in Escherichia coli. Microbiol Rev 58(3):401–65
2. Butland G et al (2005) Interaction network containing conserved and essential protein complexes in Escherichia coli. Nature 433 (7025):531–7
3. Shereda RD et al (2009) Identification of the SSB binding site on *E. coli* RecQ reveals a conserved surface for binding SSB's C terminus. J Mol Biol 386(3):612–625
4. Lu D, Keck JL (2008) Structural basis of Escherichia coli single-stranded DNA-binding protein stimulation of exonuclease I. Proc Natl Acad Sci U S A 105(27):9169–74
5. Lu D, Windsor MA, Gellman SH, Keck JL (2009) Peptide inhibitors identify roles for SSB

C-terminal residues in SSB/Exonuclease I complex formation. Biochem 48:6764–6771

6. Lu D, Bernstein DA, Satyshur KA, Keck JL (2010) Small-molecule tools for dissecting the roles of SSB/protein interactions in genome maintenance. Proc Natl Acad Sci U S A 107 (2):633–8
7. Breyer WA, Matthews BW (2000) Structure of Escherichia coli exonuclease I suggests how processivity is achieved. Nat Struct Biol 7 (12):1125–8
8. Edelhoch H (1967) Spectroscopic determination of tryptophan and tyrosine in proteins. Biochemistry 6(7):1948–54

Chapter 11

Far Western Blotting as a Rapid and Efficient Method for Detecting Interactions Between DNA Replication and DNA Repair Proteins

Brian W. Walsh, Justin S. Lenhart, Jeremy W. Schroeder and Lyle A. Simmons

Abstract

Protein–protein interactions are required for the proper function of many biological pathways. Numerous biochemical and protein blotting methods are available for probing direct and indirect interactions between two protein-binding partners. Here, we describe the methodology of far Western blotting, or immunodot blotting, as a technique for probing direct interactions between two proteins. We describe the utility of this approach as a rapid, qualitative screen for identifying novel protein-binding partners. We also describe the importance of this technique for measuring differences in interaction between wild-type and mutant forms of a known binding partner. Far Western blotting is a rapid and highly reproducible experimental approach for identifying and understanding the interaction between protein-binding partners leading to new discoveries in the function and regulation of biological pathways.

Key words: Far western blot, Immunodot blot, Protein–protein interaction, Antibodies, Antigens

1. Introduction

Immunodetection of proteins separated by sodium dodecyl sulfate (SDS) polyacrylamide gel electrophoresis (PAGE) is an extremely powerful technique which has been common in the laboratory setting for several decades (1). While this technique allows for the detection of a protein in a cell lysate, it lacks the capacity to inform the user of interactions that may have occurred between the protein of interest and other proteins in the cell. The denaturation of a protein during SDS-PAGE prevents the detection of interactions with proteins due to loss of secondary and tertiary structure. Conversely, native PAGE analysis can be very useful because it allows for the maintenance of protein structure and the detection of a

James L. Keck (ed.), *Single-Stranded DNA Binding Proteins: Methods and Protocols*, Methods in Molecular Biology, vol. 922, DOI 10.1007/978-1-62703-032-8_11,

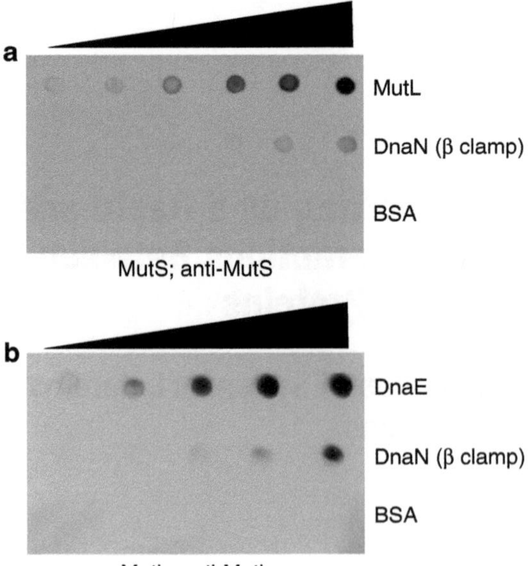

Fig. 1. Detection of direct protein–protein interactions using immunodot blotting. (**a**) Shown is a test for MutS binding partners. MutL, DnaN, and BSA were spotted in increasing amounts from 2.5 to 80 pmol of monomer on a nitrocellulose membrane. The membrane was incubated with 100 nM purified MutS followed by incubation with the primary antibody, affinity-purified anti-MutS at a 1:250 dilution. The membrane was subsequently incubated with the secondary antibody HRP-conjugated goat anti-rabbit at a 1:5,000 dilution. (**b**) Shown is a test for MutL binding partners. Proteins DnaE, DnaN, and BSA were spotted in increasing amounts from 2.5 to 80 pmol of monomer as in (**a**). The membrane was incubated with 400 nM purified MutL followed by a 1:250 dilution of the primary antibody, affinity-purified anti-MutL. The membrane was subsequently incubated with a 1:2,000 dilution of the secondary antibody HRP-conjugated goat anti-rabbit.

protein within a complex. However, it has the limitations that some protein interactions are poorly detected by this technique and that large protein complexes are difficult to resolve in a native gel (2). Far Western blotting is an effective and efficient technique used to assay interactions that occur between natively structured proteins. It can be used to specifically detect interactions between the protein of interest and any number of bait proteins immobilized on a solid support membrane, such as nitrocellulose. This is useful not only for the detection of interaction between full-length purified proteins, but also for identifying areas of interaction between purified proteins which may be truncated or have amino acid changes in putative binding interfaces (3, 4) (Fig. 1). In addition to locating sites of potential interaction, this technique can provide a measure of relative binding if the concentrations of each protein being tested are known (5). Far Western blotting can also be performed without having antibodies generated against your protein of interest through the use of functional affinity tags on the prey (capture) protein used in the assay. The protocol described here allows for the rapid screening and detection of potential protein–protein interactions and provides a method to qualitatively determine relative binding strength.

2. Materials

2.1. SDS Polyacrylamide Gel Components

1. Resolving Gel Buffer: 1.5 M Tris, pH 8.8. Add 100 mL water to a glass beaker. Add 181.81 g Tris Base to the beaker and add water to a total volume of 900 mL while mixing. Adjust pH to 8.8 with HCl, add water to a final volume of 1 L and filter sterilize.
2. Stacking gel buffer: 1.0 M Tris, pH 6.8. Add 100 mL ddH_2O to a glass beaker. Add 121.14 g Tris Base to the beaker. Then, add ddH_2O to a total volume of 900 mL while mixing. Adjust pH to 6.8 with HCl, add water to a final volume of 1 L and filter sterilize.
3. Acrylamide 40% solution. 19:1 Acrylamide : Bis-Acrylamide.
4. Ammonium persulfate: 10% (w/v) solution in ddH_2O. Prepare this solution fresh before making SDS-PAGE.
5. *N, N, N, N′*-tetramethyl-ethylenediamine (TEMED), stored at 4°C.
6. 5× SDS Electrophoresis Buffer: In a 1 L beaker, add ddH_2O to 100 mL. Add 15.1 g Tris base, 72.0 g glycine, and 5.0 g SDS. Fill to 1 L with ddH_2O. Dilute to 1× with ddH_2O for use in electrophoresis (6).
7. 6× SDS sample buffer: add 7 mL stacking gel buffer (1.0 M Tris, pH 6.8) and 3 mL of glycerol to a 15 mL conical tube. Add 1 g SDS, 1 g bromophenol blue and 0.93 g DTT to the tube. Vortex the solution to mix and aliquot into 0.5 mL quantities and store at −70°C up to 6 months (6).
8. Mini PROTEAN glass plates with 1 mm spacers and glass short plates.

2.2. Electrophoretic Transfer Components

1. Transfer Buffer, 10×: Add 30.3 g of Tris Base to 100 mL of ddH_2O in a 1 L beaker. Add 144.2 g of Glycine and fill to 1 L with ddH_2O. The solution is then filtered through a membrane with a 0.22 μm pore size. For use in electrophoresis, add 100 mL of 10× transfer buffer to 700 mL of ddH_2O. Then add 200 mL of methanol to a total volume of 1 L (6).
2. Nitrocellulose Membrane: Whatman Protran 0.22 μM cut to an appropriate size.

2.3. Affinity Purification Components

1. Ponceau stain: in 1 mL of Glacial Acetic Acid dissolve 0.5 g Ponceau S. Add ddH_2O to a final volume of 100 mL (6).
2. Phosphate Buffered Saline, pH 7.6 (PBS): Mix 87 mL of 0.2 M Na_2HPO_4 and 13 mL of 0.2 M NaH_2PO_4. Check to ensure that the pH is 7.6. Add 8.766 g of NaCl. Fill to 1 L before filter sterilizing. Final concentration used is 20 mM phosphate, 150 mM NaCl.
3. Blocking Buffer: 5% w/v dried milk solids in PBS plus 0.05% v/v Tween-20.
4. Sodium Phosphate Buffer, pH 8.0: 94.7 mL of 0.2 M Na_2HPO_4, 5.3 mL 0.2 M NaH_2PO_4 filled to 1 L with ddH_2O. Final concentration used is 20 mM phosphate.
5. Strip Buffer: 5 mM glycine, 150 mM NaCl, pH 2.4.

2.4. Immunoblotting Components

1. Blocking Buffer: 5% w/v dried milk solids in PBS plus 0.05% v/v Tween-20.
2. Protein Dilution Buffer: PBS plus 0.05% v/v Tween-20.
3. Nitrocellulose Membranes: Whatman Protran 0.22 μM cut to a usable size (about 1 cm^2 per dot). Alternatively, if a spot blot apparatus is used we recommend Bio-Rad Trans-Blot with a nitrocellulose membrane with a 0.45 μM pore size.
4. Protein-Binding Buffer: 5% w/v dried milk solids in PBS, 0.05% Tween-20, 0.2–0.4 μM prey protein.
5. Chemiluminescent Substrate: SuperSignal West Pico Chemiluminescent Substrate (Thermo Scientific #34080).

2.5. Antigens

1. Purified bait and prey proteins: purified to >95% purity as judged by SDS-PAGE.
2. Purified bovine serum albumin: 10 mg/mL (New England Biolabs) or other protein as an appropriate negative control.
3. Primary antiserum: Polyclonal antiserum is affinity purified and tested for cross reactivity with bait proteins. If primary antiserum shows no cross reactivity with other proteins, or is purchased monoclonal, affinity purification is unnecessary. Affinity-purified antiserum is used after dilution in blocking buffer (often 1:2,000). Antiserum generated by Covance is typically used in our laboratory as a primary antibody.
4. Secondary Antibody: Stabilized Peroxidase Conjugated Goat anti-Rabbit (Pierce/Thermo-scientific #32460) used at a 1:2,000 dilution in blocking buffer. If you are using primary antibodies generated in mice or other animals, use the appropriate secondary antibody.

3. Methods

3.1. Affinity Purification of Antisera

1. Purified proteins are electrophoresed on an SDS-PAGE. Typically using 10 μg of purified protein per well and 10–100 μg total purified protein per gel (see Note 1).
2. Transfer purified protein to a nitrocellulose membrane using electrophoretic transfer.
3. Stain proteins on the nitrocellulose membrane with Ponceau S, and rinse with ddH_2O to destain.
4. Cut the band of purified protein from the nitrocellulose membrane with a clean razor blade leaving as little extra space as possible surrounding the protein.
5. Block the membrane by adding blocking buffer and incubating at room temperature for 20 min with constant gentle shaking.
6. Wash the membrane twice for 2 min each time in PBS plus 0.05% Tween-20.
7. Incubate the strip of nitrocellulose binding your protein with 300 μL of antiserum for 1 h with gentle mixing at room temperature (see Note 2).
8. Following incubation, remove the antiserum (see Note 3).
9. Wash the membrane twice for 15 min each in 1–2 mL of PBS.
10. Aliquot 100 μL of sodium phosphate buffer into three separate microcentrifuge tubes.
11. Strip the membrane by adding 300 μL of strip buffer to the membrane followed by incubation.
12. Incubate at room temperature for 30 s.
13. Remove the strip buffer from the membrane and then add the buffer to one of the microcentrifuge tubes containing sodium phosphate for neutralization of the solution.
14. Repeat steps 11–13 two more times.
15. Combine all three tubes of quenched strip buffer, which now contain the affinity-purified antibodies, in order to equalize the antibody titer.
16. Divide the affinity-purified antibodies into new aliquots and store at 4°C.
17. Antibodies should be checked for titer by standard western blot methods following each purification prior to use since yields may vary between preparations (see Note 4).

3.2. Immunodot Blotting

1. Serial dilute the bait proteins to be examined for interaction in protein dilution buffer to appropriate molarity for the experiment (see Note 5).

2. Apply serial dilutions of protein samples to a precut, dry membrane leaving at least 1 cm of space between the centers of each dot of the series. Apply to the membrane either BSA or other protein which does not bind the prey protein as a negative control. Allow the membrane to dry before the next step (see Note 6).
3. Place the membrane in blocking buffer for 1 h at room temperature on an orbital shaker with gentle agitation, usually at approximately 20 rpms, to block the membrane.
4. Remove blocking buffer and replace with protein-binding buffer containing the prey protein for at least 6 h of incubation at 4°C on an orbital shaker or rocker with gentle agitation (see Note 7).
5. Following incubation with the prey protein, wash three times in blocking buffer for 5 min each with gentle rotation.
6. Incubate in blocking buffer containing affinity-purified primary antiserum overnight at 4°C on an orbital shaker or rocker at approximately 20 rpms. Alternatively, if the protein-binding step was performed at 4°C overnight primary antiserum incubation can be performed at room temperature for 4–6 h (see Note 8).
7. Remove the primary antibody solution and wash three times with blocking buffer for 5 min each on an orbital shaker or rocker at low rpms.
8. Place the blot in secondary antibody solution diluted 1:2,000 in blocking buffer. Incubate on an orbital shaker or rocker at room temperature for 2–3 h using low rpms.
9. Following incubation with secondary antibody solution, wash once with blocking buffer for 5 min. Following the first wash, use two more wash steps with protein dilution buffer for 5 min each.
10. Activate the blot by addition of 1 mL of each component of the chemiluminescent substrate. Wash the substrate over the blot for 2–5 min and remove the blot from solution.
11. Expose blot to film for an appropriate length of time, depending on the strength of your signal (see Note 9).
12. Following development of the film, examine each series. From the lowest molarity to the highest there should be an increase in signal intensity if the bait and prey proteins interact. The negative control protein should have no signal or signal increase since your antibodies should be specific to the prey protein, and your prey protein should not bind the negative control.

4. Notes

1. SDS-PAGE can be performed with various well sizes depending on the amount of protein that is to be separated. If a large amount of purified protein is to be separated, it is more convenient to use much larger wells as opposed to multiple small wells since there will be less excess nitrocellulose to remove when the membrane is cut.
2. The membrane can be cut in half prior to incubation with antisera and one half may be stored at −20°C if tightly wrapped in plastic. Additionally, when incubating with antiserum, it helps to incubate the piece of membrane in a microcentrifuge tube because this provides optimal surface-to-volume ratio for contact of the protein with the antiserum.
3. The depleted antiserum may still contain a significant amount of antibodies and may be used for affinity purification again if the antiserum is in short supply.
4. Affinity-purified antibodies are stored at 4°C in an effort to prevent loss of function of the antibodies due to multiple freeze–thaw cycles. Sodium azide may also be added to 1 mM to prevent contamination.
5. 10 pmol of monomeric bait protein seems to be a good starting point for detecting interactions. If interactions are hypothesized to be stronger, then the amount of protein used should be decreased to 2.5 pmol or lower.
6. Bio-Rad sells a dot-blot apparatus (Bio-Dot Apparatus), which allows for the exact spacing between protein samples (Bio-Rad #170-6545). When using this apparatus, we use a nitrocellulose membrane that has 0.45 μM pore size and the membrane is pre-wet in PBS prior to use. Rather than using the force of a vacuum, the protein is allowed to bind the membrane by gravity filtration. The vacuum is used only when placing the membrane onto and removing the membrane from the Bio-Dot apparatus.
7. Membranes can be incubated with as little as 4 mL of protein-binding buffer with prey protein but 6–7 mL seems to yield the best results. Incubation can also be done at room temperature for only 3 h.
8. Our lab generally uses affinity-purified primary antiserum diluted 1:250 in blocking buffer. It should be noted that the titer of affinity-purified antibody can vary between preparations and thus the dilution used may need to be adjusted for each experiment. We discard affinity-purified primary antiserum after one use.

9. A good starting point for exposure time with film is usually 1 min, but often the exposure requires 5–10 min. If a satisfactory exposure time is not achieved before loss of the luminescent signal occurs, the blot can be washed and activated a second time by washing with protein dilution buffer for 1 min. The blot can then be reactivated by repeating the application of chemiluminescent substrate. This seems to work best if the chemiluminescent substrate is applied in the dark room to prevent the early loss of signal.

Acknowledgements

We wish to thank Nick Bolz, Eileen Brandes, and Gabriella Szewczyk for comments on the manuscript. We also thank Dr. Andrew Klocko for his work in helping to establish protein-blotting assays in our laboratory. This work was supported by start-up funds from the University of Michigan and the Department of Molecular, Cellular and Developmental Biology. This work was also supported by grant MCB1050948 from the National Science Foundation and a grant from the Wendy Will Case Cancer fund to L.A.S. An NIH genetics training grant to the University of Michigan supported J.W.S in this work (T32GM007544).

References

1. Towbin H, Staehelin T, Gordon J (1979) Electrophoretic transfer of proteins from polyacrylamide gels to nitrocellulose sheets: procedure and some applications. Proc Natl Acad Sci U S A 76(9):4350–4
2. Wittig I, Schagger H (2005) Advantages and limitations of clear-native PAGE. Proteomics 5 (17):4338–46
3. Simmons LA et al (2008) Beta clamp directs localization of mismatch repair in Bacillus subtilis. Molecular Cell 29(3):291–301
4. Klocko AD et al (2011) Mismatch repair causes the dynamic release of an essential DNA polymerase from the replication fork. Molecular Microbiol 82(3):648–63
5. Ohba T et al (1998) Dot far-western blot analysis of relative binding affinities of the Src homology 3 domains of Efs and its related proteins. Anal Biochem 262(2):185–92
6. Gallagher SR, Wiley EA (2008) Current protocols essential laboratory techniques. Wiley, Hoboken

Chapter 12

Methods for Analysis of SSB–Protein Interactions by SPR

Asher N. Page and Nicholas P. George

Abstract

Surface plasmon resonance (SPR) is a widely employed technique for studying protein–protein interactions. Here, we describe a method for the analysis of single-stranded DNA binding protein (SSB)–heterologous protein interactions by SPR. This method avoids several pitfalls often associated with SPR, particularly difficulties in immobilizing the protein while still allowing for facile regeneration of the sensor chip surface for subsequent experiments. Essentially, the method entails immobilizing a biotinylated single-stranded DNA oligo onto the chip surface, which is then bound by SSB prior to analyte addition to the SSB-coated chip. This allows for rapid qualitative and detailed quantitative analysis of both equilibrium and kinetic parameters of the SSB–protein interaction.

Key words: MgsA, Surface plasmon resonance, Kinetics, Equilibrium binding

1. Introduction

Since the initial development of SPR as a technology for analyzing molecular interactions (1, 2), the method has gained wide use in molecular biology due to its ability to measure protein–protein interactions in real time and without the use of labels. SPR-based technologies utilize a biosensor chip that converts binding of two macromolecules into a signal that is monitored in real time by the user. The chip is composed of a sensor surface upon which a ligand is immobilized, and four flow cells are sealed to the surface through which an analyte is flowed. As analyte binds to the ligand, the accumulation of mass near the sensor surface alters the local refractive index. This change in refractive index is measured in real time and displayed as response units (RUs). A plot of RUs versus time is termed a sensorgram and records analyte association with and dissociation from the ligand. Analysis of the sensorgram allows the user to decipher both equilibrium and kinetic binding data regarding the interaction.

James L. Keck (ed.), *Single-Stranded DNA Binding Proteins: Methods and Protocols*, Methods in Molecular Biology, vol. 922, DOI 10.1007/978-1-62703-032-8_12,

A drawback of using SPR to study protein–protein interactions is that one of the proteins must be immobilized onto the chip sensor surface. This is often performed by covalently coupling the protein via a functional linker onto the chip surface, which can yield a nonhomogenous sensor surface and interfere with the interaction between the proteins. Direct covalent immobilization of SSB to the sensor surface has been successfully utilized in other studies (3). In the method described below, SSB is not directly coupled to the chip, but rather bound to single-stranded DNA (ssDNA) that is immobilized onto the chip via a biotin–streptavidin interaction (4–6). This method of capture allows SSB to be presented in a more biologically relevant manner and allows for easier regeneration of the chip for repeated experiments with SSB variants. The SSB interacting protein (SIP), the analyte, is then injected into the flow cell and binding is monitored. After equilibrium is reached, buffer alone is flowed through the flow cell and the dissociation of the SIP from the SSB–ssDNA complex is followed as the RUs return to a baseline reading. Injecting increasing concentrations of SIP and determining the equilibrium RUs allows for relatively simple determination of the binding constant for the SSB–SIP interaction. Furthermore, nonlinear curve fitting of the associative and dissociative phases of the sensorgram can generate binding kinetics. Thus, one can utilize this method to obtain accurate equilibrium and kinetic binding data while avoiding several of the pitfalls often associated with SPR.

2. Materials

1. Biacore Sensor Chip SA (GE Healthcare) (see Note 1).
2. Biotinylated dT_{35} oligonucleotide.
3. TBS50: 20 mM Tris–HCl, pH 7.5, 50 mM NaCl, 3 mM EDTA, 0.01% Tween 20°K.
4. TBS200: 20 mM Tris–HCl, pH 7.5, 200 mM NaCl, 3 mM EDTA, 0.01% Tween 20°K.
5. 6 M Guanidinium hydrochloride.

3. Methods

Solution preparation

1. Prepare all necessary buffers for experiment. TBS50 and TBS200 are needed for dialysis so several liters of both buffers should be prepared.

2. Filter all buffers through 0.2 μm filter and thoroughly degas at room temperature (see Note 2).

Immobilize biotinylated dT_{35} onto sensor chip

3. Remove Biacore Sensor Chip SA from packaging and dock onto instrument.
4. Dilute biotinylated dT_{35} DNA into TBS200 to a final concentration of 100–400 nM.
5. Pretreat all four flow cells with three 1 min pulses of 1 M NaCl at 100 μl/min prior to injecting DNA.
6. Do not inject DNA into flow cell 1. Under most circumstances, flow cell 1 should be reserved as a no-ligand control to be used for baseline subtraction.
7. Inject DNA onto the chip surface in flow cell 2, 3, or 4 (experimental flow cell) at flow rate of 5 μl/min while monitoring response units (RUs). Stop the flow when the RUs reach $\approx$ 300 (see Note 3).
8. Rinse the chip surface (control and experimental flow cells) with TBS200 at a flow rate of 100 μl/min and ensure that the RUs remain stable.

Immobilize SSB onto Sensor chip

9. Dialyze SSB or SSB variant against TBS200 buffer (see Notes 4 and 5).
10. Dialyze the SIP against TBS50.
11. Remove SSB and SIP from dialysis. Centrifuge in microfuge at 10,000 x g for 10 min to get rid of any precipitated sample or particulates. Determine concentration of both protein samples.
12. Dilute SSB or SSB variant to 200 nM with TBS50.
13. Equilibrate the experimental flow cell with TBS50 at 100 μl/min until the RUs stabilize.
14. Load the experimental flow cell with 200 nM SSB at 5–50 μL/min until saturation. Rinse experimental flow cell with TBS50 until the RUs stabilize (see Notes 6 and 7).

Inject SIP onto sensor surface

15. Make working dilutions of SIP (diluted in TBS50). We have found that using concentrations from 50 to 1,500 nM yields a sufficient binding curve (see Note 8).
16. Load SIP dilutions into reaction vials, cap vials, and transfer to SPR instrument (see Note 9).
17. Inject all samples through the control flow cell followed by the experimental flow cell.

18. First, inject the least concentrated SIP dilution onto the chip sensor surface at a flow rate of 90 μl/min for approximately 30 s (see Note 10). An increase in RUs indicates that the SIP is accumulating on the sensor surface.
19. After the injection reaches equilibrium, flow TBS50 at the same flow rate for ≈5 min to completely dissociate the SIP from the sensor surface. The RUs should return to the baseline reading of only DNA and SSB (see Note 11).
20. Sequentially inject all the SIP dilutions, from least to most concentrated.

Remove SSB from sensor surface

21. After completing the experiment you may remove the SSB from the DNA in order to reuse the flow cell with another SSB variant or to repeat the experiment with fresh SSB.
22. Rinse the chip sensor surface by injecting 6 M Guanidinium hydrochloride at 100 μl/min for 15 s to remove the SSB from the ssDNA.
23. Equilibrate the flow cell with TBS50 as in step 13. The flow cell can now be reused for another experiment.

Data analysis

24. Measure the equilibrium RUs for each injection in both the control and experimental flow cell. Subtract the control RUs (flow cell 1) from the experimental value to determine the actual level of SIP binding.
25. Plot the equilibrium RU (after subtracting control) against the concentration of SIP to generate a binding curve (5, 7).
26. To determine the K_D for the interaction, fit the data with a nonlinear curve using a suitable program such as GraphPad Prism.
27. Fit the association and dissociation phases of the sensorgram with nonlinear least squares fits to determine the k_{on} and k_{off} rates (see Note 12). Several suitable programs exist for this type of analysis, including BIAevaluation (supplied by BIA-core), GraphPad Prism, or SCRUBBER-2 (available from the Center for Biomolecular Interaction Analysis at http://www.cores.utah.edu/Interaction/index.php). Omit the first few seconds of the associative and dissociative phases if they are particularly noisy. This should be repeated for all the injections (see Note 13).

4. Notes

1. The surface of the Biacore Sensor Chip SA consists of a carboxymethylated dextran pre-immobilized with streptavidin, which allows for the capture of biotinylated ligands.
2. The BIAcore system utilizes small flow cells that may be disrupted or damaged by particulates or air bubbles. Therefore, it is important to thoroughly degas all buffers.
3. It is important to load as little DNA as possible while maintaining a stable signal over background. If the concentration of DNA is too high, mass transport limitations may affect equilibrium and kinetic measurements (6, 8). A survey of the published literature demonstrates that groups have achieved success in similar experiments loading biotinylated oligo to ≈300–1,000 RUs (3, 4, 6).
4. SSB is dialyzed into high salt (200 mM NaCl) to aid in solubility. Immediately prior to loading onto the sensor surface, SSB is diluted with TBS50 to lower the salt concentration.
5. Any SSB variant that retains DNA binding can be substituted for wild-type SSB to examine the effect on the SSB–protein interaction.
6. SSB-tethering to DNA and the biacore surface is limited to conditions under which SSB association with the biotinylated ssDNA is essentially irreversible. Under other conditions (particularly high salt, detergent, etc.), SSB associations are predicted to be less stable. With all experiments it is important that potential SSB dissociation from the SPR chip surface be measured through (at least) the lifetime of the experiment.
7. A slight variation on this technique has been used to measure the DNA binding affinity of SSBs (8, 9).
8. The concentrations used must be altered depending on the dissociation constant (K_D) of the interaction. It may be useful to perform a preliminary experiment over a wide range of SIP concentrations to gain an idea of appropriate concentrations to use in order to generate a complete binding curve.
9. It is helpful to tap the bottoms of the vials to dislodge any air bubbles from the samples.
10. The exact flow rates and times required for association and dissociation will need to be empirically determined for each experiment. During the association phase it is important to ensure that the injections reach equilibrium, evidenced by reaching stable RUs. During the dissociation phase the RUs must return to a baseline reading indicating that the SIP has dissociated.

11. An assumption is made that the level of immobilized SSB remains constant throughout the experiment. To ensure this, compare the RU generated by an identical injection at the beginning and end of the experiment.

12. To ensure that mass transport is not limiting and causing an underestimation of binding kinetics it is important to repeat the experiment in other flow cells with different levels of immobilized DNA and SSB (vary by at least twofold). If the measured rate constants are not similar between these experiments then mass transport is a limiting factor. Increasing the flow rate and lowering the level of immobilized DNA and SSB can reduce the effect of mass transport issues. However, all measured rate constants must be considered lower estimates.

13. A "global analysis" may also be attempted in which the associative and dissociative phases for all injections are fitted simultaneously (10).

References

1. Jonsson U, Fagerstam L, Ivarsson B, Johnsson B, Karlsson R, Lundh K, Lofas S, Persson B, Roos H, Ronnberg I, Sjolander S, Stenberg E, Stahlberg R, Urbaniczky C, Ostlin H, Malmqvist M (1991) Real-time biospecific interaction analysis using surface-plasmon resonance and a sensor chip technology. BioTechniques 11:620–627
2. Fagerstam LG, Frostell-Karlsson A, Karlsson R, Persson B, Ronnberg I (1992) Biospecific interaction analysis using surface plasmon resonance detection applied to kinetic, binding site and concentration analysis. J Chromatogr 597:397–410
3. Witte G, Urbanke C, Curth U (2003) DNA polymerase III chi subunit ties single-stranded DNA binding protein to the bacterial replication machinery. Nucleic Acids Res 31:4434–4440
4. Purnapatre K, Handa P, Venkatesh J, Varshney U (1999) Differential effects of single-stranded DNA binding proteins (SSBs) on uracil DNA glycosylases (UDGs) from Escherichia coli and mycobacteria. Nucleic Acids Res 27:3487–3492
5. Page AN, George NP, Marceau AH, Cox MM, Keck JL (2011) Structure and biochemical activities of Escherichia coli MgsA. J Biol Chem. 286:12075–12085.
6. Kelman Z, Yuzhakov A, Andjelkovic J, O'Donnell M (1998) Devoted to the lagging strand-the subunit of DNA polymerase III holoenzyme contacts SSB to promote processive elongation and sliding clamp assembly. EMBO J 17:2436–2449
7. Nicholson MW, Barclay AN, Singer MS, Rosen SD, van der Merwe PA (1998) Affinity and kinetic analysis of L-selectin (CD62L) binding to glycosylation-dependent cell-adhesion molecule-1. J Biol Chem 273:763–770
8. Ehn M, Nilsson P, Uhlen M, Hober S (2001) Overexpression, rapid isolation, and biochemical characterization of Escherichia coli single-stranded DNA-binding protein. Protein Expression Purif 22:120–127
9. Scaltriti E, Tegoni M, Rivetti C, Launay H, Masson JY, Magadan AH, Tremblay D, Moineau S, Ramoni R, Lichiere J, Campanacci V, Cambillau C, Ortiz-Lombardia M (2009) Structure and function of phage p2 ORF34 (p2), a new type of single-stranded DNA binding protein. Mol Microbiol 73:1156–1170
10. Morton TA, Myszka DG, Chaiken IM (1995) Interpreting complex binding kinetics from optical biosensors: a comparison of analysis by linearization, the integrated rate equation, and numerical integration. Anal Biochem 227:176–185

Chapter 13

Use of Native Gels to Measure Protein Binding to SSB

Jin Inoue and Tsutomu Mikawa

Abstract

We describe a procedure to detect protein binding to SSB by polyacrylamide gel electrophoresis under non-denaturing conditions. As an example, we show the interaction of *Thermus thermophilus* (*Tth*) SSB with its cognate RecO protein. The interaction is detected as decay of the band corresponding to SSB by addition of RecO. We also demonstrate analysis of the RecO–RecR interaction as another example of this method.

Key words: Native-PAGE, Protein–protein interaction, Electrophoresis, SSB-binding protein, RecO, RecR, Mutational analysis

1. Introduction

1.1. Background

Bacterial SSB can interact with several proteins involved in DNA metabolism, and SSB acts as a hub protein in the DNA metabolism pathway (1, 2). Therefore, analysis of the interaction of SSB with other proteins is important for understanding the mechanism of DNA metabolism. At present, many methods and techniques are used to analyze protein–protein interactions. However, most of these methods require special instruments that are generally expensive. In addition, large amounts of protein are often required. In this chapter, we describe a simple method, which only employs polyacrylamide gel electrophoresis (PAGE) under non-denaturing conditions and needs only small amounts of proteins, to detect protein–protein interactions. This technique, native-PAGE, is suitable for the analysis of the subunit composition of protein complexes since the proteins retain their ternary and quaternary structure under electrophoresis (3). However, this method can also be used for monitoring protein–protein interactions.

In sodium dodecyl sulfate (SDS)-PAGE, all proteins migrate into the gel since the electric charge of each protein is canceled by binding to negatively charged SDS (4, 5). In contrast, in

James L. Keck (ed.), *Single-Stranded DNA Binding Proteins: Methods and Protocols*, Methods in Molecular Biology, vol. 922, DOI 10.1007/978-1-62703-032-8_13, © Springer Science+Business Media, LLC 2012

native-PAGE, the direction of migration and the mobility of each protein in the gel depend only on the protein's own charge. In our procedure, negatively charged proteins such as SSB enter the gel and form a band. In contrast, positively charged proteins do not migrate into the gel since these proteins move to the opposite side of the gel. If these two proteins interact with each other, decay of intensity of the band corresponding to SSB and/or appearance of the band corresponding to SSB–protein complex is observed.

Here, we describe a procedure to detect the interaction of SSB and RecO proteins using this method. In addition, we show the interaction between RecR and RecO as another example of this procedure.

1.2. Sample Result 1: Interaction Between SSB and RecO

SSB, which is a negatively charged protein, migrates into the gel and forms a band (Fig. 1a, lane 1). Since RecO is positively charged, like many proteins that interact with SSB, RecO does not migrate into the gel (Fig. 1a, lane 2). By this method, the SSB–RecO interaction is detected as the decay of intensity of the band corresponding to SSB with the addition of RecO (Fig. 1a, lanes 4–9). When a large amount of lysozyme, which is also a positively charged protein, was added to SSB as a control, decay was not observed, although there were some disturbances in the electrophoresis (Fig. 1a, lane 10). These results indicate that RecO specifically interacts with SSB. Thus, this method is useful for the easy detection of interactions between SSB and other proteins. In addition, this method is useful for screening mutant proteins that lack the ability to bind SSB. Based on the results of NMR measurements that suggest the SSB-binding site of RecO (2), some candidates for RecO mutants have been prepared by single amino acid substitution. Then, to examine which mutants have reduced ability to bind SSB, the native-PAGE method was employed (Fig. 1b). The result indicates that the R127E mutation drastically decreases the binding affinity of RecO for SSB (Fig. 1b, lane 5), whereas other mutants (Fig. 1b, lanes 7, 9, and 11) and the wild-type protein (Fig. 1b, lane 3) interact with SSB normally. The R127A RecO mutant also showed decreased binding affinity for SSB (data not shown, see Fig. 1c). To confirm the validity of this native-PAGE method, we also examined the interaction between SSB and R127A using surface plasmon resonance analysis (Fig. 1c). SSB was immobilized to the sensor chip, and the wild-type protein and R127A RecO mutant were injected. The wild-type protein showed an increase in the response unit by binding to immobilized SSB and gave normal sensorgrams (Fig. 1c left panel). In contrast, the R127A mutant did not give normal sensorgrams (Fig. 1c right panel), indicating that the R127A mutant lacks the ability to bind SSB. Thus, we have shown that the native-PAGE method is a simple and useful way to detect SSB–protein interactions using small amounts of protein.

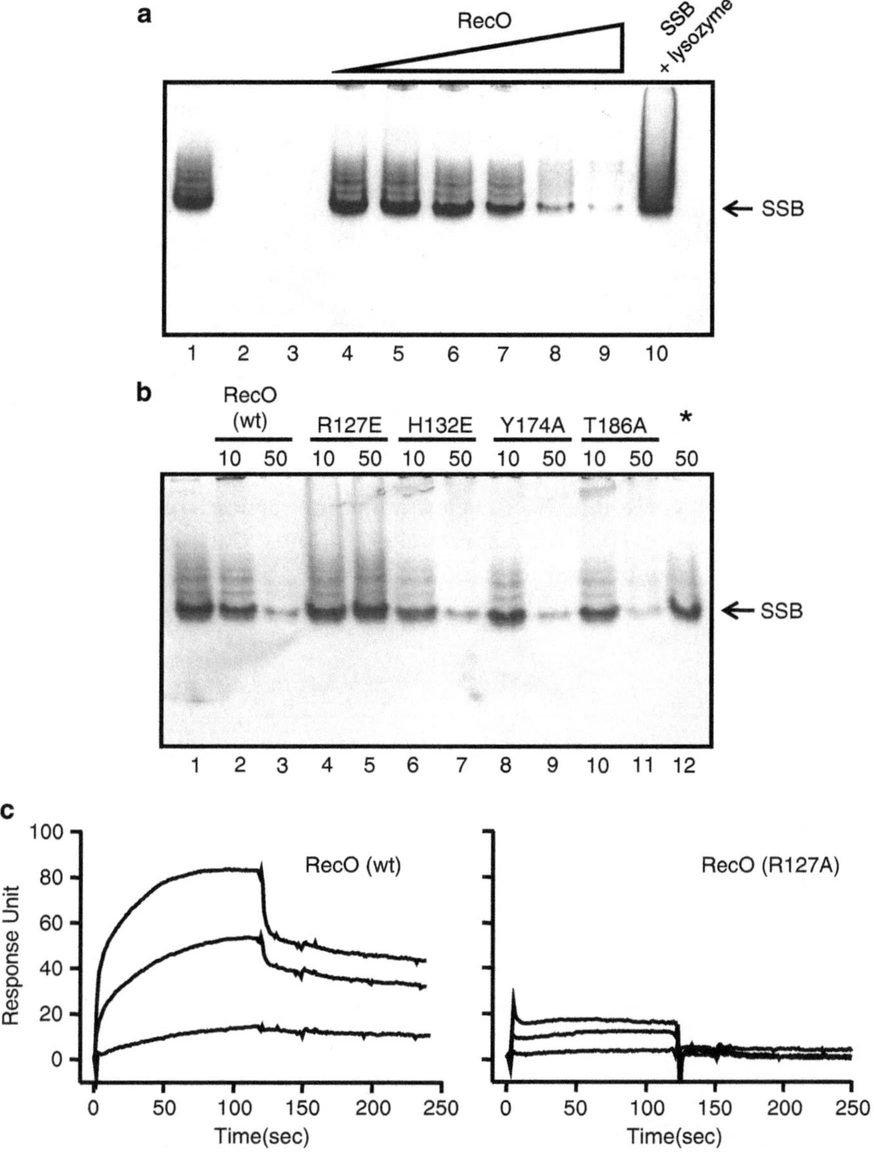

Fig. 1. Analysis of the interaction of SSB with RecO. (**a**) SSB (10 μM) was incubated with 1, 2, 5, 10, 20, or 50 μM RecO. Lane 1, SSB alone (10 μM); Lane 2, RecO alone (10 μM); Lane 3, lysozyme alone (10 μM). SSB (10 μM) was also incubated with 50 μM lysozyme as the control (lane 10). (**b**) Analysis of the interaction of SSB with RecO mutants by native-PAGE. SSB (10 μM) was incubated with 10 or 50 μM wild-type or mutant RecO. Lane 1, SSB alone (10 μM). SSB (10 μM) was also incubated with 50 μM cytochrome C as the control (*asterisk*). (**c**) Analysis of the interaction of SSB with RecO by surface plasmon resonance analysis. Wild-type RecO (*left*) or mutant RecO (*right*), 0.1, 0.5, or 1 μM (40 μl), was applied to an SSB-immobilized CM5 sensor chip. The response measured on a reference surface was subtracted from the response obtained from the SSB surface.

1.3. Sample Result 2: Interaction Between RecR and RecO

RecR forms a band in a similar way to SSB since RecR is also a negatively charged protein (Fig. 2a, lane 1). When RecO is added to RecR, the band corresponding to RecR decays and that corresponding to the RecR–RecO complex can be observed (Fig. 2a, lanes 3–5). The advantage of this native-PAGE method is that qualitative data can be easily obtained concerning protein–protein interactions. However, this method can also detect subtle differences in binding affinity of protein–protein interaction. For example, in our study, one result generated by native-PAGE indicated that the affinity of R127A RecO mutant for RecR seemed to be higher than that of wild-type RecO, since the degree of decay of the band corresponding to RecR caused by addition of the R127A mutant was slightly higher than that caused by addition of wild-type RecO (Fig. 2a). Surface plasmon resonance analysis supported this result, in which the dissociation constant of the RecR–wild-type RecO or RecR–R127A RecO mutant interaction were 6.6×10^{-8} M and 4×10^{-8} M, respectively (2). Thus, our native-PAGE method can also be used for simply comparing binding affinities in protein–protein interactions.

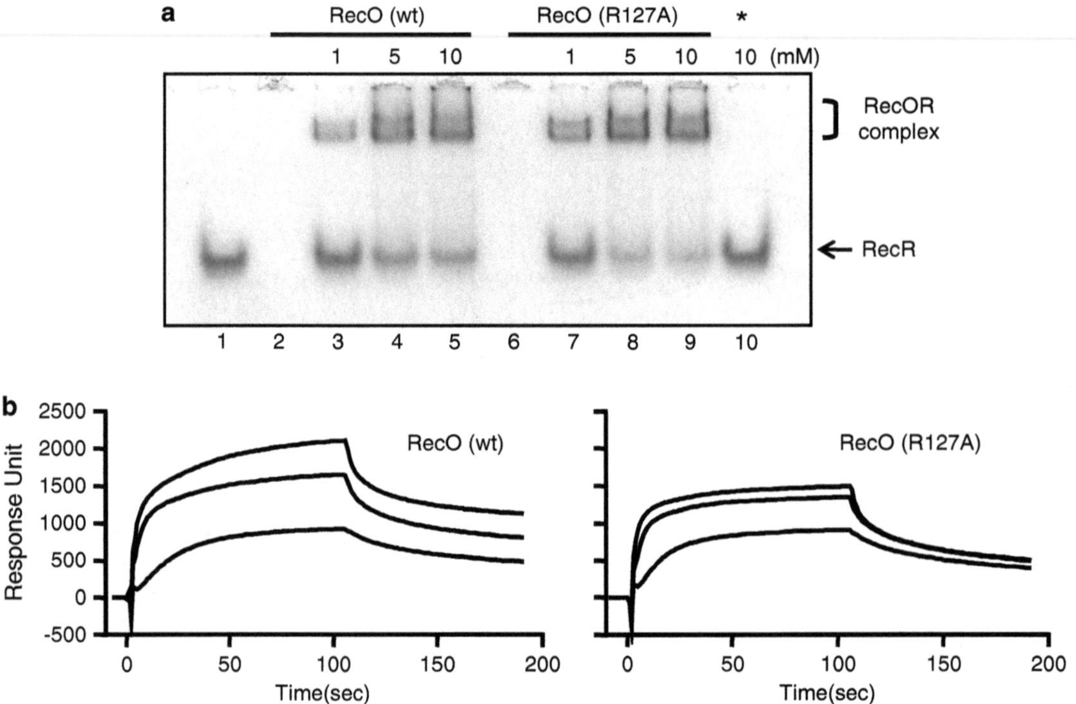

Fig. 2. Analysis of the interaction of RecR with RecO. (**a**) Analysis of the interaction of RecR with mutant RecO (R127A) by native-PAGE. RecR (5 μM) was incubated with 1, 5, or 10 μM wild-type or mutant RecO. Lane 1, RecR alone (5 μM). RecR (5 μM) was also incubated with 10 μM cytochrome C as the control (*asterisk*). (**b**) Analysis of the interaction of RecR with RecO by surface plasmon resonance analysis. Wild-type RecO (*left*) or mutant RecO (*right*), 0.1, 0.5, or 1 μM (40 μl), was applied to a RecR-immobilized CM5 sensor chip. The response measured on a reference surface was subtracted from the response obtained from the RecR surface.

2. Materials

Prepare all solutions using ultrapure water at room temperature and store reagents at room temperature (unless indicated otherwise).

1. 1 M Tris–HCl, pH 6.8 or pH 7.5: Add approximately 40 mL water to a 50 mL glass beaker. Weigh 6.06 g Tris and transfer to the beaker. Mix with a magnetic stirrer and adjust pH with HCl. Transfer to a graduated cylinder and make up to 50 mL with water.
2. 0.5 M Ethylenediamine tetraacetate (EDTA), pH 8.0: Weigh 9.3 g *di*-sodium dihydrogen ethylenediamine tetraacetate dihydrate and prepare a 50 mL solution as in item 1 (see Note 1).
3. Resolving gel buffer: 1.5 M Tris–HCl, pH 8.8. Weigh 9.09 g Tris and prepare a 50 mL solution as in item 1.
4. Acrylamide/bis solution (30%; 29:1 Acrylamide:Bis): Weigh 14.5 g acrylamide monomer and 0.5 g Bis (cross-linker) and transfer to a graduated cylinder. Add water and stir until dissolved. Make up to 50 mL with water. Store at 4°C in a dark bottle (see Notes 2 and 3).
5. Ammonium persulfate (APS): 10% solution in water. Weigh 1 g APS and transfer to a 15-mL polypropylene tube. Add water and stir until dissolved. Make up to 10 mL with water. Store at 4°C (see Note 3).
6. *N*,*N*,*N*,*N*′-tetramethyl ethylenediamine (TEMED): Store at 4°C.
7. Running buffer: 25 mM Tris, 192 mM glycine. Weigh 3 g Tris and 14.4 g glycine and make up to 1 L with water.
8. Sample buffer (5×): 125 mM Tris–HCl, pH 6.8, 50% glycerol, 0.5 mg/mL bromophenol blue (BPB). Add 5 mL glycerol and 1.25 mL 1 M Tris–HCl, pH 6.8 to a 15-mL polypropylene tube. Make up to 10 mL with water and mix well using a vortex mixer. Add 5 mg BPB to the solution and mix well again (see Note 4).
9. Reaction buffer (10×): 500 mM Tris–HCl, pH 7.5, 20 mM EDTA. Add 5 mL 1 M Tris–HCl, pH 7.5 and 0.4 mL 500 mM EDTA to a 15-mL polypropylene tube. Make up to 10 mL with water.
10. CBB staining solution: 0.25% Coomassie brilliant blue R-250, 5% methanol, 7.5% acetic acid. Add 100 mL water to a 200-mL plastic bottle. Add 10 mL methanol and 15 mL acetic acid and make up to 200 mL with water. Add 0.5 g CBB R-250 to the solution and mix well.

11. Destaining solution: 25% methanol, 7.5% acetic acid. Add 100 mL water to a 200-mL plastic bottle. Add 50 mL methanol and 15 mL acetic acid and make up to 200 mL with water.

3. Methods

Perform all the following steps at room temperature (unless indicated otherwise).

3.1. Preparation of a 10% Polyacrylamide Gel

1. Clean glass plates (short plate: 10 cm × 7.4 cm, and plate with spacer: 10 cm × 8.3 cm × 1 mm) using 70% ethanol and cleaning tissue, and assemble the glass plates using a gel casting kit (see Note 5).
2. Mix 1.5 mL of resolving buffer, 2 mL of 30% acrylamide/bis solution, 2.5 mL of water, and 60 μL of APS in a 15-mL polypropylene tube. Stand the tube on ice for a few minutes (see Note 6). Add 6 μL of TEMED and stir the solution using a vortex mixer.
3. Pour the solution into the space between the glass plates quickly until the space is filled. If air bubbles are introduced, remove the air bubbles by tapping.
4. Insert a 12-well gel comb without introducing air bubbles.
5. Stand the gel for approximately 30 min (see Note 7).
6. After gelation, remove the gel comb and wrap the gel with plastic wrap. Store at 4°C until use (see Note 8).

3.2. Sample Preparation

The following samples are for the Subheading 1.2

1. Prepare *Thermus thermophilus* SSB and RecO as described previously (6, 7) (see Note 9).
2. Prepare lysozyme at a high concentration (several hundred μM) in 1× reaction buffer containing 50% glycerol (see Note 10).
3. Prepare reaction mixtures (10 μL) containing 10 μM SSB and 1, 2, 5, 10, 20, or 50 μM RecO in 1× reaction buffer. As control experiments, prepare reaction mixtures containing 10 μM SSB, RecO, or lysozyme alone.
4. For a negative control, prepare a reaction mixture containing 10 μM SSB and 50 μM lysozyme.
5. Incubate the reaction mixtures at 50°C for 5 min (reaction mixtures containing lysozyme should be incubated at 37°C, see Note 11).
6. Add 2.5 μL of sample buffer to the samples and mix well (see Note 12).

3.3. Electrophoresis

1. Assemble the gel cassette using the PAGE system. Pour the running buffer into the PAGE system until the electrodes of the PAGE system are completely submerged. Clean and wash the loading wells using a syringe.
2. Load the samples gently into the loading wells (10 μL/well).
3. Run at a constant current of 10 mA for 1.5 h (see Note 13).
4. After electrophoresis, carefully detach the gel from the glass plates and transfer to a Tupperware container.
5. Add CBB staining solution until the gel is completely submerged. Heat the solution using a microwave oven until almost boiling. Agitate the container gently for 10 min.
6. Remove the CBB staining solution completely. Rinse the gel with water a few times. Add destaining solution until the gel is completely submerged. Add 1 piece of tissue (see Note 14). Heat the solution using a microwave oven until almost boiling. Agitate gently until bands are clearly visible (see Note 15).
7. After destaining, remove the destaining solution and rinse the gel with water three times for 20 min each time.
8. To preserve the gel, place the gel on filter paper, cover with cellophane, and desiccate using a gel dryer.

4. Notes

1. To dissolve EDTA, the pH must be adjusted using NaOH.
2. Acrylamide and its solution are harmful. When handling these reagents, disposable gloves and a mask should be worn, and the acrylamide powder must be carefully weighed, taking care not to scatter any powder.
3. This solution can be stored for about 6 months.
4. To dissolve BPB easily, the BPB should be added to the buffer last.
5. Glass plates purchased from many suppliers can be used. The volume of gel solution in the next step (3.1.2) must be adjusted according to the size of the glass plates.
6. If the solution is not sufficiently chilled, gelation may begin before pouring is complete.
7. If room temperature is below 25°C, the waiting time should be extended. Check the state of gelation by observing the small amount of gel solution remaining in the 15-mL tube.
8. The gel can be stored for up to 1 month. However, the freshness of the gel affects the resolution of PAGE.

9. Other SSBs and proteins that interact with the SSBs can also be used.
10. For a negative control experiment, a positively charged protein, lysozyme, is used instead of RecO. Cytochrome C can also be used.
11. Samples should be incubated at an appropriate temperature. Since we used proteins from a thermophile, we used an incubation temperature of 50°C.
12. Do not heat the samples.
13. For proteins that are not thermostable, electrophoresis should be performed at 4°C. Check the temperature of the running buffer at intervals during electrophoresis. The use of a large volume of running buffer may prevent an excessive increase in temperature.
14. Addition of a piece of tissue facilitates destaining.
15. Exchange of destaining solution will also facilitate destaining.

References

1. Costes A, Lecointe F, McGovern S, Quevillon-Cheruel S, Polard P (2010) The C-terminal domain of the bacterial SSB protein acts as a DNA maintenance hub at active chromosome replication forks. PLoS Genet 6:e1001238
2. Inoue J, Nagae T, Mishima M, Ito Y, Shibata T, Mikawa T (2011) A mechanism for single-stranded DNA-binding protein (SSB) displacement from sindle-stranded DNA upon SSB-RecO interaction. J Biol Chem 286:6720–6732
3. Wittig I, Schagger H (2005) Advantages and limitations of clear-native PAGE. Proteomics 5:4338–4346
4. Weber K, Osborn M (1969) The reliability of molecular weight determinations by dodecyl sulfate-polyacrylamide gel electrophoresis. J Biol Chem 244:4406–4412
5. Laemmli UK (1970) Cleavage of structural proteins during the assembly of the head of bacteriophage T4. Nature 227:680–685
6. Inoue J, Shigemori Y, Mikawa T (2006) Improvement of rolling circle amplification (RCA) efficiency and accuracy using *Thermus thermophilus* SSB mutant protein. Nucleic Acids Res 34:e69
7. Inoue J, Honda M, Ikawa S, Shibata T, Mikawa T (2008) The process of displacing the single-stranded DNA-binding protein from single-stranded DNA by RecO and RecR proteins. Nucleic Acids Res 36:94–109

Chapter 14

Identification of Small Molecules That Disrupt SSB–Protein Interactions Using a High-Throughput Screen

Douglas A. Bernstein

Abstract

Bacterial single-stranded DNA-binding proteins (SSBs) recruit a diverse array of genome maintenance enzymes to their sites of action through direct protein interactions. The essential nature of these SSB–protein interactions makes inhibitors that block SSB-partner complex formation valuable biochemical tools and attractive potential antibacterial agents. However, many of these protein–protein interactions are weak and not amenable to the high-throughput nature of small molecule screens. Here I describe a high-throughput screen to identify small molecules that inhibit the interaction between Exonuclease I (ExoI) and the final 10 amino acids of the SSB C-terminal tail (SSB-Ct). The strength of the binding between ExoI and the SSB-Ct tail is fundamental to the interaction's utility in the high-throughput screen.

Key words: Single-stranded binding protein, Small molecule screen, Protein–protein interaction, Antibiotic target, High-throughput screen

1. Introduction

Although DNA unwinding processes are obligatory, they also present intrinsic risks to cells. Single-stranded DNA (ssDNA) is sensitive to damage and can self-associate in turn creating structural impediments to genome maintenance reactions. To protect and stabilize unwound ssDNA, cells have evolved specialized SSBs that bind DNA with high affinity and in a sequence-independent manner (1–3). Most bacterial SSBs function as homotetramers, with each subunit containing an N-terminal DNA-binding/oligomerization (OB) domain and a C-terminal tail segment (2, 4). While the C-terminal tail segment can vary in length, the SSB-Ct is highly conserved among bacteria. All known protein–protein interactions between bacterial SSBs and their binding partners are mediated through the SSB-Ct (5–15). This structural arrangement is distinct

James L. Keck (ed.), *Single-Stranded DNA Binding Proteins: Methods and Protocols*, Methods in Molecular Biology, vol. 922,
DOI 10.1007/978-1-62703-032-8_14,

from the major eukaryotic SSB (Replication Protein A), which functions as a heterotrimer and lacks the C-terminal tail element found in bacterial SSBs (16). Interaction between the SSB-Ct and cellular genome maintenance machinery is essential in *Escherichia coli*, and given the conservation of the SSB-Ct, such interactions are likely to be common among bacteria (17–21).

One well-characterized protein-binding partner of *E. coli* SSB is ExoI (5, 22). The X-ray crystal structure of *E. coli* ExoI bound to a peptide comprising the SSB-Ct sequence has provided a molecular model of SSB–protein interactions (4). The SSB-Ct sequence includes highly conserved acidic and hydrophobic segments (Asp-Asp-Asp-Ile-Pro-Phe in *E. coli* SSB), and in this structure the C-terminal-most Phe of the SSB-Ct sequence forms a critical contact with ExoI in which the Phe side chain is enveloped in a hydrophobic pocket and its α-carboxyl group is bound by an Arg side chain from ExoI. Intimate recognition of the SSB-Ct Phe appears to be a conserved feature in other SSB–protein interactions (5, 11). Roles for the acidic SSB-Ct residues in mediating interaction with ExoI have also been identified (4) and the SSB-Ct tail is essential in vivo (23). The identification of this binding scheme has raised a number of questions as to the conservation of SSB-Ct binding sites among its many binding partners and the consequences of inhibiting interactions with SSB in reconstituted systems and in cells. To begin to answer these questions, a unique set of small molecule biochemical tools for probing the roles of SSB–protein interactions were developed. These results have been published (24).

2. Materials

2.1. In House

1. pLysS plasmid (Novagen).
2. pET15b plasmid (New England Biolabs).
3. Ampicillin.
4. Chloramphenicol.
5. Isopropyl β-D-thiogalactopyran.
6. Tris.
7. Imidazole.
8. β-mercaptoethanol.
9. Glycerol.
10. NaCl.
11. Bovine serum albumin.

12. $MgCl_2$.
13. Centrifuge (Beckman Coulter Avanti J-25).
14. Rotor (Beckman JLA-10.5).
15. Sonicator (cell disruptor Heat systems-Ultrasonicas W 200 R).
16. Thrombin.
17. Ni-NTA Agarose (Qiagen).
18. Bio-Rad 20 ml plastic columns.
19. AKTA FPLC.
20. Hi prep 16/60 sephacryl/S-300 high resolution (GE healthcare).
21. Cary 100 Bio UV-visible spectrophotometer.
22. Guanidine–HCl.
23. Panvera Beacon 2000 FP system.
24. Glass tubes (Kimble Chase Borosilicate glass culture tubes 6×50 mm).
25. Dimethyl Sufoxide.
26. SSB-Ct peptide (sequence: Trp-Met-Asp-Phe-Asp-Asp-Asp-Ile-Pro-Phe).
27. F-SSB-Ct (SSB-Ct with a N-terminal fluorescein).

2.2. Small Molecule Screening Materials

1. Corning Costar 384-well black plates.
2. Biomek FX Liquid handling system.
3. Tecan Safire II.
4. Biomek FX tips 20 μL and 250 μL.

2.3. Buffers

1. Lysis buffer: 20 mM Tris–HCl, pH 8.0, 20 mM imidazole, pH 8.0, 300 mM NaCl, 1 mM β-mercaptoethanol, 10% glycerol.
2. Elution buffer: 20 mM Tris–HCl, pH 8.0, 100 mM imidazole, pH 8.0, 300 mM NaCl, 1 mM β-mercaptoethanol, 10% glycerol.
3. Dialysis buffer: 20 mM Tris–HCl, pH 8.0, 300 mM NaCl, 1 mM β-mercaptoethanol, 10% glycerol.
4. Binding buffer: 20 mM Tris–HCl, pH 8.0, 100 mM NaCl, 1 mM $MgCl_2$, 1 mM β-mercaptoethanol, 0.1 g/L bovine serum albumin, 4% (v/v) glycerol, 3.33% (v/v) DMSO.
5. Master Mix buffer: 20 mM Tris–HCl, pH 8.0, 100 mM NaCl, 1 mM $MgCl_2$, 1 mM β-mercaptoethanol.
6. Storage Buffer: 20 mM Tris–HCI, pH 8.0, 300 mM NaCl, 1 mM β-mercaptoethanol, 40% glycerol.

3. Methods

3.1. Exonuclease I Protein Purification

1. Amplify *Escherichia coli exoI* from strain DH5α using PCR with primers specific for the 5′ and 3′ ends of the protein-coding region of the gene. The primers incorporate *Nde*I (5′) and *Bam*HI (3′) restriction sites to allow subcloning into the pET15b bacterial over-expression plasmid, allowing inducible over-expression of ExoI protein that incorporates an N-terminal hexahistidine Ni^{2+}-affinity purification tag and a thrombin cleavage site (20 residues total). Sequence the recombinant *exoI* gene to confirm the proper DNA sequence for the gene.
2. Grow 8 L of *Escherichia coli* BL21(DE3) cells transformed with pLysS (Novagen) and the ExoI overexpression plasmid at 37°C in Luria–Bertani medium supplemented with 100 μg/ml ampicillin and 25 μg/ml chloramphenicol. Induce protein expression in cells at early logarithmic phase (OD_{600} of 0.3–0.4) by the addition of 1 mM isopropyl β-D-thiogalactopyranoside and harvest by centrifugation.
3. Resuspend cells in lysis buffer and lyse by sonication at 100% power for 3 × 2 min on ice with a 1 min rest on ice in between each interval. Perform all subsequent purification steps at 4°C. Pour a fresh 2 ml Ni^{2+}–NTA column into a plastic column and wash with 10 column volumes of lysis buffer. Load soluble lysate onto the column and wash with lysis buffer until protein is undetectable in the eluent. Elute His-tagged protein with elution buffer.
4. Dialyze the eluent against lysis buffer, digest with thrombin at room temperature for 1 h to remove the His-tag (a Gly-Ser-His sequence remains on the N-terminus of ExoI) and pass over a Ni^{2+}–NTA column to remove *E. coli* proteins that fortuitously bind to Ni^{2+}–NTA resin (see Note1).
5. Dialyze ExoI into dialysis buffer and load onto a Sephacryl S-300 column (Pharmacia) equilibrated in the same buffer. Identify highly purified fractions using polyacrylamide gel electrophoresis (PAGE), pool and concentrate to >1 mg/ml prior to storage at 4°C. Determine protein concentration by measuring A_{280} in 6.0 M guanidine–HCl (see Note 2).

3.2. High-Throughput Inhibitor Screen

1. Incubate 1 μM *E. coli* ExoI with 10 nM F-SSB-Ct in binding buffer. After 5 min, add 0–200 μM unlabeled SSB-Ct at room temperature and incubate for an additional 5 min at room temperature. Measure fluorescence polarization (FP) at 25°C using a Panvera Beacon 2000 FP system with 490 nm excitation and 535 nm emission wavelengths with three replicates. As the concentration of unlabeled peptide increases,

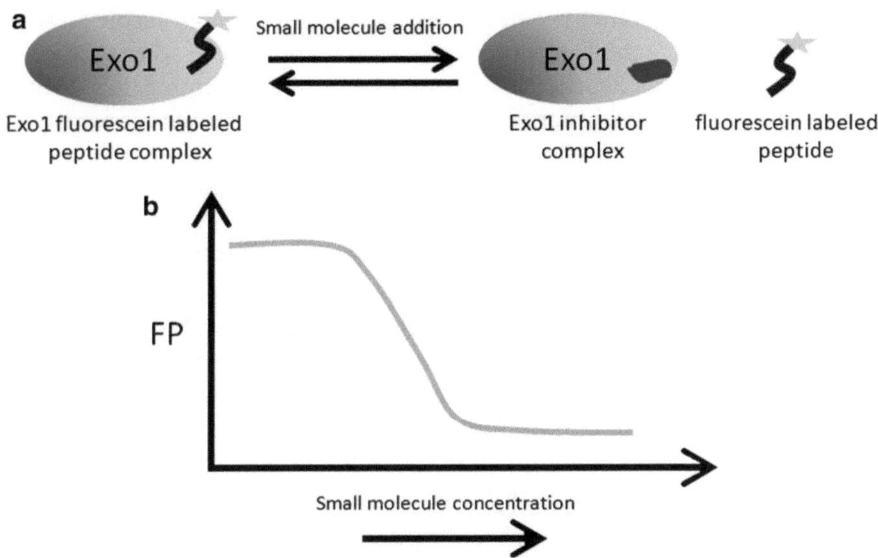

Fig. 1. Experimental design of small molecule screen. (**a**) Schematic model depicting the effects of small molecule competing with fluorescein-labeled SSB-Ct peptide for binding ExoI. (**b**) Drop in FP values due to increased dissociation of fluorescein-labeled SSB-Ct from ExoI.

dissociation of the fluorescein-labeled peptide increases, causing FP values to drop to near unbound peptide levels (see Notes 3–5).

2. The unlabeled peptides ability to compete away bound fluorescein-labeled peptide suggests that the framework of the proposed assay had the potential to identify small molecules that decreased fluorescein-labeled peptide binding (Fig. 1a, b) (see Notes 4–7).
3. To determine the optimal concentration of compound library for the high-throughput screen, perform a small-scale screen of one plate from each chemical library at (1 μL of 1 mM or 100 μM stock in 100% DMSO) to 30 μL of ExoI/F-SSB-Ct. (see Note 8–11).
4. Incubate a mixture of 1 μM *E. coli* ExoI and 10 nM F-SSB-Ct peptide in master mix buffer for 5 min at room temperature. Use a liquid handling system (e.g., Biomek FX) to add small molecule to ExoI/F-SSB-Ct mixture in black-walled 384-well plates. Measure the FP of each reaction using a Tecan Saphire II. Include control reactions using SSB-Ct peptide (positive control, 1 μL of 2 mM stock in 100% DMSO) or DMSO (negative control, 1 μL of 100% DMSO) on each plate (see Notes 12 and 13). From this small-scale screen, determine the number of significant inhibitors that are likely to be identified per 384-well plate of small molecules.

5. Once the best concentration of small molecule is established for the high-throughput screen, screen compounds from the Chemical Diversity, Maybridge, Chembridge, or other chemical libraries in an analogous fashion (see Notes 14–16).

3.3. Identification of Small Molecule Hits

1. Measure mean fluorescence for each plate; compounds that lower FP values more than two standard deviations of the mean fluorescence are considered hits.
2. Remove any small molecules that had been previously identified as fluorescent, fluorescence quenching, or were adjacent to highly fluorescent compounds from consideration. Also, screening facilities may have databases that provide information on compounds that had been previously identified as hits in multiple FP-based screens. These may also be removed from consideration (see Note 17).
3. Upon completion, repeat FP inhibition tests with a practical number of compounds (30–100) by hand.

3.4. Retesting Hits at Home

1. Incubate 1 μM *E. coli* ExoI with 10 nM F-SSB-Ct (SSB-Ct with an N-terminal fluorescein) and 0–200 μM small molecule at room temperature for 5 min in binding buffer. Measure FP at 25°C with 490 nm excitation and 535 nm emission wavelengths for three replicates (see Note 18).
2. This should allow you to reduce the number of "hits" from the high-throughput screen to a smaller number (~4–10) for subsequent mechanism of action studies (see Note 15).

4. Notes

1. If ExoI protein is greater than 95% pure after step 4 of the purification protocol, further purification using size exclusion is not needed for the FP assays herein described. However, a more thorough purification of ExoI may be needed for further biochemical characterization of the inhibitor compounds not described in this chapter.
2. If purified ExoI is not likely to be used within 1 week of purification for a high-throughput assay, it should be dialyzed into storage buffer and stored at −20°C. ExoI can be stored at −20°C for up to 2 months.
3. ExoI's stability at room temperature is advantageous, as the high-throughput liquid handling systems we had access to operate at these temperatures.
4. While testing to determine if the fluorescent peptide can be dissociated from the protein by unlabeled peptide, the FP

values may never drop to unbound FP values since small amounts of unspecific binding may occur between the fluorescent peptide and ExoI.

5. Stock concentrations of fluorescent peptide are often times extremely high. Dilution of the peptide from these stock concentrations before adding it to the peptide-ExoI master mix allows for more consistent allocation of the peptide.
6. The only difference between the buffer master mix for the screen and the binding buffer is 3.33% DMSO in the binding buffer. This is the same amount of DMSO that is added to each reaction with the addition of small molecule.
7. Most small molecule libraries are stored in DMSO, thus the development of the assay should take into consideration the final concentration of DMSO that will be present after addition of the small molecule.
8. DMSO and unlabeled peptide in DMSO controls placed in multiple wells or an entire plate can help gauge the reproducibility of inhibition and how evenly fluorescent peptide is distributed to each well.
9. Depending upon the scale (number of compounds) to be tested in the small molecule screen, the amount of enzyme and fluorescently labeled peptide master-mix required will vary. Making a single master mix allows for more consistent data collection, but after the peptide is bound to ExoI, I have found it cannot be kept for longer than 1–2 h before inhibition with wild type peptide becomes inconsistent.
10. ExoI binding strongly to the SSB-Ct and the stability of the complex in 3.33% DMSO at room temperature made it an excellent candidate to perform a high-throughput screen at our small molecule facility. Developing an assay that is compatible with conditions present in the screening facility which your assay will be performed is crucial.
11. Inconsistent readings of the edge wells by a plate reader can make data interpretation difficult. We did not collect any data along the first or last columns of each plate since we knew these readings would be less reproducible.
12. The criteria by which our assay was deemed suitable for a high-throughput screen was based on the screening centers previous experience that any more than one to two significant hits per 384-well plate would make it difficult for us to perform adequately focused follow up experiments. The small-scale small molecule screen allowed us to determine the concentration of the small molecule compound library necessary to selectively inhibit peptide binding at roughly this frequency.

13. For our assay, black-walled plates worked best to shield adjoining wells from fluorescent small molecules in the screen. Highly fluorescent compounds still influenced measurements in adjacent wells.
14. We found that addition of 1 μL of 1 mM small molecule to 30 μL of master mix worked well in our screen giving a hit rate of roughly 0.1–1%. However, if our hit rate had been considerably higher, decreased small molecule concentration or alteration of our binding assay would have been necessary in order to ensure practical levels of stringency for the screen.
15. Not all small molecules in all small molecule libraries are available for purchase. Before undertaking a small molecule screen, consideration of how identified hits will be further examined is important.
16. Many liquid handling robotics systems are equipped to use 1,536-well and 384-well plates. While there are obvious advantages to the higher throughput the 1,536-well plates afford, the relative ease in which we could purify ExoI, purchase peptide, and the size of the libraries we intended to screen made the 384-well platform sufficient for our needs.
17. If screening with libraries that have been previously characterized, the researcher is able to disregard small molecules that produce a significant FP value change but have also been shown to alter FP readings in prior FP-based assays. This significantly streamlines downstream hit identification and verification.
18. The ability to quickly retest hits at home is essential. We found ~80% of our hits had differences between the readings given on the Tecan and Panvera FP detection systems. Potential reasons for these differences are differences in the way the small molecules were preserved for screening, differences in screening technique (glass tube versus plastic plate), and the number of freeze thaw cycles the chemicals had undergone.

References

1. Meyer RR, Laine PS (1990) The single-stranded DNA-binding protein of Escherichia coli. Microbiol Rev 54(4):342–380
2. Shereda RD et al (2008) SSB as an organizer/mobilizer of genome maintenance complexes. Crit Rev Biochem Mol Biol 43(5):289–318
3. Lohman TM, Ferrari ME (1994) Escherichia coli single-stranded DNA-binding protein: multiple DNA-binding modes and cooperativities. Annu Rev Biochem 63:527–570
4. Lu D, Keck JL (2008) Structural basis of Escherichia coli single-stranded DNA-binding protein stimulation of exonuclease I. Proc Natl Acad Sci USA 105(27):9169–9174
5. Lu D et al (2009) Peptide inhibitors identify roles for SSB C-terminal residues in SSB/exonuclease I complex formation. Biochemistry 48 (29):6764–6771
6. Buss JA, Kimura Y, Bianco PR (2008) RecG interacts directly with SSB: implications for

stalled replication fork regression. Nucleic Acids Res 36(22):7029–7042

7. Cadman CJ, McGlynn P (2004) PriA helicase and SSB interact physically and functionally. Nucleic Acids Res 32(21):6378–6387
8. Genschel J, Curth U, Urbanke C (2000) Interaction of E. coli single-stranded DNA binding protein (SSB) with exonuclease I. The carboxy-terminus of SSB is the recognition site for the nuclease. Biol Chem 381(3):183–192
9. Glover BP, McHenry CS (1998) The chi psi subunits of DNA polymerase III holoenzyme bind to single-stranded DNA-binding protein (SSB) and facilitate replication of an SSB-coated template. J Biol Chem 273 (36):23476–23484
10. Kelman Z et al (1998) Devoted to the lagging strand-the subunit of DNA polymerase III holoenzyme contacts SSB to promote processive elongation and sliding clamp assembly. EMBO J 17(8):2436–2449
11. Shereda RD et al (2009) Identification of the SSB binding site on E. coli RecQ reveals a conserved surface for binding SSB's C terminus. J Mol Biol 386(3):612–625
12. Yuzhakov A, Kelman Z, O'Donnell M (1999) Trading places on DNA – a three-point switch underlies primer handoff from primase to the replicative DNA polymerase. Cell 96 (1):153–163
13. Suski C, Marians KJ (2008) Resolution of converging replication forks by RecQ and topoisomerase III. Mol Cell 30(6):779–789
14. Shereda RD, Bernstein DA, Keck JL (2007) A central role for SSB in Escherichia coli RecQ DNA helicase function. J Biol Chem 282 (26):19247–19258
15. Arad G et al (2008) Single-stranded DNA-binding protein recruits DNA polymerase V to primer termini on RecA-coated DNA. J Biol Chem 283(13):8274–8282
16. Wold MS (1997) Replication protein A: a heterotrimeric, single-stranded DNA-binding protein required for eukaryotic DNA metabolism. Annu Rev Biochem 66:61–92
17. Wang TC, Smith KC (1982) Effects of the ssb-1 and ssb-113 mutations on survival and DNA repair in UV-irradiated delta uvrB strains of Escherichia coli K-12. J Bacteriol 151 (1):186–192
18. Vales LD, Chase JW, Murphy JB (1980) Effect of ssbA1 and lexC113 mutations on lambda prophage induction, bacteriophage growth, and cell survival. J Bacteriol 143(2):887–896
19. Greenberg J et al (1974) exrB: a malB-linked gene in Escherichia coli B involved in sensitivity to radiation and filament formation. Genet Res 23(2):175–184
20. Greenberg J, Donch J (1974) Sensitivity to elevated temperatures in exrB strains of Escherichia coli. Mutat Res 25(3):403–405
21. Meyer RR et al (1980) A temperature-sensitive single-stranded DNA-binding protein from Escherichia coli. J Biol Chem 255 (7):2897–2901
22. Sandigursky M et al (1996) Protein-protein interactions between the Escherichia coli single-stranded DNA-binding protein and exonuclease I. Radiat Res 145(5):619–623
23. Curth U et al (1996) In vitro and in vivo function of the C-terminus of Escherichia coli single-stranded DNA binding protein. Nucleic Acids Res 24(14):2706–2711
24. Lu D et al (2010) Small-molecule tools for dissecting the roles of SSB/protein interactions in genome maintenance. Proc Natl Acad Sci USA 107(2):633–638

Chapter 15

Detection of Posttranslational Modifications of Replication Protein A

Cathy S. Hass, Ran Chen, and Marc S. Wold

Abstract

Replication Protein A (RPA) is a single-strand DNA-binding protein that is found in all eukaryotes. RPA is subjected to multiple posttranslational modifications including serine- and threonine-phosphorylation, poly-ADP ribosylation, and SUMOylation. These modifications are believed to regulate RPA activity through modulating interactions with DNA and partner proteins. This article describes two methods used to detect posttranslational modified RPA: immunofluorescence and immmuoblotting.

Key words: DNA replication, DNA Repair, Recombination, Immunoblotting, Immunofluorescence, Protein phosphorylation

1. Introduction

Replication protein A (RPA) is a conserved single-stranded DNA-binding protein that is found in all eukaryotic cells (1, 2). RPA binds ssDNA intermediates in DNA replication, DNA repair, and recombination and is involved in damage recognition in the cellular response to DNA damage. RPA is a stable complex composed of three subunits of approximately 70-, 32-, and 14-kDa (referred to as RPA1, RPA2, and RPA3, respectively). RPA is subjected to multiple posttranslational modifications. These include phosphorylation of serine and threonine residues, poly-ADP ribosylation, and SUMOylation (1, 3, 4). Although limited phosphorylation of RPA is observed in S and M phases, posttranslational modification of RPA primarily occurs after DNA damage (4, 5). These modifications have been mapped to both the RPA1 and RPA2 subunits. The N-terminus of RPA2 and the C-terminal domain of RPA1 are hyperphosphorylated and SUMOylated (respectively) after cells are exposed to DNA damage (4, 5). The role(s) of posttranslational modifications of RPA are not well understood but are thought to modulate RPA activity and help coordinate the cellular DNA damage response (5–7).

James L. Keck (ed.), *Single-Stranded DNA Binding Proteins: Methods and Protocols*, Methods in Molecular Biology, vol. 922, DOI 10.1007/978-1-62703-032-8_15,

Immunofluorescence microscopy and immunoblotting are two powerful techniques for identifying posttranslationally modified proteins (see also Chapter 16 in this volume). Antibodies to specific posttranslational modifications can be used to detect modified proteins. There are also a number of antibodies that interact with specific phosphorylated residues in RPA that are commercially available (8, 9). This allows phosphorylation of specific sites on RPA2 to be monitored either by immunoblotting or immunofluorescence. The methods for detecting these modifications of RPA will be described below.

1.1. Detection of Posttranslationally Modified RPA by Immunofluorescence

Immunofluorescence can be used to examine the localization of proteins throughout the cell and to determine localization of specific forms of proteins in response to different cell conditions. RPA is normally localized in the nucleus of cells and shows diffuse nuclear staining by immunofluorescence (10). When cells are exposed to DNA damage, RPA localizes to sites of DNA damage resulting in a punctate staining pattern (11, 12). This localized RPA is tightly associated with chromatin and can be visualized by immunofluorescence after detergent extraction (13). RPA is also phosphorylated and SUMOylated after DNA damage. Phospho-residue-specific antibodies to RPA can be used to visualize chromatin-associated modified RPA.

1.2. Detection of Posttranslationally Modified RPA by Immunoblotting

The mobility of RPA2 and RPA1 is reduced when phosphorylated or SUMOylated, respectively. Antibodies to native RPA can be used to detect these changes in mobility after separation by SDS-PAGE using immunoblotting. Depending on the number of phosphorylated residues, the change in mobility observed after phosphorylation of RPA2 can be small. To obtain optimal separation of the different phosphorylated forms of RPA, it is necessary to separate RPA on high percentage (12–13%) polyacrylamide gels (14, 15). If it is desired to analyze multiple subunits of RPA simultaneously, the size distribution of RPA subunits (70-, 32-, and 14-kDa) calls for full size, gradient gels. Gradient gels can be purchased commercially or poured in house. The protocol below includes a description for pouring gradient SDS-PAGE gels. Phosphorylated RPA can also be detected by immunoblotting with specific anti-phospho-residue RPA2 antibody.

2. Materials

All the buffers and solutions are made in ultrapure deionized water. Buffers for immunofluorescence staining are stored at 4°C unless otherwise noted. Procedures are carried out at room temperature unless otherwise noted.

2.1. Cell Culture Components

1. 6-Well cell culture polystyrene plates.
2. 22 × 22 Cover glass.
3. Dulbecco's Modified Eagle Medium (DMEM) plus 10% Calf Serum.
4. HeLa cells.
5. DNA-damaging agent: camptothecin, dissolved at 14.4 mM in DMSO to make stock solution.

2.2. Immunofluoresence Staining Components

1. CSK buffer: 10 mM HEPES (diluted from 1 M stock at pH 7.8), 300 mM sucrose, 100 mM NaCl, and 3 mM $MgCl_2$ in sterile water. To make 1 L of CSK buffer, weigh 102.7 g of sucrose, 5.84 g of NaCl, and 0.61 g of $MgCl_2$ and add to 800 mL of water. Add 10 mL of 1 M stock of HEPES. Bring volume to 1 L with water.
2. 0.5% TritonX-100 CSK: 0.5% TritonX-100, 10 mM HEPES, 300 mM sucrose, 100 mM NaCl, and 3 mM $MgCl_2$ in sterile water. To make 1 L of 0.5% TritonX-100 CSK, add 5 mL TritonX-100 to 1 L of CSK buffer.
3. Fixing solution: 10% Neutral Buffered Formalin.
4. Phosphate Buffered Saline (PBS) buffer: 137 mM NaCl, 2.7 mM KCl, 4.3 mM $Na_2HPO_4{\cdot}7H_2O$, 1.4 mM KH_2PO_4. To make 1 L of PBS buffer, weigh 8 g NaCl, 0.2 g KCl, 0.062 g $Na_2HPO_4{\cdot}7H_2O$, and 0.168 g KH_2PO_4 to 800 mL of water. Bring volume to 1 L with water.
5. 0.5% NP-40/PBS: 0.5% NP-40 in PBS. To make 500 mL of 0.5% NP-40/PBS, add 2.5 mL NP-40 to 500 mL PBS.
6. Blocking solution: 5% calf serum in PBS. To make 500 mL blocking solution, add 25 mL calf serum to 500 mL PBS.
7. Antibody dilution buffer: 0.1% TritonX-100 in PBS. To make 100 mL antibody dilution buffer, add 0.1 mL TritonX-100 to 100 mL PBS.
8. Primary antibody: Phospho RPA32 (S33) Antibody (Bethyl Laboratories, Inc., Montgomery, TX); RPA2 antibody (71-9A) (16).
9. Secondary antibody: FITC-conjugated donkey anti-rabbit (Jackson ImmunoResearch), goat anti-mouse Texas Red (Invitrogen).
10. DAPI: 1 μg/μL 4′,6-diamidino-2-phenylindole.
11. Precleaned Micro Slides 1 × 3 × 1.0 mm.
12. Mounting media: ProLong Antifade Kit (Invitrogen).
13. Fluorescence microscope.

2.3. SDS Polyacrylamide Gel Components

1. 4× Upper Tris (Stacking Buffer): 0.5 M Tris, pH 6.8, 0.4% SDS. To make 1 L of 4× Upper Tris, weigh 60.5 g of Tris Base and 4 g of SDS, and add them to 800 mL of water. Add water to bring volume to 900 mL. Mix and adjust pH with HCl. Bring volume to 1 L with water. Store at room temperature.
2. 4× Lower Tris (Resolving Buffer): 1.5 M Tris pH 8.8, 0.4% SDS. To make 1 L of 4× Lower Tris, weigh 182 g of Tris Base and 4 g of SDS and add them to 800 mL of water. Add water to bring volume to 900 mL. Mix and adjust pH with HCl. Bring volume to 1 L with water. Store at room temperature.
3. 50% glycerol: To make 100 mL of solution, mix 50 mL of glycerol with 50 mL of water.
4. 40% Acrylamide/Bis Solution, 19:1 (Bio-Rad). Store at 4°C.
5. 10% Ammonium persulfate (10% APS): Add 0.1 g of ammonium persulfate to 800 μL of water. Add water to bring volume to 1 mL. Make fresh 10% APS every 3–4 days. Store 10% APS at 4°C.
6. TEMED. Store at 4°C.
7. Water-saturated N-butanol. To make the water-saturated n-butanol, mix equal volume of water and n-butanol in a container and vortex vigorously. The top layer will be water-saturated n-butanol and the bottom layer will be n-butanol-saturated water.
8. Precision plus protein standards.
9. 1× SDS-PAGE running buffer: 0.1% SDS, 0.025 M Tris Base, 0.192 M Glycine. To make 1 L of 10× SDS-PAGE running buffer stock (1% SDS, 0.25 M Tris Base, 1.92 M Glycine), weigh 30 g Tris Base, 144 g glycine, 10 g SDS, mix first with 600 mL of water, and bring the volume to 1 L with water. No pH adjustment needed. To make 1 L of 1× SDS-PAGE running buffer, mix 100 mL of 10× stock with 900 mL of water.
10. 30 mL gradient maker.
11. Full-sized SDS-PAGE gel apparatus. Protocol describes making gels for Hoefer SE600 series gel apparatus but can be adjusted to any commercial gel apparatus.

2.4. Immunoblotting Components

1. 3× SDS sample buffer (0.188 M Tris, pH 6.8 at 25°C, 6% SDS, 30% Glycerol, 2.14 M β-mercaptoethanol, 0.03% bromophenol blue). To make 10 mL of 3× SDS sample buffer, mix 3.75 mL 4× Upper Tris, 0.6 g SDS, 3.75 mL of 80% Glycerol, 1.5 mL of β-mercaptoethanol and 300 μL of 1% bromophenol blue.
2. Transfer Buffer: 25 mM Tris Base, 192 mM Glycine, 0.04% SDS, 20% Methanol. To make 1 L of transfer buffer, weigh 3 g of Tris Base, 14.4 g of Glycine, 0.4 g of SDS, mix with 200 mL

of methanol and 600 mL of water. Add water to bring volume to 1 L. Store at room temperature.

3. 10× Tris Buffered Saline (TBS): 1.5 M NaCl, 0.2 M Tris Base, pH 7.6. To prepare for 1 L of 10× TBS, weigh 24.2 g Tris Base, 88 g NaCl, and add them to 800 mL of water. Adjust the pH to 7.6 with HCl. Bring the volume to 1 L with water. Store at room temperature.
4. Wash buffer (TBS/T): 1× TBS, 0.1% Tween-20. To make 1 L of TBS/T, mix 1 mL of Tween-20 with 1 L of 1× TBS. Store at room temperature.
5. Blocking buffer: 1× TBS, 0.1% Tween-20 with 5% nonfat dry milk powder. To make 150 mL blocking buffer, add 15 mL 10× TBS to 135 mL of water and mix well. Add 7.5 g nonfat dry milk powder and 150 μL Tween-20 while stirring.
6. Electroblotting apparatus. Procedure described for Semi Dry Electroblotting system (Theromo scientific).
7. 3 MW Blotting paper, 0.33 mm (MDSCI).
8. Nitrocellulose membranes, 0.45 μM (Bio-Rad).
9. SuperSignal West Pico Chemiluminescent Substrate (Thermo).
10. Plastic wrap.
11. X-ray film.
12. Primary antibody: Phospho RPA32 (S33) antibody (Bethyl Laboratories, Inc., Montgomery, TX); RPA2 antibody (71-9A) (16).
13. Secondary antibody: HRP-conjugated anti-mouse or HRP-conjugated anti-rabbit antibody (cell signaling).

3. Methods

3.1. Cell Culture

1. Sterilize coverslips in 70% ethanol in six-well tissue culture plates under UV light in tissue culture hood for 5 min.
2. Wash coverslips and wells with sterile PBS, twice. Leave coverslips on bottom of wells.
3. Seed wells with 5×10^5 cells. Determine number of cells per mL using a hemocytometer. Dilute with growth media to obtain 2.5×10^5 cells/mL. Add 2 mL of diluted cells per well.
4. Grow cells on coverslips in six-well tissue culture plates in DMEM with 10% calf serum at 37°C with 5% CO_2 for 24 h prior to starting experiments.
5. Treat cells as desired for experiment. This can involve transfecting siRNA or expression plasmids and/or treatment with drugs or

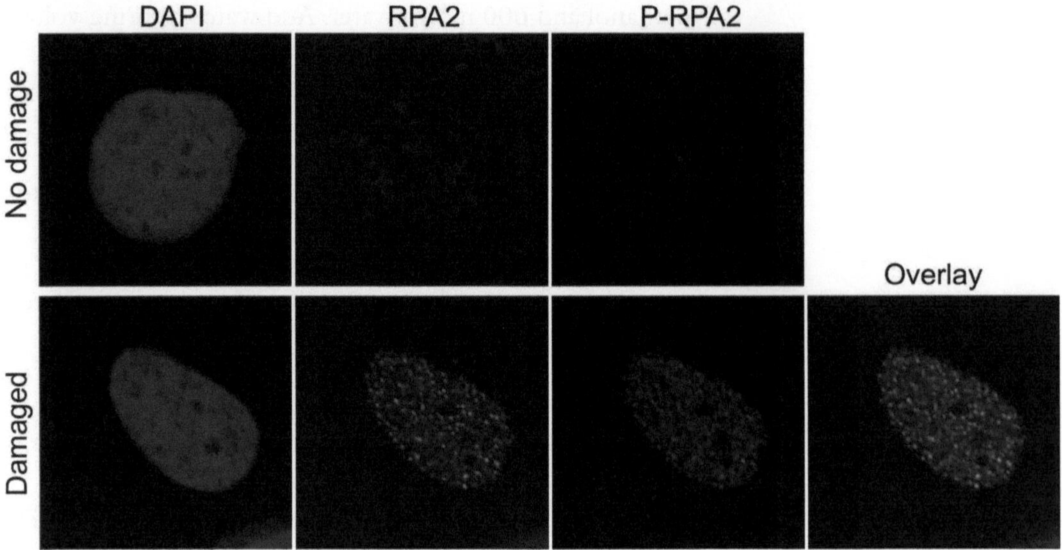

Fig. 1. Immunofluorescence of extracted HeLa cells. HeLa cells grown on coverslips were exposed, where indicated, to 20 μM DNA-damaging agent, camptothecin, an inhibitor of Topoisomerase II, for 4 h to cause single-strand and double-strand breaks throughout the nucleus. Cells were then fixed and extracted to leave only chromatin-associated proteins. Nuclei were then stained for DNA (DAPI), treated with mouse RPA2 antibody (RPA2; 1:1,000) or rabbit anti-phospho RPA32 (S33) antibody (P-RPA2; 1:1,000). The antibodies were detected with secondary antibodies and visualized using confocal microscopy.

DNA-damaging agents. In the example shown in Fig. 1, DNA damage was induced by treatment with 20 μM camptothecin for 4 h. Stock solution of camptothecin in DMSO was diluted 1:720 into cell growth media to obtain 20 μM concentration. Then 2 mL of the camptothecin containing media was added to each well.

3.2. Immunofluorescence Staining

1. Wash coverslips with 1 mL per well of CSK buffer for 2 min, twice.
2. Extract nonchromatin bound RPA with 1 mL per well of 0.5% TritonX-100 CSK for 5 min (see Immunofluorescence Note 1(a)).
3. Fix cells with 1 mL per well 10% Neutral Buffered Formalin for 20–30 min.
4. Wash coverslips with 1 mL per well PBS for 2 min, twice.
5. Treat cells with 1 mL per well 0.5% NP-40/PBS for 5 min for second extraction treatment (see Immunofluorescence Note 1(a)).
6. Wash coverslips with 1 mL per well PBS for 2 min, twice.
7. Block with 1 mL per well of 5% serum in PBS for 30–60 min (see Immunofluorescence Note 1(b)).

8. Incubate coverslips in 1 mL per well of primary antibody diluted in 0.1% Triton PBS overnight at 4°C (see Immunofluorescence Note 1(c)).
9. Wash coverslips with 1 mL per well PBS for 2 min, twice.
10. Incubate coverslips in 1 mL per well of secondary antibody diluted in 0.1% Triton PBS for 1–2 h at room temperature (see Immunofluorescence Note 1(c)).
11. Wash coverslips with 1 mL per well PBS for 2 min, twice.
12. Incubate coverslips in 1 mL per well of DAPI diluted to 1 μg/μ L in PBS for 5 min.
13. Wash coverslips with 1 mL per well PBS for 2 min, twice.
14. Mount coverslips onto slides. Remove coverslips from wells with tweezers and place cell-side up on edge of glass slide. Add approx 30 μL of mounting media (Invitrogen ProLong Antifade Kit) on each coverslip. Invert coverslip (cell-side down) on the center of the glass slide. Seal edges with clear nail polish.
15. Collect images on fluorescence or confocal microscope. Figure 1 shows an example of HeLa nuclei that were stained with antibodies to native RPA2 and phosphoserine 33 in RPA2. Images collected on a confocal microscope.

3.3. Pouring 8–14% Gradient Gel

1. Clean and assemble gel plates in pouring stand as recommended by manufacturer.
2. Set up dry gradient maker on a magnetic stir plate with tubing leading to top of assembled gel apparatus. (Make sure the gradient maker sits above assembled gel plates.) The gradient maker has two chambers for two different concentration acrylamide solutions. The high concentration will be in the chamber (with outlet) and should have a small stir bar in it.
3. The following recipe is for a 1 mm thick 18 × 16 cm gel. Volumes can be adjusted proportionally for different size or thickness gels. Make 8% and 14% acrylamide solutions for running gel. There is no need to degas solutions. The amount of TEMED and APS should cause the gel to polymerize in 1 h. see Gradient Gel Note 2.

	H_2O	4× Lower Tris	50% Glycerol	40% Acril/Bis		TEMED	10% APS
8%	5.67 mL	2.7 mL	0.21 mL	2.18 mL	Mix	6.7 μL	13.3 μL
14%	3.25 mL	2.7 mL	1 mL	3.75 mL	Mix	6.7 μL	13.3 μL

4. Make sure the valves are in the closed position. Pour 8% acrylamide solution into the chamber of the gradient maker

with a single valve. Briefly open and close valve between chambers to fill the valve with 8% solution and to make sure valve is not blocked with air bubbles. Pour 14% acrylamide solution into the chamber with the outlet to the gel apparatus. Start stir bar and open valve at outlet to start flow into gel apparatus. If solution does not start flowing immediately, provide some positive pressure by gently pressing thumb over top chamber or by gentle suction with a syringe attached to the other end of the tubing. Once the 14% acrylamide solution is flowing through the tubing, open valve between chambers. Adjust the rate of stirring so that there is good mixing but little aeration. Schlieren lines should be observed from mixing the 8% solution with the 14% solution. Let the gradient pour until the surface is around 0.5 cm from the maximum depth of the comb or until the gradient is finished. Immediately rinse out the gradient maker with water. Then cover the gel with water saturated n-butanol. Let the gel polymerize. This should take around 1 h and is indicated by the appearance of a second interface below the butanol/aqueous boundary.

5. Pick up the gel sandwich and pour off the butanol and unpolymerized acrylamide. Rinse the gel several times (>3) with water to remove any residual butanol.
6. Make 5% stacking gel. The amount of TEMED and APS should cause the stacking gel to polymerize in 15 min. Pipette stacking gel solution on top of running gel and insert comb.

	H_2O	4× Upper Tris	50% glycerol	40% Acril/Bis		TEMED	10% APS
5%	4.2 mL	1.7 mL	0 mL	0.8 mL	Mix	13.3 μL	27 μL

7. After stacking gel is polymerized, gently remove comb and rinse out wells. Rinse at least two times with water and at least one time with 1× SDS-PAGE running buffer. Shake all liquid out after each wash. Fill wells with 1× SDS-PAGE running buffer in preparation for loading samples.

3.4. Sample Preparation and Electrophoresis

1. Mix protein samples with 3× SDS sample buffer to make the gel samples. For 1 mm gel, samples should be 15–25 μL with 2/3 final volume of combined water and protein sample and 1/3 volume of the 3× SDS sample buffer. Samples should contain 1–3 μg of purified protein or 20–30 μg of cell lysate/extracts.
2. Heat gel samples at 90°C (or boil) for 2 min.
3. Spin the tubes momentarily to get entire volume into bottom of the tube.
4. Apply the protein standards and samples to the bottoms of the wells through the buffer using gel loading tips.

5. Place loaded gel in gel running apparatus and fill both buffer chambers with 1× SDS-PAGE running buffer. For Hoefer SE600 series gel apparatus, the lower chamber holds 3–3.5 L and upper chamber holds 500 mL 1× SDS-PAGE buffer. Make sure no air bubbles are trapped under the gel. Buffer can be used for up to four gels before replacing.
6. Attach to power supply and apply current. Negative electrode is attached to top of gel. For optimal separation, run the gel at constant power at 30–40 W for 1–1.5 h. At this power, gel will get very warm and must be submerged in stirring 1× SDS-PAGE buffer to keep thermal stress constant. Alternatively, gel can be separated overnight under constant voltage at 55 V. Tracking dye should be run to bottom of gel.

3.5. Immunoblotting Electrotransfer

Either wet or semidry transfers can be used. Follow manufacturer's recommended protocol for set up and transfer. The following procedure is for a semidry electroblotting apparatus (Thermo Scientific).

1. Remove gel from apparatus. Separate gel plates with plastic spatula. Prepare gel for transfer by cutting off stacking gel with a razor blade. Running gel is then transferred to electroblotting apparatus. It is easiest to do this in two steps. First peel gel off plate by holding the plate upside down over a tray containing water (to prevent gel from sticking to tray). Gently peel one corner of the gel off the glass plate and let gravity pull gel into tray. The gel can then be transferred to the blotting apparatus when ready (step 3). Alternatively, the gel can be lifted carefully from bottom (14% section) and placed directly on blotting paper.
2. Cut the nitrocellulose membrane to the size of the gel and soak in water. Also cut six pieces of 20 cm × 20 cm blotting paper. (Blotting paper should just cover surface of electroblotting apparatus.) Soak blotting paper in the transfer buffer.
3. Arrange a gel-membrane sandwich by laying three layers of presoaked blotting paper on the bottom surface of the semidry apparatus. Place the gel on top of blotting papers, followed by the presoaked nitrocellulose membrane, and another three layers of presoaked blotting papers on the top (see Immunoblotting Note 3(a)).
4. Put the lid over the semidry plate, finger-tighten three knobs on the lid, and connect to a high current power supply. The orientation of the sandwich should be as follows: negative electrode-gel-nitrocellulose-positive electrode.
5. Run the transfer at 500 mA and 20 V for 1 h. Transfer of prestained protein standards to the membrane indicates successful transfer.

3.6. Immunoblotting

1. After transfer, wash the nitrocellulose membrane three times for 5 min each with TBS/T at room temperature with constant rocking.
2. Block the nitrocellulose membrane with blocking buffer for 1 h at room temperature with constant rocking.
3. Wash the nitrocellulose membrane three times for 5 min each with TBS/T with constant rocking
4. Incubate the nitrocellulose membrane in the blocking buffer with diluted primary antibody (1:1,000 for antibodies shown in Fig. 2). The incubation occurs at 4°C overnight with gentle rocking (see Immunoblotting Note 3(b)).

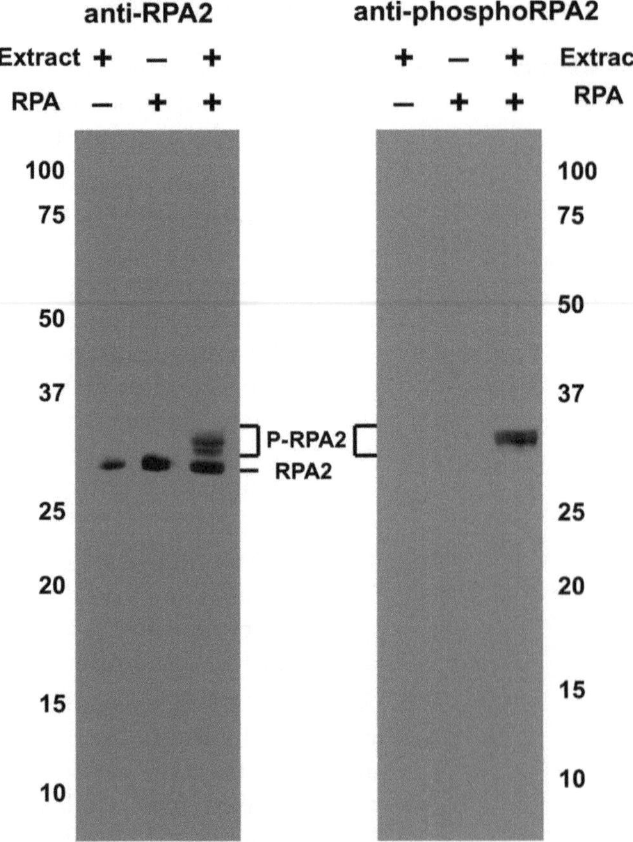

Fig. 2. Immuoblot of phosphorylated RPA. Protein samples containing either 75 μg HeLa cell extract protein or 300 ng purified RPA or both were incubated under conditions that allow efficient hyperphosphorylation of RPA (14). Identical reactions were separated on an 8–14 % gradient gel and transferred to nitrocellulose. Membrane was then incubated with a 1:15,000 dilution mouse anti-RPA2 antibody (*left panel*) or a 1:10,000 dilution of rabbit anti-phospho RPA32 (S33) antibody (*right panel*). Positions of molecular mass markers are indicated on either side of each panel. Phosphorylated forms of RPA2 (lower mobility bands) were only observed in the complete reaction (*third lane*). Note that the RPA2 antibody detects all forms of RPA2 including multiple phosphorylated forms while the phospho-serine33-specific antibody detects only a subset of low-mobility, phosphorylated forms. (Unphosphorylated endogenous RPA2 is observed in the extract-only lanes).

5. Wash the membrane three times for 5 min each with TBS/T with gentle rocking
6. Incubate membrane with diluted HRP-conjugated secondary antibody (anti-mouse IgG HRP was diluted 1:20,000, anti-rabbit IgG HRP was diluted in 1:15,000) in TBS/T with gentle rocking for 1–2 h at room temperature.
7. Wash the membrane three times for 5 min each with TBS/T with gentle rocking
8. Incubate the membrane with 1:1 mixture of luminol and peroxide solutions from Chemiluminescent Substrate kit for 5 min at room temperature. Mixture should be prepared right before using.
9. Drain the membrane by tapping one corner of membrane on tissue paper to remove excess developing solution (do not let dry), wrap the membrane in plastic wrap and expose to X-ray film (see Immunoblotting Note 3(c)).

4. Notes

1. Immunofluorescence
   (a) Extraction steps should not be done if total RPA is to be visualized.
   (b) Optimal blocking solution will vary depending on the source of primary and secondary antibodies. We find that bovine serum generally is effective. However, background can usually be reduced by blocking with serum from the animal that is the source of the secondary antibody.
   (c) Optimal dilution will need to be determined for each antibody empirically. Usual ranges are 1:100–1:1,000 for commercial primary antibodies and higher dilutions for more concentrated antibody stocks. Secondary antibodies are usually used at dilutions of at least 1:1,000. Can use any antibody specific for RPA or a modified residue.
2. Gradient Gel
   (a) Acrylamide is a neurotoxin. Wear personal protective equipment.
3. Immunoblotting
   (a) Make sure there are no air bubbles between gel and nitrocellulose membrane. These will prevent protein transfer.
   (b) Optimal dilution will need to be determined for each antibody empirically. Usual ranges are 1:1,000–1:10,000 for most

commercial primary antibodies and higher dilutions for more concentrated antibody stocks. Secondary antibodies are usually used at dilutions of 1:1, 000–1:20,000. Any antibody specific for RPA or a modified residue can be used.

(c) An initial 1-min exposure should indicate the proper exposure time.

Acknowledgment

Thanks to all the current and past members of the Wold lab who have contributed to optimizing these protocols.

References

1. Wold MS (1997) Replication protein A: a heterotrimeric, single-stranded DNA-binding protein required for eukaryotic DNA metabolism. Annu Rev Biochem 66:61–92
2. Oakley GG, Patrick SM (2010) Replication protein A: directing traffic at the intersection of replication and repair. Front Biosci 15:883–900
3. Eki T, Hurwitz J (1991) Influence of poly (ADP-ribose) polymerase on the enzymatic synthesis of SV40 DNA. J Biol Chem 266:3087–3100
4. Dou H, Huang C, Singh M, Carpenter PB, Yeh ET (2010) Regulation of DNA repair through DeSUMOylation and SUMOylation of replication protein A complex. Mol Cell 39:333–345
5. Binz SK, Sheehan AM, Wold MS (2004) Replication protein A phosphorylation and the cellular response to DNA damage. DNA Repair (Amst) 3:1015–1024
6. Zou Y, Liu Y, Wu X, Shell SM (2006) Functions of human replication protein A (RPA): from DNA replication to DNA damage and stress responses. J Cell Physiol 208:267–273
7. Anantha RW, Borowiec JA (2009) Mitotic crisis: the unmasking of a novel role for RPA. Cell Cycle 8:12903–12908
8. Block WD, Yu Y, Lees-Miller SP (2004) Phosphatidyl inositol 3-kinase-like serine/threonine protein kinases (PIKKs) are required for DNA damage-induced phosphorylation of the 32 kDa subunit of replication protein A at threonine 21. Nucleic Acids Res 32:997–1005
9. Anantha RW, Vassin VM, Borowiec JA (2007) Sequential and synergistic modification of human RPA stimulates chromosomal DNA repair. J Biol Chem 282:35910–35923
10. Kenny MK, Schlegel U, Furneaux H, Hurwitz J (1990) The role of human single-stranded DNA binding protein and its individual subunits in simian virus 40 DNA replication. J Biol Chem 265:7693–7700
11. Golub EI, Gupta RC, Haaf T, Wold MS, Radding CM (1998) Interaction of human Rad51 recombination protein with single-stranded DNA binding protein, RPA. Nucleic Acids Res 26:5388–5393
12. Vassin VM, Wold MS, Borowiec JA (2004) Replication protein A (RPA) phosphorylation prevents RPA association with replication centers. Mol Cell Biol 24:1930–1943
13. Binz SK, Dickson AM, Haring SJ, Wold MS (2006) Functional assays for replication protein A (RPA). Methods Enzymol 409:11–38
14. Henricksen LA, Umbricht CB, Wold MS (1994) Recombinant replication protein A: expression, complex formation, and functional characterization. J Biol Chem 269:11121–11132
15. Zernik-Kobak M, Vasunia K, Connelly M, Anderson CW, Dixon K (1997) Sites of UV-induced phosphorylation of the p34 subunit of replication protein A from HeLa cells. J Biol Chem 272:23896–23904
16. Erdile LF, Wold MS, Kelly TJ (1990) The primary structure of the 32-kDa subunit of human replication protein A. J Biol Chem 265:3177–3182

Chapter 16

Detecting Posttranslational Modifications of Bacterial SSB Proteins

Dusica Vujaklija and Boris Macek

Abstract

Posttranslational modifications of single-stranded DNA-binding proteins (SSBs), which are essential proteins in DNA metabolism, have been reported from prokaryotic to eukaryotic systems. While eukaryotic SSBs are regulated by phosphorylation on serine and threonine residues, bacterial SSB proteins are also phosphorylated on tyrosine residues. This was initially observed during a systematic search for global phosphotyrosine-containing proteins in *Streptomyces*, complex life cycle bacteria that support mycelial growth and spore formation. Tyrosine phosphorylation was further confirmed on SSB proteins from the spore-forming bacterium *Bacillus subtilis* and in the simpler prokaryote, *Escherichia coli*. However, a thorough study of this modification and its cognate kinase has been performed only on SSB proteins from *Bacillus subtilis*. It was shown that phosphorylation of *B. subtilis* SSB (SsbA) significantly increases binding affinity with single-stranded DNA in vitro. Mass spectrometry analysis of SsbA identified Tyr82 as the phosphorylation site. Analyses of the resolved and predicted crystal structures of SSB proteins from *B. subtilis*, *E. coli*, and *S. coelicolor* revealed that the Tyr phosphorylation site occupies similar positions in all three structures. Our results indicate that tyrosine phosphorylation of bacterial SSBs is a conserved modification in taxonomically distant bacteria.

Key words: Detection of phosphorylated SSBs, Purification of His-tagged SSB proteins, Western blotting, Identification of phosphorylation sites by mass spectrometry

1. Introduction

Protein phosphorylation on Serine/Threonine/Tyrosine residues is crucial in regulating cellular activities in all living organisms. This type of protein modification was recently identified in bacteria. Moreover, the number of proteins known to be phosphorylated on Ser/Thr/Tyr in bacteria has grown based on recently developed proteomics approaches using phosphopeptide enrichment and high-accuracy mass spectrometry (1–4). The newly identified phosphorylated proteins strongly support Ser/Thr/Tyr protein phosphorylation as a posttranslational modification that is widely

James L. Keck (ed.), *Single-Stranded DNA Binding Proteins: Methods and Protocols*, Methods in Molecular Biology, vol. 922, DOI 10.1007/978-1-62703-032-8_16, © Springer Science+Business Media, LLC 2012

distributed among bacteria in proteins that participate in a variety of cellular processes. Research of tyrosine phosphorylation in bacteria is not as advanced as Ser/Thr phosphorylation; however, the increasing number of identified tyrosine phosphorylated proteins from bacteria highlights the prominent role of this modification in the regulation of bacterial physiology (5).

Tyrosine phosphorylation of SSB proteins from taxonomically distant bacteria has been discovered (6). Phosphorylated SSB has been isolated from whole cell extracts utilizing phospho-specific antibodies and immunoaffinity purification. A monoclonal antibody 4G10 (anti-phosphotyrosine antibody) produced by hybridoma cells that is purified and covalently coupled to Affi-gel 10 is very useful in this separation technique. The efficiency of the immunochromatographic technique can be readily followed using western blot. MS sequencing of fractions with detected tyrosine phosphorylated proteins allowed us to identify bacterial SSB as a target of bacterial tyrosine kinase (6). Although this procedure is labor-intensive and time-consuming it remains applicable to proteins that are phosphorylated to a very low extent. Notably, only one additional case of SSB phosphorylation was reported applying gel-free and site-specific analysis in different bacteria (7).

Using western blot analysis we have detected tyrosine phosphorylation of SSB proteins from three distantly related bacteria. Mass spectrometry analysis of SSB from *B. subtilis* identified Tyr82 as a phosphorylation site (6). The position of the phosphorylated tyrosine residue is highly conserved in SSBs from Gram-positive bacteria, but is often absent in SSBs from Gram-negative organisms. In spite of this, it has been known that bacterial SSB proteins share common structural features and predominantly exist as homotetramers (8). This structural conservation is shown schematically in (Fig. 1). As depicted, Tyr82 in the predicted structure of *B. subtilis* SsbA monomer, Tyr98 in the crystal structures of *E. coli* SSB (9) and Tyr88 in *S. coelicolor* SSB (8) occupy similar positions.

We have demonstrated that phosphorylation of *B. subtilis* SsbA is linked with increased single-stranded DNA-binding affinity by almost 200-fold in vitro, and this reaction is affected antagonistically by tyrosine kinase and phosphatase (6). We have also observed that under DNA-damaging conditions, *B. subtilis* cells undergo a reduction in the level of SSB phosphorylation thus lowering its affinity for ssDNA. The biological significance of tyrosine phosphorylation on DNA metabolism in *B. subtilis* cells was reported recently. The authors showed that *B. subtilis* deficient for a protein tyrosine kinase exhibits impaired DNA replication, with accumulation of extra chromosomes in a subpopulation of cells (10). Moreover, in the same strain, SsbA intracellular localization was affected, with less DNA-associated foci than in the wild-type strain (11).

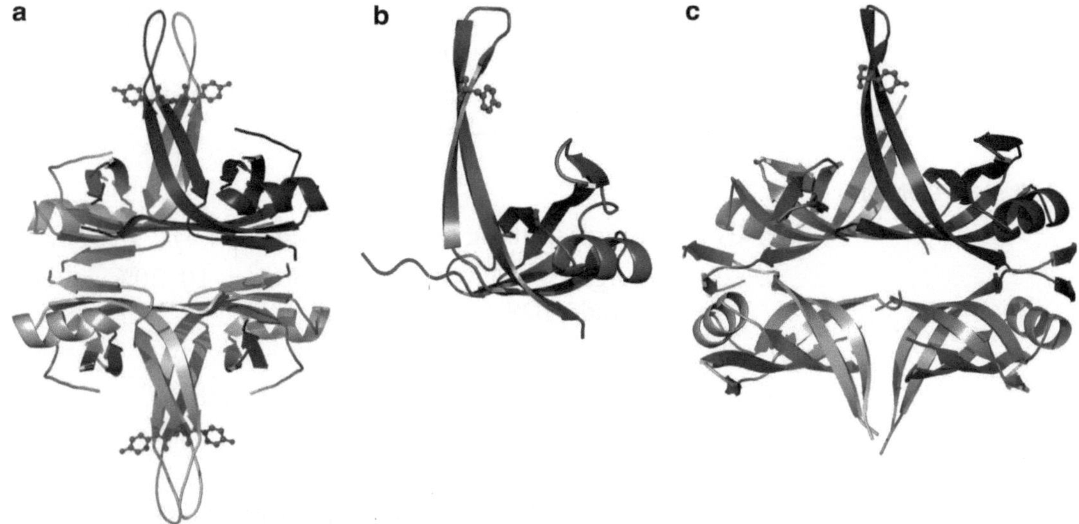

Fig. 1. Ribbon representation of the SSB structures with tyrosines (*presented in ball-and-stick model*) from: *E. coli* tetramer (**a**), *B. subtilis* predicted monomer, (**b**) and *S. coelicolor* tetramer (**c**). All figures were prepared using: *PyMOL*, PDB codes (*E.coli*: 1kaw; *S.coelicolor*: 3eiv) and the PHYRE protein fold recognition server for predicting the structure of *B. subtilis* Ssb monomer (by Z. Stefanic, Rudjer Boskovic Institute).

In this chapter, we describe standard methods to detect posttranslational modification of bacterial SSB proteins in heterologous and homologous host, as well as optimized LC-MS analysis for identification of phosphorylated tyrosine residue in SSB.

2. Materials

All solutions and solvents should be prepared with high-purity deionized Milli-Q water of the resistivity 18.2 MΩcm (Millipore Q-Gard 2 cartridge).

2.1. Preparative Protein Synthesis and Purification of His-Tagged SSB Proteins

1. DNase I lyophilized powder (see Note 1).
2. Ni-NTA resin.
3. Ampicillin stock solution: 100 mg/mL H_2O, sterilize through a 0.2 μm filter, store in aliquots at −20°C.
4. 1 M IPTG stock solution: 238 mg/mL H_2O, sterilize through a 0.2 μm filter, store in aliquots at −20°C.
5. PD-10 Desalting Columns.
6. LB medium: 10 g/L tryptone, 5 g/L yeast extract, 10 g/L NaCl.
7. Lysis buffer: 50 mM Tris–HCl, pH 7.5, 150 mM NaCl, 10% glycerol (see Note 2).

8. Wash buffer: 50 mM Tris–HCl, pH 7.5, 150 mM NaCl, 10% glycerol, 30 mM imidazole.
9. Elution buffer: 50 mM Tris–HCl, pH 7.5, 150 mM NaCl, 10% glycerol, 300 mM imidazole.

2.2. Western Blotting

1. BSA.
2. PVDF or nitrocellulose membrane (see Note 3).
3. 12% prepared SDS-PAGE gels or precast 10–20% gradient polyacrylamide SDS gels.
4. Alkaline phosphatase.
5. anti-P-Tyr peroxidase conjugates (Sigma-Aldrich).
6. Wash buffer (TTBS): 20 mM Tris–HCl, pH 7.5, 150 mM NaCl, 0.1% Tween 20 (see Note 4).
7. Blocking buffer: TTBS supplemented with 3% BSA.
8. Colorimetric AEC-staining kit or Chemiluminescent detection kit, ECL plus western blot detection.

2.3. Determination of Phosphorylation Sites by Mass Spectrometry

Denaturation, reduction, and alkylation buffers can be frozen as stock solutions at −20°C.

2.3.1. In-solution Protein Digestion

1. Denaturation buffer: 6 M urea, 2 M thiourea, 1% *n*-octylglucoside (w/v) in 10 mM HEPES, pH 8.0.
2. Reduction buffer (stock solution): 1 M dithiothreitol, 50 mM ammonium bicarbonate.
3. Alkylation buffer (stock solution): 550 mM iodoacetamide, 50 mM ammonium bicarbonate.
4. Protease: Trypsin, sequencing grade, modified.

2.3.2. Titanium Oxide (TiO_2) Chromatography

1. Loading solution: 30 mg/mL 2,5 dihydrobenzoic acid, 80% acetonitrile in water.
2. Washing solution I: 30% acetonitrile/3% trifluoroacetic acid.
3. Washing solution II: 80% acetonitrile/0.1% trifluoroacetic acid.
4. Elution solution: 40% NH_4OH (aq, 25% NH_3), 60% acetonitrile (pH > 10.5).
5. Beads: Titansphere TiO_2 (10 μm; GL Sciences).
6. C8 microcolumns (see Note 5).

2.3.3. Liquid Chromatography–Mass Spectrometry

1. HPLC solvent A: 0.5% acetic acid (Fluka) in water (see Note 6).
2. HPLC solvent B: 0.5% acetic acid, 80% acetonitrile in water.
3. HPLC loading solvent: 1% trifluoroacetic acid, 2% acetonitrile in water.
4. HPLC columns: Reversed phase C18 nano-HPLC column (15 cm × 75 μm ID) (New Objective).

3. Methods

Detection of posttranslational modification of bacterial SSB proteins in vivo

Follow general cloning strategy for DNA manipulations

1. Design PCR primers with apropriate restriction sites for cloning and amplification of the selected gene encoding SSB from the respective genomic DNA (see Note 7).
2. Amplify and ligate the target DNA into expression vector for homologous or heterologous gene expression in appropriate bacterial host. Various bacterial protein expression systems can be used for this purpose (see Note 8).
3. Transform ligation product into bacterial host according to the protocol recommended by supplier of expression system.
4. Isolate plasmids from the resulting bacterial colonies and confirm the correct clone by sequencing recombinant DNA fragment.

3.1. Preparative Protein Synthesis and Purification of His-Tagged SSB Proteins

1. Inoculate 10 mL of LB broth containing 100 μg/mL ampicillin and grow *E. coli* at 37°C with vigorous shaking.
2. Inoculate 500 mL culture (1:100) with non-induced overnight culture and grow the cells at 37°C with vigorous shaking until OD_{600} reaches 0.7 (see Note 9).
3. Induce expression by adding IPTG to a final concentration of 1 mM.
4. Incubate cell culture for additional 3–4 h.
5. Harvest the cells by centrifugation at 5,000 × *g* for 15 min (see Note 10).
6. Resuspend cells in Lysis buffer (5 mL) and add DNase to 50 μg/mL.
7. Disrupt the cells with sonication on ice. Apply four 30 s bursts with cooling in between and use the sonicator with macrotip.
8. Centrifuge lysate at 10,000 × *g* for 15 min at 4°C to remove cell debris and transfer supernatant to a fresh tube.
9. Add 1 mL of 50% slurry of Ni-NTA resin into the gravity flow column and pre-equilibrate the resin with 10 mL of Lysis buffer (see Note 11).
10. Run supernatant twice over Ni-NTA resin and save flow through.
11. Wash column with 10 mL of ice-cold Lysis buffer.
12. Wash column with 10 mL of ice-cold Wash buffer.

13. Elute protein with Elution buffer; collect 1 mL fractions (see Note 12).
14. Pool the fractions containing the highest protein content (see Note 13).
15. To exchange buffer use PD-10 Desalting Columns as recommended by supplier and store the proteins at −80°C in aliquots (see Note 14).

3.2. Western Blotting

The western blot (the protein immunoblot) is the most common method used for detecting the phosphorylation state of a protein. Detection of posttranslational modification of purified 6xHis-tagged SSBs by immunoblotting assay was performed in several steps as described (6).

1. Run two SDS-PAGE gels in parallel with 50–100 ng of SSB protein per lane (see Note 15) and transfer proteins to PVDF membranes or nitrocellulose (see Note 16) according to standard protocols.
2. Incubate one membrane for 1–2 h at room temperature in Blocking buffer and in control experiment, incubate the second membrane in the same buffer with addition of alkaline phosphatase (1 U/mL) to dephosphorylate P-proteins (Fig. 2).
3. Briefly wash both membranes in the same buffer without BSA (Wash buffer).
4. Incubate the membranes for 1 h with anti-P-Tyr peroxidase conjugates in the Blocking buffer (1:40,000) (see Note 17).
5. Wash the membrane for 30–45 min with three or more changes of Wash buffer.
6. Use colorimetric AEC-staining kit (Fig. 2), or chemiluminescent detection kit, ECL plus western blotting detection system (Fig. 3), to visualize protein bands (see Note 18).

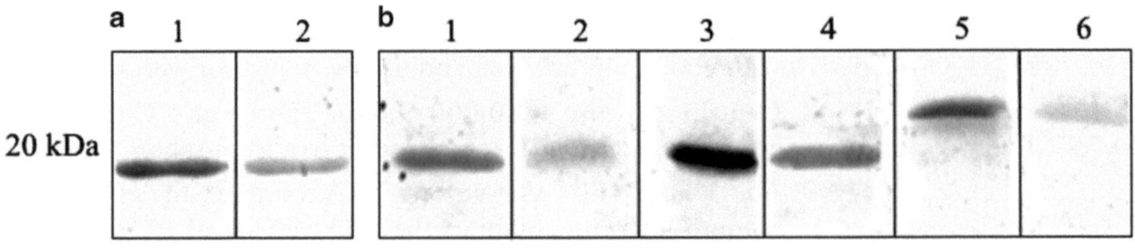

Fig. 2. Phosphorylation of bacterial SSBs in heterologous and homologous host. (**a**) Phosphorylation of *B. subtilis* His-tagged SsbA isolated from homologous host has been detected by western blotting (lane 1); dephosphorylation by alkaline phosphatase during imunoblotting assay is shown in lane 2. (**b**) Lanes 1, 3, and 5 show phosphorylated SSBs from *B. subtilis*, *E. coli*, and *S. coelicolor*, respectively. All proteins were synthesized and isolated from *E. coli*. Dephosphorylation of these proteins by alkaline phosphatase is shown in lanes 2, 4, and 6. Purification and western blotting are performed as described and colorimetric AEC-staining kit has been applied to visualize protein bands. This figure is modified from ref. (6).

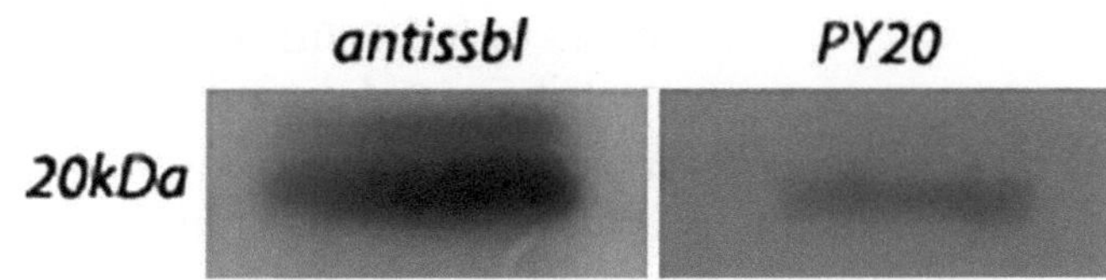

Fig. 3. In vivo phosphorylation of *S. coelicolor* SSB-L. His-tagged protein is isolated from homologous host, 2× 100 ng of protein was run on SDS-PAGE and electroblotted on PVDF membrane. The membrane was cut into two strips and western blotting was performed with polyclonal anti-SSB-L and PY20 antibody separately. The proteins were visualized with chemiluminescent detection kit. Membranes treated with anti-SSB-L and for PY20 were exposed 1 min and 7 min, respectively.

3.3. Determination of Phosphorylation Sites by Mass Spectrometry

3.3.1. In-solution Protein Digestion

1. Dissolve protein sample (50 μg of purified SSB protein, (6) in 100 μl denaturation buffer) (see Note 19).
2. Add reduction buffer to the sample to final concentration of 1 mM DTT; incubate 1 h at room temperature (see Note 20).
3. Add alkylation buffer to the sample to final concentration of 5.5 mM IAA; incubate 1 h at room temperature in the dark.
4. Dilute sample with 4 volumes of water (see Note 21).
5. Check pH (should be 8.0); adjust if necessary with very low volume of 1 M Tris–HCl, pH 8.0.
6. Add 1 μg trypsin per 100 μg sample protein and incubate overnight at room temperature. When needed double digestion with LysC/trypsine might be performed (see Notes 22 and 23).

3.3.2. TiO_2 Chromatography

Protein phosphorylation is often of a very low level and substoichiometric; therefore, enrichment of phosphopeptides from mixture of their unmodified and often highly abundant counterparts is essential. This can be achieved by several approaches. Enrichment of SSB peptides was originally performed using IMAC beads (6). Here, we describe a novel and generic method for phosphopeptide enrichment based on TiO_2 chromatography.

1. Stop the trypsin digestion by acidifying to pH 2–3 using TFA.
2. Add the loading solution to the sample (1:6) (see Note 24).
3. Weigh 5 mg of the TiO_2 beads (see Note 25).
4. Add 100 μl of loading solution to the TiO_2 beads and shake for 10 min at room temperature.
5. Add aliquots containing 5 mg of the TiO_2 beads slurry into each sample.
6. Incubate for at least 30 min (end-over-end rotation) at room temperature, centrifuge, and discard supernatant.

7. Wash with 1.5 mL of washing solution I, shake vigorously, centrifuge, and discard supernatant.
8. Wash with 1.5 mL of washing solution II, shake vigorously, centrifuge, and discard supernatant.
9. Transfer the TiO_2 beads into dedicated C8 microcolumns.
10. Elute phosphopeptides from each microcolumn with 3× 100 μl of elution solution and check pH of the last eluate—it should be >10; if it is not, elute with additional 100 μl of elution solution and then pool all eluates (see Note 26).
11. Dry the eluates to ~5 μl without heating and make sure they do not run dry (see Note 27).
12. Add HPLC loading solvent to 10 μl (final concentration of 1% acetonitrile and ~0.5% trifluoroacetic acid)—the samples are now ready for nanoLC-MS.

3.3.3. LC-MS

Liquid chromatography

1. The liquid chromatography (LC) part of the analytical LC-MS system described here consists of a Thermo EasyLC system (Thermo Fisher Scientific) comprising nanoflow pumps and a thermostated micro-autosampler with a 20 μl injection loop. Other systems could be substituted.
2. Use a nano-HPLC column packed in a fused-silica emitter (New Objective, 75-μm inner diameter with a 5–8 μm laser-pulled tip).
3. Connect the C18 RP HPLC column directly to the outlet of 6-port valve of the HPLC autosampler through a 20 cm long, 25-μm inner diameter-fused silica transfer line (Composite Metals) and a micro Tee-connector (Upchurch) ("liquid junction" connection) (see Note 28).
4. Load 2–5 μl of the phosphopeptide mixture using the HPLC autosampler onto the packed emitter at a fixed maximum pressure of 280 bar ("Intelliflow" feature) using 2% of HPLC solvent B.
5. After loading, reduce the flow-rate to 200 nl/min and increase the HPLC solvent B content to 10%.
6. Separate and elute the bound peptides with a 90 min linear gradient from 10 to 30% of HPLC solvent B. Wash out hydrophobic peptides by linearly increasing the HPLC solvent B content to 80% over 15 min.

Mass spectrometry

1. All mass spectrometric experiments discussed here are performed on an LTQ orbitrap "XL" mass spectrometer (Thermo Fisher Scientific) connected to an EasyLC nanoflow system

(Thermo Fisher Scientific) via a nano-electrospray LC-MS interface (Thermo Fisher Scientific) (see Note 29). Other instruments can also be used.

2. Operate the mass spectrometer in the data-dependent mode to automatically switch between MS and MS/MS using the Tune and Xcalibur software package.
3. Use the following settings in the "Tune" acquisition software:
   (a) FT full scan: accumulation target value 1E6; max. fill time 1,000 ms
   (b) FT MS^n: accumulation target value 5E4; max. fill time 500 ms
   (c) IT MS^n: accumulation target value 5E3; max. fill time 150 ms
4. For accurate mass measurements enable the "Lock mass" option in both MS and MS/MS mode in the Xcalibur software (12). Use the background polydimethylcyclosiloxane (PCM) ions generated from ambient air (e.g. $m/z = 445.120025$) for internal recalibration in real time. For single SIM scan injections of the lock mass ion into the C-trap set the lock mass "ion gain" at 10% of the target value of the full mass spectrum. If the fragment ion measurements are performed in the orbitrap, use the PCM ion at m/z 429.088735 (PCM with neutral methane loss).
5. In the "Xcalibur Instrument Setup" create a data-dependent acquisition method in which full scan MS spectra, typically in the m/z range from 300 to 1,800, are acquired by the orbitrap detector with resolution $R = 60{,}000$ (defined at m/z 400).
6. For high-accuracy and full mass range measurements of fragment ions, set the data-dependent MS^2 of the three most intense multiply-charged ions ($z \geq 2$) to be measured in the FT analyzer (orbitrap) at the resolution of $R = 15{,}000$ and enable the HCD option (Higher-energy C-trap dissociation). Set the first mass in mass range to $m/z = 80$. This will allow for low mass reporter ions such as immonium ions to be identified (13) (see Note 30).
7. For fast-scanning and high-sensitivity but low-resolution measurements of fragment ions, set the data-dependent MS^2 of the five most intense multiply-charged ions ($z \geq 2$) to be measured in the linear ion trap. Enable the preview mode for FTMS master scans to perform data-dependent MS^2 in-parallel with the full scan in the orbitrap. Set the fragmentation mode to CID and enable the multi-stage activation fragmentation option by which the neutral loss species at 97.97, 48.99, or 32.66 m/z below the precursor ion will be successively activated for 30 ms each (*pseudo* MS^3) (14) (see Note 31).

8. Standard acquisition method settings:
   (a) Electrospray voltage, 2.4 kV (see Note 32).
   (b) No sheath and auxiliary gas flow
   (c) Ion transfer (heated) capillary temperature, 150°C
   (d) Collision gas pressure, 1.3 mTorr
   (e) Dynamic exclusion of up to 500 precursor ions for 60 s upon MS/MS; exclusion mass width of 10 ppm
   (f) Normalized collision energy using wide-band activation mode; 35% for both CID and HCD
   (g) Ion selection thresholds: 1,000 counts for CID and 10,000 counts for HCD
   (h) Activation $q = 0.25$; Activation time = 30 ms

4. Notes

1. We use DNase in the powder form; a stock solution could also be prepared and stored according to supplier's instruction. Repeated freeze–thaw cycles should be avoided.
2. To reduce binding of nonspecific proteins add 10 mM imidazole to Lysis buffer and increase NaCl to 300 mM.
3. We perform the standard anti-phosphotyrosine western blot analysis of SSB proteins using nitrocellulose membrane, chemiluminescent detection kit, and PY20 antibody.
4. Different buffer composition had been used for previous immunobloting assay (6). Washing buffer (WB) contained 25 mM Tris–HCl, pH 8.3, 192 mM glycine, 0.05% Tween 20 while Blotting buffer was made adding 3% BSA to WB.
5. Instead of commercial C8 micorcolumns, "stage-tip" columns could be prepared by placing a ~1 mm^2 piece of Empore C8 material (Empore C8 Disk—Varian) into a 200 µl (yellow) pipette tip for each fraction, as described previously (see refs. 15, 16).
6. Acetic acid is the optimal ion-pairing reagent for the column packing material used here; an alternative is 0.1% formic acid. Do not use TFA in LC-MS solvents, as it may interfere with electrospray ionization (TFA is added only into the sample prior to loading).
7. A PCR primer may carry a 5′ tag, for instance hexahistidine, in order to facilitate protein purification. Such recombinant protein tagged at N- or C-terminus could be easily purified with Ni^{2+}-affinity chromatography in one-step purification procedure from crude cell lysate. It has been known that conserved C-terminal domain of SSB interacts with various proteins

(see ref. 17); therefore in previous work SSBs were tagged with poly-His at N-terminus (see ref. 6).

8. Vector pQE-30 (Qiagen) and *E. coli* NM522 were used for expression SSB-encoding genes of *B. subtilis*, *S. coelicolor*, and *E. coli* (6). DNA fragments were inserted between *Bam*HI and *Pst*I sites in the correct reading frame to allow synthesis of the six histidines at the N-terminus of each SSB. Vector map can be found in Qiagen Handbook & Protocols (The QIAexpressionist™).
9. Optimal conditions for SSB synthesis should be established at small-scale culture prior to preparative purification. Follow the same protocol as described for large-scale and apply time-course analysis varying the concentration of IPTG and growth temperature.
10. At this step cell pellet can be stored at −20°C.
11. Keep supernatant and all solution ice-cold and work quickly.
12. Proteins usually elute in the second and third fractions.
13. We usually apply quick quantitative assay for determination of the protein in the fractions: 5 μl of each fraction mix with 20 μl of Bradford reagent (Biorad) onto a strip of parafilm. The fractions with the protein will change reagent color from brown to blue in proportion the amount of protein.
14. In standard protocol, we exchange Elution buffer to Lysis buffer, aliquot the eluates and store the proteins at −80°C.
15. For routine electrophoresis we prepare SDS-PAGE (12%) with Mini PROTEAN 3 System glass plates (BioRad).
16. To evaluate the transfer efficiency, membrane can be stained after the transfer with Ponceau S or prestained protein marker for determination of protein mass can be loaded in the right and left lanes of the protein samples.
17. (a) The optimal dilution of antibody must be determined experimentally; We usually dilute PY20 antibody 1:10, 000–40,000. (b) The membrane can also be incubated overnight at 4°C in solution containing primary antibody. (c) If primary antibody is not labeled with peroxidase, add one more step in the procedure. After primary antibody, do some brief washing as described and add secondary antibody that are labeled with enzyme and directed against the primary antibody. After a 1 h incubation in blotting solution proceed with the washing and detection steps as described.
18. We recommend checking both detection methods.
19. Keep the salt content at minimum during cell lysis and protein extraction. High salt content may interfere with TiO_2 chromatography. Protein precipitation with acetone or chloroform/ methanol may be applied to remove salt.

20. Do not heat-up the samples because high concentrations of urea will lead to carbamylation of free amino groups.
21. To keep the salt concentration low, use pure water rather than ammonium bicarbonate buffer to dilute the sample prior to trypsin incubation.
22. In case of double digestion, after step 3 adjust pH to 8.0 if necessary, and add 1 μg of lysyl endopeptidase LysC per 100 μg protein, incubate for 3 h at room temperature and proceed with step 4.
23. In the procedure described previously (see ref. 6), the trypsin digest was separated into 10 fractions on a Source™ 15RPC ST 4.6/100 column (Amersham Pharmacia Biotech); however, the current method is optimized and this step can be omitted.
24. This is particularly important to increase the binding specificity in the bacterial samples with very little phosphorylation (e.g. prokaryotic cell lysates); when possible it should be avoided to decrease DHB contamination during LC-MS. Alternatively, it is possible to use lactic acid as a competitive binder instead of DHB (see ref. 18).
25. If sample is fractionated prior to TiO_2 chromatography use the same amounts of TiO_2 beads for each fraction.
26. To quickly neutralize the high pH upon elution, we recommend eluting the samples into a small volume (~50 μl) of the HPLC loading solvent.
27. Desalting of the samples on either a microcolumn or a HPLC pre-column may lead to a loss of hydrophilic phosphopeptides (e.g. shorter or multiply phosphorylated peptides).
28. There is no pre-column or split in this LC-MS setup.
29. The timing between the MS and the LC system should be controlled with a standard double contact closure cable.
30. This gives a total scan cycle time (full scan + 3 MS^2 events) of up to 5 s on an LTQ Orbitrap XL, but only about 2 s on the newer generation LTQ Orbitrap Velos.
31. Multistage activation (MSA) produces information-rich spectra, where many of the fragment ions show pronounced neutral loss of phosphoric acid (e.g. −97.97, −48.99, or −32.66 for singly-, doubly-, or triply charged fragment ions, respectively). This information is very useful in validation of the phosphopeptide spectra. Note that this feature is useful for analysis of other labile modifications, such as O-glycosylation (the MSA values have to be adjusted for the corresponding modification).
32. Source settings have to be optimized for the emitter and nano-LC-MS setup.

Acknowledgement

This work was supported by Croatian Ministry of Science, Education and Sport (Project 098-0982913-2877 to DV) and Juniorprofessoren-Programm of the Baden-Württemberg Stiftung, SFB766 of the Deutsche Forschungsgemeinschaft and PRIME-XS (to BM).

References

1. Macek B, Gnad F, Soufi B, Kumar C, Olsen JV, Mijakovic I, Mann M (2008) Phosphoproteome analysis of *E. coli* reveals evolutionary conservation of bacterial Ser/Thr/Tyr phosphorylation. Mol Cell Proteomics 7:299–307
2. Macek B, Mijakovic I, Olsen JV, Gnad F, Kumar C, Jensen PR, Mann M (2007) The serine/threonine/tyrosine phosphoproteome of the model bacterium *Bacillus subtilis*. Mol Cell Proteomics 6:697–707
3. Parker JL, Jones AM, Serazetdinova L, Saalbach G, Bibb MJ, Naldrett MJ (2010) Analysis of the phosphoproteome of the multicellular bacterium *Streptomyces coelicolor* A3(2) by protein/peptide fractionation, phosphopeptide enrichment and high-accuracy mass spectrometry. Proteomics 10:2486–2497
4. Gnad F, Forner F, Zielinska DF, Birney E, Gunawardena J, Mann M (2010) Evolutionary constraints of phosphorylation in eukaryotes, prokaryotes, and mitochondria. Mol Cell Proteomics 9:2642–2653
5. Bechet E, Guiral S, Torres S, Mijakovic I, Cozzone AJ, Grangeasse C (2009) Tyrosine-kinases in bacteria: from a matter of controversy to the status of key regulatory enzymes. Amino Acids 37:499–507
6. Mijakovic I, Petranović D, Macek B, Cepo T, Mann M, Davies J, Jensen PR, Vujaklija D (2006) Bacterial single-stranded DNA-binding proteins are phosphorylated on tyrosine. Nucleic Acids Res 34:1588–1596
7. Sun X, Ge F, Xiao Chuan-Le, Yin X-F, Ge R, Zhang L-H, He Qing-Yu (2010) Phosphoproteomic analysis reveals the multiple roles of phosphorylation in pathogenic bacterium *Streptococcus pneumonia*. J Proteome Res 9:275–282
8. Stefanic Z, Vujaklija D, Luic M (2009) Structure of the single-stranded DNA-binding protein from *Streptomyces coelicolor*. Acta Crystallogr D Biol Crystallogr D65:974–979
9. Raghunathan S, Ricard CS, Lohman TM, Waksman G (1997) Crystal structure of the homo-tetrameric DNA binding domain of *Escherichia coli* single-stranded DNA-binding protein determined by multiwavelength x-ray diffraction on the selenomethionyl protein at 2.9-A resolution. Proc Natl Acad Sci USA 94:6652–6657
10. Jers C, Pedersen MM, Paspaliari DK, Schütz W, Johnsson C, Soufi B, Macek B, Jensen PR, Mijakovic I (2010) Bacillus subtilis BY-kinase PtkA controls enzyme activity and localization of its protein substrates. Mol Microbiol 77:287–299
11. Petranovic D, Michelsen O, Zahradka K, Silva C, Petranovic M, Jensen PR, Mijakovic I (2007) *Bacillus subtilis* strain deficient for the protein-tyrosine kinase PtkA exhibits impaired DNA replication. Mol Microbiol 63:1797–1805
12. Olsen JV, de Godoy L, Li MG, Macek B, Mortensen P, Pesch R, Makarov A, Lange O, Horning S, Mann M (2005) Parts per million mass accuracy on an Orbitrap mass spectrometer via lock mass injection into a C-trap. Mol Cell Proteomics 4:2010–2021
13. Olsen JV, Macek B, Lange O, Makarov A, Horning S, Mann M (2007) Higher-energy C-trap dissociation for peptide modification analysis. Nat Methods 4:709–712
14. Schroeder M, Shabanowitz JJ, Schwartz JC, Hunt DF, Coon JJ (2004) A neutral loss activation method for improved phosphopeptide sequence analysis by quadrupole ion trap mass spectrometry. Anal Chem 76:3590–3598
15. Rappsilber J, Ishihama Y, Mann M (2003) Stop and go extraction tips for matrix-assisted laser desorption/ionization, nanoelectrospray, and LC/MS sample pretreatment in proteomics. Anal Chem 75:663–670
16. Rappsilber J, Mann M, Ishihama Y (2007) Protocol for micro-purification, enrichment,

pre-fractionation and storage of peptides for proteomics using StageTips. Nat Protoc 2:1896–1906

17. Shereda RD, Kozlov AG, Lohman TM, Cox MM, Keck JL (2008) SSB as an organizer/ mobilizer of genome maintenance complexes. Crit Rev Biochem Mol Biol 43:289–318

18. Sugiyama N, Masuda T, Shinoda K, Nakamura A, Tomita M, Ishihama Y (2007) Phosphopeptide enrichment by aliphatic hydroxy acid-modified metal oxide chromatography for nano-LC-MS/MS in proteomics applications. Mol Cell Proteomics 6:1103–1109

Chapter 17

Fluorescent SSB as a Reagentless Biosensor for Single-Stranded DNA

Katy Hedgethorne and Martin R. Webb

Abstract

Helicases are an important and much studied group of enzymes that generally couple ATP hydrolysis to the separation of strands of base-paired nucleic acids. Studying their biochemistry at different levels of organization requires assays that measure the progress of the reaction in different ways. One such method makes use of the single-stranded DNA-binding protein (SSB) from *Escherichia coli*. This is used as a protein framework to produce a "reagentless biosensor," making use of its tight and specific binding of single-stranded DNA. The attachment of a fluorophore to this protein produces a signal in response to that binding. Thus the (G26C)SSB, labeled with a diethylaminocoumarin, gives a ~5-fold fluorescence increase on binding to single-stranded DNA and this can be used to assay the progress of helicase action along double-stranded DNA. A protocol for this is described along with a variant that can be used to follow the unwinding on a single molecule scale.

Key words: Reagentless biosensor, Helicase, Fluorescence, Kinetics, TIRFM

1. Introduction

Helicases are energy-transducing enzymes that are involved in many types of DNA or RNA manipulation (1, 2). There is a wide variety of proteins within this family, for example involved in nucleic acid packaging, replication, recombination, and repair. Often the helicase is part of a larger protein complex that performs several parts of such manipulations simultaneously. Because of this wide variety and the importance of these processes for the cell, there needs to be a range of different types of assay to study them, both to understand the mechanisms by which they operate and to examine the control of nucleic acid manipulations. The example, described here, is a reagentless biosensor for single-stranded DNA (ssDNA) that can measure the rate and extent of helicase activity on double-stranded DNA (dsDNA) in real time (3).

James L. Keck (ed.), *Single-Stranded DNA Binding Proteins: Methods and Protocols*, Methods in Molecular Biology, vol. 922, DOI 10.1007/978-1-62703-032-8_17,

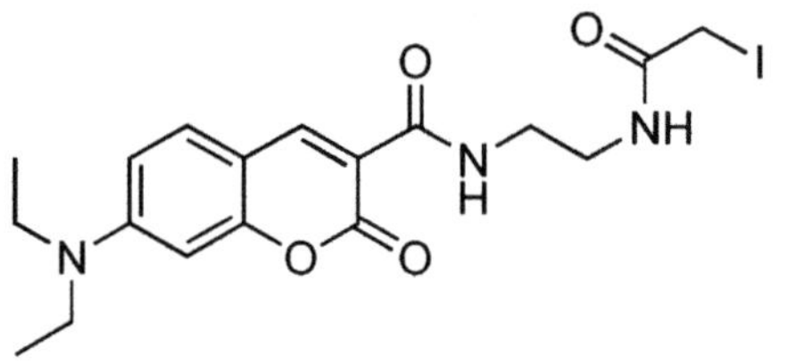

Fig. 1. Structure of the fluorescent label, IDCC.

Fluorescent reagentless biosensors form a class of probe molecules that provide a signal to give concentrations of a specific target analyte molecule. In particular, there is a class of such biosensors, based on a protein framework that is labeled with a fluorophore. Examples of fluorophore-protein adducts acting as such biosensors include ones for phosphate biosensors (4, 5), ADP (6), and certain other small molecules, such as sugars, amino acids, and metals ions (7–11).

The protein's ligand-binding site provides potentially a high degree of specificity, which may be important in applying the biosensor to assay specific molecules, particularly in the presence of similar ones. Such processes as conformation changes, resulting from ligand binding, can provide a coupling mechanism between interaction of ligand and the fluorophore response. This requires appropriate positioning of the fluorophore on the protein surface, for example, to be able to respond to, but not interfere with, ligand binding. It also requires the fluorophore to have the property of changing its fluorescence when its environment changes. Thus some fluorophores that are inherently bright, with a high fluorescence quantum yield under most conditions (for example, Cy3B), may not be suitable.

In the work described here, a diethylaminocoumarin (Fig. 1) is a suitable fluorophore with a fluorescence that can increase by more than an order of magnitude in response to changes around it (12, 13). It also has the practical advantage that its excitation maximum wavelength closely coincides with a strong Hg emission line at 436 nm, so that in assay formats such as stopped flow, the fluorescence of the coumarin is enhanced relative to background. Indeed, the mechanism in this case, as has been shown in other uses, involves the formation of a coplanar system of the rings and the dialkylamino moiety (14, 15). In this state there is high fluorescence quantum yield. If the diethylamino group is not so constrained, the fluorescence is lower. Understanding such mechanisms, as well as having defined protein structures, helps in developing reagentless biosensors for particular targets.

Fluorescence has been found to be a useful signal for bulk solution measurements because of its rapid response and high sensitivity. A protein framework provides advantages of a specific binding site, although the ability to manipulate the precise sequence potentially allows alteration of that specificity. Often ligand binding and associated conformation changes are rapid, at least allowing responses on the timescale of likely measurements, for example, using stopped-flow fluorescence with reaction times of a few milliseconds. An additional advantage of reagentless biosensors is that they usually target the product of the enzyme reaction being studied and this minimizes interference as the reaction itself is not compromised. This is not the case if substrates or the enzyme itself are modified, for example by a fluorophore. In those cases the modification can affect the properties of the enzyme system, such as rate or affinity. For example, fluorescent ATP analogues are widely used to study ATPases and kinases, but the properties of the analogues may differ greatly from unmodified ATP or ADP (16–18).

1.1. ssDNA Biosensor Properties

The protein framework is the single-stranded DNA-binding protein (SSB) from *Escherichia coli*. This is a ~20 kDa protein that acts as a tetramer, normally binds 65–70 bases of ssDNA at physiological ionic strength, with the DNA wrapping around the tetramer in a surface channel, as revealed by X-ray crystallography (19–21). This binding is rapid (322 μM^{-1} s^{-1} for the labeled protein) at 20 °C and tight (dissociation constant ~3 nM) (3, 22). At concentrations well above this dissociation constant the SSB binds to all ssDNA. At low ionic strength or high concentration of SSB, the binding mode changes to ~35 bases per tetramer. This change in binding can be significant in planning measurements using SSB and is discussed further, below.

After surveying several single cysteine mutations as sites for attaching one fluorophore per monomer, and several different types of fluorescent labels, the best combination of G26C and IDCC (*N*-[2-(iodoacetamido)ethyl]-7-diethylaminocoumarin-3-carboxamide) (23) (Fig. 1) was chosen so the adduct DCC-SSB acts as the ssDNA biosensor. It gives ~5-fold fluorescence increase on binding to ssDNA with little or no shift in the excitation or emission spectra, fairly normal for diethylaminocoumarin. SSB has no significant conformation change apparent on DNA binding (19, 24) and so the basis of the fluorescence change is likely to be the alteration in coumarin environment by the presence of DNA itself. The labeling does, indeed, cause ~30-fold decrease in affinity of the SSB for ssDNA, although the binding is still tight enough to be quantitative in most conditions (22). The preparation of DCC-SSB is described below.

1.2. Considerations in Using DCC-SSB

The particular focus of this chapter is the use of DCC-SSB to assay dsDNA unwinding by helicases in real-time, continuous assays (25). In such assays, DCC-SSB needs to be present in sufficient concentration to bind ssDNA rapidly as it is formed by the helicase and allow fluorescence increase to be followed with time. Such assays might be considered as being of two types. One is a multi-turnover, steady-state assay with the helicase concentration much lower than DNA. The unwinding can be followed over several minutes. More incisively, a single-turnover assay has all DNA unwound at the same time by the helicase present in relatively high concentration. This type of assay tends to be richer in information and will be described experimentally below. However, the assay timescale depends on the length of DNA unwound and the speed of unwinding. The former depends on processivity, which varies between helicases from a few base pairs to many thousands. The translocation rates of helicases also vary widely from $<1\ s^{-1}$ to $>1{,}000\ s^{-1}$. This means that the assay may be on the millisecond to second timescale and stopped-flow fluorescence may be the format of choice.

This section considers the factors that are important in developing a successful assay using DCC-SSB. It also outlines some of the pitfalls of using fluorescent reagentless biosensors, in general, and this fluorescent probe, in particular. Many of these considerations can be accommodated practically by performing an initial titration of DNA into a solution of SSB under the assay conditions and this will be described.

Firstly, there are several factors that will determine what concentration of DCC-SSB should be present. At any likely set of conditions, the probe binds to all of the DNA, so that an excess of DCC-SSB is required over the number of binding sites in terms of 65 bases per site. Typically, a 4- to 5-fold excess seems a suitable compromise between high background and other factors, described below. In addition, as for all coupled assays, the rate of fluorescence response should be considered so that it is significantly larger than the rate of the helicase reaction to ensure that the observed rate is not limited by the SSB but is purely the rate of the helicase reaction. One way to consider this is that the time it takes the helicase to move 65 bases reveals one complete binding site for the SSB. The rate of binding at the particular concentration of DCC-SSB, as described (22), should give a time constant at least fivefold more rapid. Given the rapid binding, this is unlikely to be a problem for most assays. Finally, inner filter affects may only start to be significant if the absorbance of the coumarin is >0.1 for the pathlength of the assay cell. In that case, absorbance of the exciting light in the near side of the cell may reduce the emission from the far side. This would require DCC-SSB tetramers at >500 nM for 1 cm pathlength. However, because the absorbance of the coumarin is unaltered by DNA binding, in practice at least several fold higher DCC-SSB can be used, as long as nothing else absorbs at the same wavelength.

Low salt concentration, high ratio of SSB to DNA and high concentration of SSB, favor 35-base binding and potentially extra SSB will bind onto the DNA during the time course of the assay. This "extra" binding is relatively slow (22) and is accompanied by a fluorescence decrease, but can be avoided or at least minimized for most likely conditions by keeping the total concentration of DCC-SSB below 500 nM. An example of this phenomenon has been described (26): in that case the extra binding did not affect the observed time course during unwinding.

As with any fluorescence assay, the signal may vary with conditions such as salt concentrations and temperature, so that a calibration, if required, needs to be at the same conditions that will be used in the assay. A titration, measuring fluorescence at various levels of protein saturation by DNA is shown in Fig. 3. Figure 3a is a titration using a standard spectrofluorimeter, while Fig. 3b shows a titration done under the same conditions as the helicase activity assay, shown in Fig. 4. This assay shows the unwinding of two different lengths of plasmid by the combination of initiator protein RepD and helicase PcrA. The assay allows the unwinding rate to be measured as a function of DNA length (25).

1.3. Use of Cy3B-SSB as Probe in a TIRFM Assay

Although the main focus of this chapter is the use of the ssDNA biosensor in bulk kinetic assays, a version of the probe has also been used to measure unwinding by a single helicase in a surface-attached assay using Total Internal Reflectance Fluorescence Microscopy (TIRFM) (27). Although diethylaminocoumarin has significantly favorable properties for bulk assays, that is not so for single molecule visualization. It is not a particularly bright fluorophore, as it has a low extinction coefficient (~45,000 M^{-1} cm^{-1}) and is easily photobleached under the high-intensity light conditions of a single molecule assay. So far, there is not a labeled SSB that successfully combines the necessary fluorescence characteristics for single molecule sensitivity and a large fluorescence change. However, the surface attachment of the helicase·DNA complex, combined with the high affinity of SSB for ssDNA, means that a fluorophore with a high quantum yield, namely Cy3B, can be used as a label for SSB even though it gives rise to only a small increase in fluorescence on binding ssDNA. The evanescent wave excitation of TIRFM means that the unwinding is observed as a single fluorescent spot for each complex that grows in intensity as Cy3B-SSB is recruited to the complex in response to release of the ssDNA during unwinding. However, the bulk solution, containing unbound and, therefore, mobile Cy3B-SSB, does not greatly contribute to background fluorescence. An advantage of this single molecule technique is that many events can be monitored as separate spots in a single field of view. It also means that there can be a more-or-less direct comparison of single molecule and bulk solution kinetic measurements of unwinding.

Helicase assays using this format have been described for three helicases, AddAB, RecBCD, and PcrA (27) with either DNA or protein attachment to the surface. These demonstrate the assay at very different unwinding rates depending on the particular helicase, but give similar average rates of unwinding to assays under the same solution conditions in bulk solution using DCC-SSB. The methodology of this technique and experimental aspects have been described in a previous Methods in Molecular Biology (28).

2. Materials

Prepare all solutions with double distilled water and analytical grade reagents.

2.1. SSB Protein Preparation

1. Glycerol stock of BL21(DE3) *E. coli*, transformed with pET22b vector, containing G26C SSB (3).
2. L. agar/ampicillin plates.
3. L. broth.
4. 10% ampicillin.
5. IPTG.
6. 1 M dithiothreitol (DTT).
7. 1 M Tris–HCl, pH 7.5.
8. 1 M Tris–HCl, pH 8.3.
9. 5 M NaCl.
10. 0.5 M EDTA.
11. Sucrose.
12. Glycerol.
13. 12% polyacrylamide gels.
14. Complete, EDTA-free protease inhibitor cocktail tablet (Roche).
15. 10% polyethyleneimine, pH 6.9.
16. Glass homogenizer.
17. $(NH_4)_2SO_4$ solid (biochemistry or enzyme grade).
18. 0.2 μm polyethersulfone syringe membrane filter.
19. 5 mL HiTrap Heparin column (GE Healthcare).
20. SnakeSkin pleated dialysis tubing (Thermo Scientific).
21. Centrifuge for 15 mL to 1 L volumes.
22. Preparative ultracentrifuge for ~70 mL volumes.
23. Chromatography fraction collector, equipped with absorbance detection and pump.

2.2. Labeling

1. 1 M DTT.
2. 1 M Tris–HCl, pH 7.5.

3. 5 M NaCl.
4. 0.5 M EDTA.
5. Glycerol.
6. PD10 desalting gel filtration column, 5 ml (GE Healthcare).
7. N-[2-(iodoacetamido)ethyl]-7-diethylaminocoumarin-3-carboamide (Invitrogen, U.S.A. or Synchem, Germany). Stored as a 20 mM solution in dry dimethylformamide at −80 °C, protected from light.
8. 100 mM sodium 2-mercaptoethane-sulfate (MESNA).
9. P4 gel filtration column (Biorad; 1 × 30 cm).
10. 0.2 μm polyethersulfone syringe filter membrane.
11. Amicon Ultracentrifugal filter.

2.3. Characterization

1. Spectrophotometer.
2. Fluorescence spectrophotometer.
3. Oligonucleotide dT_{70}.

2.4. Helicase Assay

1. Stopped-flow fluorimeter (HiTech, TgK Ltd), equipped with Xe-Hg lamp.
2. Plasmid, containing *ori*D double-stranded origin of replication (29).
3. PcrA helicase and RepD initiation factor (25).
4. ATP.
5. Assay buffer: 50 mM Tris–HCl, pH 7.5, 100 mM KCl, 1 mM EDTA, 10 mM $MgCl_2$, 10% ethanediol.

3. Methods

3.1. Preparation of G26C SSB

3.1.1. Expression in E. coli

1. Streak two, pre-warmed L. agar/ampicillin plates with the glycerol stock of the G26C SSB plasmid in BL21(DE3) *E. coli* cells.
2. Incubate overnight at 37 °C.
3. To 100 mL L. broth in a sterile 250 mL flask, add 100 μL 10% ampicillin and inoculate with two to three single colonies from the streaked L. agar plates.
4. Incubate overnight at 37 °C with shaking.
5. Add 500 μL 10% ampicillin to each of 4× 500 mL L. broth in 2 L flasks and warm to 37 °C.
6. Inoculate each flask with 5 mL of the overnight cell culture.

7. Incubate at 37 °C with shaking.
8. Measure the absorbance at 595 nm of two flasks at 30 min intervals, using L. broth as a reference.
9. Make up 10 mL of 200 mM IPTG by dissolving 0.5 g IPTG in water, and filter the solution.
10. When the absorbance of the cultures reaches 0.5 cm^{-1}, add 2.5 mL 200 mM IPTG to each flask to induce protein expression (see Note 1).
11. Incubate the flasks for a further 3 h.
12. Following the 3 h incubation, centrifuge the cells in 2× 1 L centrifuge bottles at 3,500 × *g* for 30 min at 4 °C.
13. Prepare 50 mL cell storage buffer by mixing 2.5 mL 1 M Tris–HCl, 2 mL 5 M NaCl, and 100 μL 0.5 M EDTA, making up to 50 mL with water and then dissolving 5 g sucrose into the buffer.
14. Return the supernatant to the 2 L flasks and decontaminate.
15. Resuspend the harvested cells in 50 mL cell storage buffer and transfer 25 mL aliquots to 2× 50 mL tubes. Quick-freeze on dry ice and store at −80 °C.

3.1.2. SSB Protein Purification

Prepare the following buffers:

1. 50 mL First Resuspension Buffer: 50 mM Tris–HCl pH 8.3, 400 mM NaCl, 1 mM EDTA, 20% glycerol
2. 50 mL Second Resuspension Buffer: 50 mM Tris–HCl, pH 8.3, 200 mM NaCl,1 mM EDTA, 20% glycerol
3. 500 mL Column Buffer A: 50 mM Tris–HCl, pH 8.3, 1 mM EDTA, 20% glycerol
4. 500 mL Column Buffer B: 50 mM Tris–HCl. pH 8.3, 1 mM EDTA, 20% glycerol, 1 M NaCl
5. 2 L Dialysis buffer: 50 mM Tris–HCl, pH 8.3, 1 mM EDTA, 50% glycerol, 500 mM NaCl

All procedures should be carried out at 4 °C.

3.1.3. Protein Extraction (see Note 2)

1. Thaw stored cells in a beaker of cold water.
2. Add one Complete EDTA-free protease inhibitor tablet.
3. Sonicate the cells in 4× 20 s bursts on ice using a medium-sized probe.
4. Centrifuge down the cell debris at 38,000 × *g* for 20 min at 4 °C.

3.1.4. Polyethyleneimine Precipitation

1. Measure the volume of supernatant.
2. Slowly add 10% polyethyleneimine, pH 6.9, while stirring, to give 0.4% final concentration and continue to stir gently for a further 15 min.
3. Centrifuge at 11,000 × *g* at 4 °C for 20 min.
4. Keep the supernatant on ice.
5. Resuspend the pellet in 50 mL First Resuspension Buffer + 1 mM DTT (see Note 3) and homogenize using a glass homogenizer.
6. Stir for 30 min to redissolve any SSB in the pellet.
7. Centrifuge at 11,000 × *g* for 20 min at 4 °C.

3.1.5. Ammonium Sulfate Precipitation

1. Measure the volume of supernatant.
2. Add $(NH_4)_2SO_4$ solid slowly to the stirred supernatant to 150 g/L and stir for a further 30 min.
3. Centrifuge at 14,000 × *g* for 20 min at 4 °C.
4. Keep the supernatant on ice.
5. Resuspend the pellet in 50 mL Second Resuspension Buffer + 1 mM DTT until dissolved completely.
6. Centrifuge at 38,000 × *g* for 20 min at 4 °C.
7. Store the supernatant on ice, overnight if necessary.

3.1.6. Heparin Column Chromatography (see Note 4)

1. Attach a 5 mL HiTrap heparin column to a chromatography system.
2. Wash out preservative from the column and pre-equilibrate at 2 mL/min with 50 mL Column Buffer A, containing 1 mM DTT.
3. Filter the stored supernatant using a 0.2 μm polyethersulfone syringe filter membrane.
4. To ensure the protein binds to the heparin column, load the supernatant at 2 mL/min so that 16% supernatant and 84% Column Buffer A are pre-mixed and loaded onto the column. Collect 10 mL fractions while the protein is loading.
5. Once the supernatant is loaded, wash the column with 15 mL Column Buffer A.
6. Start a gradient 0–100% Column Buffer B in 150 mL, collecting 5 mL fractions and following the absorbance at 280 nm.

3.1.7. Dialysis

1. Run a gel of the fractions surrounding and including the elution peak.
2. Pool fractions containing SSB and use dialysis tubing to dialyze against 2 L dialysis buffer + 1 mM DTT overnight.

3. Measure the absorbance spectrum from 220 to 320 nm of a 1:50 dilution using dialysis buffer as a reference.
4. Calculate the protein concentration using the absorbance at 280 nm and the SSB extinction coefficient, 27,880 M^{-1} cm^{-1}.
5. Quick freeze 1.5 mL aliquots and store at −80 °C.

3.2. Labeling with IDCC

3.2.1. Labeling

1. Make up 250 mL labeling buffer by mixing 5 mL 1 M Tris–HCl, pH 7.5, 25 mL 5 M NaCl, 500 μL 0.5 M EDTA, 50 mL glycerol, and water to make the volume 250 mL. Mix until glycerol is dissolved completely and filter the buffer under vacuum.
2. Incubate 3 mg (G26C)SSB with 1.6 μL 1 M DTT for 20 min at room temperature.
3. Meanwhile, pre-equilibrate a PD10 column by passing 25 mL filtered, labeling buffer through the column and discard the eluate.
4. Apply the SSB to the column and allow the sample to completely enter the packed bed before adding labeling buffer to make up the total applied volume to 2.5 mL. Discard the eluate.
5. Elute the protein with 5 mL buffer, collecting ten 0.5 mL fractions.
6. Use Bradford reaction to identify fractions containing protein. To 10 μL Bradford reagent add 1 μL of the collected fraction and mix thoroughly. A color change to blue indicates the presence of protein in the solution (see Note 5).
7. Using a spectrophotometer, measure the concentration of those fractions containing protein by absorbance of a 1/10 dilution of each fraction at 280 nm and pool two to three fractions to make a solution such that the protein concentration is approximately 100 μM (see Note 6).
8. Add a twofold molar excess of IDCC and incubate, protected from light, for 2 h at room temperature with end-over-end stirring.
9. Add a tenfold molar excess of MESNA over protein to react with any remaining IDCC and incubate for a further 30 min.

3.2.2. Purification

1. Pre-equilibrate a P4 gel filtration column with labeling buffer by passing 60 mL filtered buffer through the column.
2. Pass the labeled protein solution through a 0.2 μm syringe filter membrane.
3. Allow equilibration buffer to drain into the column bed before applying the sample to the column.

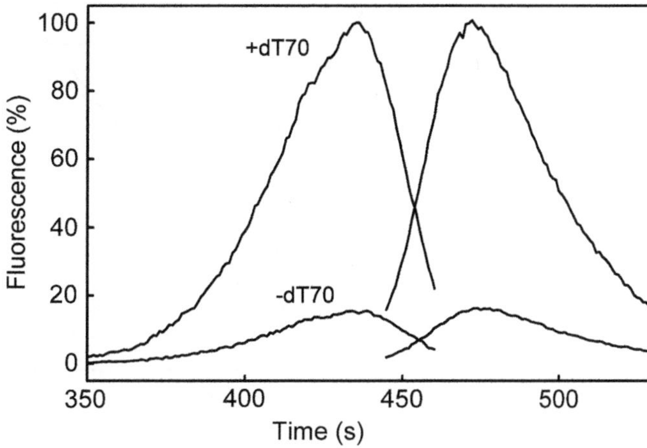

Fig. 2. Fluorescent spectral changes of DCC-SSB in response to ssDNA binding. Excitation and emission spectra are shown in the presence and absence of dT_{70} as described in the text.

4. Allow the sample to enter the column bed completely before applying additional buffer and attaching the column to a buffer reservoir.
5. Collect 0.5 mL fractions and pool the first colored peak.
6. Concentrate the labeled protein using an Amicon Ultra centrifugal filter and centrifuging for ~15 min at 2,600 × *g*.
7. Aliquot into opaque tubes and quick freeze on dry ice for storage at −80 °C.

3.2.3. Characterization

1. Measure the absorbance spectrum of a 1/50 dilution from 220 to 520 nm. Calculate the protein concentration taking the extinction coefficient for the coumarin to be 44,800 M^{-1} cm^{-1} at 430 nm (see Note 7).
2. Measure fluorescence excitation and emission spectra of a 250 nM solution of DCC-G26C SSB tetramer in 25 mM Tris–HCl, pH 7.5, 200 mM NaCl at 20 °C. Excite at 435 nm and record the emission spectrum. Then, with the fluorescence emission at 475 nm, record the excitation spectrum. An example of spectra is shown in Fig. 2.
3. Add an excess of dT_{70} (~600 nM) and repeat the excitation and emission scans. Comparison of the two spectra allows the determination of the fluorescence change of the DCC-SSB upon DNA binding.
4. Perform a fluorescence titration to characterize the DNA binding of the labeled SSB. To 200 μL of 250 nM DCC-SSB tetramer in 25 mM Tris–HCl, pH 7.5, 200 mM NaCl at 20 °C, add 4 pmol aliquots dT_{70}. The titration should show a linear increase in

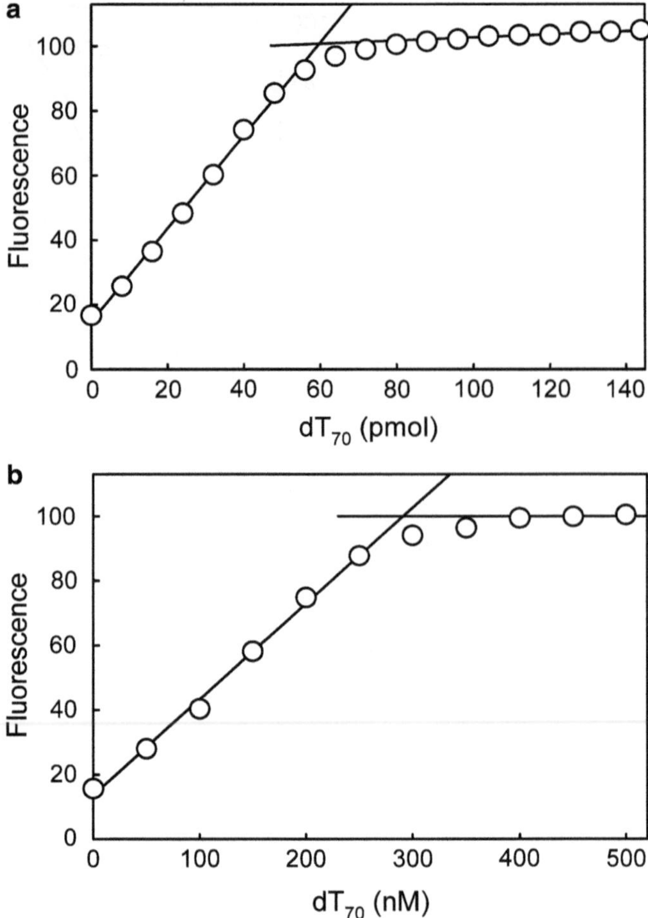

Fig. 3. Titration of ssDNA (dT_{70}) into a solution of DCC-SSB. (**a**) A titration in a cuvette using a spectrofluorimeter using 200 μL solution of 250 nM DCC-SSB tetramers (50 pmol). Solution conditions are described in the text. The additions are given in picomoles to allow for dilution by sequential additions. The fluorescence is corrected for dilution. (**b**) Titration in a stopped-flow fluorimeter under conditions of the helicase assay in Fig. 4. The DCC-SSB was at 250 nM tetramer. In each assay the data were fit to two linear regions. The intercept gives the effective capacity of DCC-SSB. The intercept is similar when poly(dT) is used.

fluorescence up to 50 pmol, at which point the SSB is saturated with DNA and there is no further increase in fluorescence. An example titration is shown in Fig. 3 (see Note 8).

3.3. Example of Helicase Assay (see Note 9)

1. Make up a reactant solution containing 2 mM ATP and 200 nM DCC-SSB tetramer in assay buffer and add to a stopped flow syringe.
2. In the second syringe, pre-incubate 1 nM pCERoriD plasmid with 4 nM RepD monomer for 30 s at 30 °C in assay buffer.
3. Add 200 nM PcrA and 200 nM DCC-SSB tetramer to the second syringe and mix.

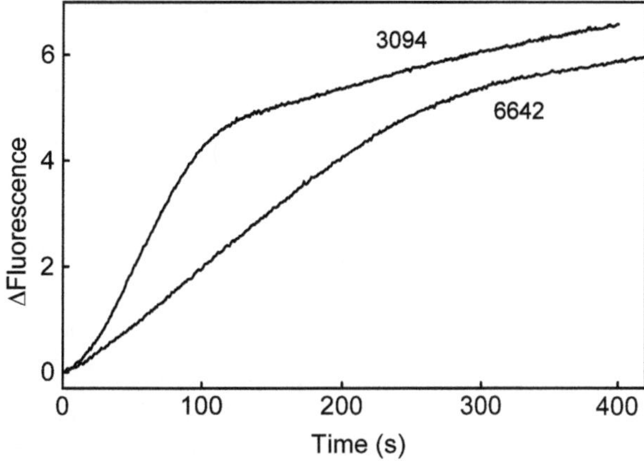

Fig. 4. Helicase assay using DCC-SSB, using two different lengths of plasmid (3,094 or 6,642 bp), containing the double-stranded origin of replication, *ori*D (25, 29). 0.5 nM plasmid, 2 nM RepD, 100 nM PcrA were mixed with 1 mM ATP in a stopped-flow instrument in the presence of 200 nM DCC-SSB tetramer. Solution conditions are described in the text and concentrations are those in the mixing chamber.

4. Rapidly mix the two solutions in the stopped-flow instrument and follow the fluorescence change for ~400–450 s by exciting at 436 nm and using a 455 nm glass cut-off filter on the emitted light. Example traces are shown in Fig. 4 using two different lengths of plasmid DNA.

4. Notes

1. To check for successful induction of protein expression, an SDS-PAGE can be used. Before addition of IPTG, take 1 mL samples from each flask, spin down in a benchtop centrifuge at 10,000 × *g* for 3 min and discard the supernatant. Quick-freeze the cell pellet and store at −80 °C. Repeat this following the 3 h incubation and run the samples on a SDS-polyacrylamide gel. An enhancement of the band at ~20 kDa indicates successful induction of protein expression.
2. To follow the progress of the protein purification, a 50 μL sample of supernatant can be taken following each centrifugation step and kept aside on ice. These samples can then be run on an SDS-polyacrylamide gel. A band should be seen at ~20 kDa in all samples except for polyethyleneimine and $(NH_4)_2SO_4$ supernatants.
3. 1 mM DTT should be added to all buffers from first resuspension buffer onwards immediately before using.
4. The protein is pure following the polyethyleneimine and $(NH_4)_2SO_4$ precipitations, so the chromatography step may not be needed. It does, however, result in a more concentrated

protein solution and this is useful both for subsequent labeling and for storage of the protein. If the heparin column is not used, continue directly on to dialyze the supernatant.

5. Protein usually elutes in fractions 3–5.
6. If starting with a dilute solution of protein, it may need to be concentrated at this point using an Amicon Ultra centrifugal filter to produce 1–1.5 mL of 100 μM protein solution.
7. To measure the total protein concentration, use the absorbance at 280 nm, where the extinction coefficient of SSB is 27,880 M^{-1} cm^{-1}. The extinction coefficient of IDCC at 280 nm is 7,470 M^{-1} cm^{-1} and this must also be taken into account. This calculation allows an estimation of the percentage of successfully labeled protein in the sample.
8. The titration is described using a length of DNA similar to the length of the SSB tetramer-binding site. In this way, one dT_{70} binds to each tetramer. A titration may be preferred under the same conditions as will be used in a helicase assay. An example of such a titration in the stopped-flow format is shown in Fig. 3.
9. The described assay has a complete plasmid unwound and at this point there is a break in the fluorescence time course (Fig. 4). This provides a calibration of the signal. In some cases, it is necessary to calibrate the signal, using known amounts of ssDNA. This can be either a full titration, as in Fig. 3b, or by using two or three different concentrations to quantify the fluorescence response, equivalent to the first few points of Fig. 4b.

Acknowledgements

We would like to thank our coworkers, who have been involved in development and application of this fluorescent SSB and are coauthors of publications cited here. We thank the Medical Research Council, UK, for support (ref. U117512742).

References

1. Singleton MR, Dillingham MS, Wigley DB (2007) Structure and mechanism of helicases and nucleic acid translocases. Annu Rev Biochem 76:23–50
2. Pyle AM (2008) Translocation and unwinding mechanisms of RNA and DNA helicases. Annu Rev Biophys 37:317–336
3. Dillingham MS, Tibbles KL, Hunter JL, Bell JC, Kowalczykowski SC, Webb MR (2008) Fluorescent single-stranded DNA binding protein as a probe for sensitive, real time assays of helicase activity. Biophys J 95:3330–3339
4. Brune M, Hunter JL, Corrie JET, Webb MR (1994) Direct, real-time measurement of rapid inorganic phosphate release using a novel fluorescent probe and its application to actomyosin subfragment 1 ATPase. Biochemistry 33:8262–8271

5. Okoh MP, Hunter JL, Corrie JET, Webb MR (2006) A biosensor for inorganic phosphate using a rhodamine-labeled phosphate binding protein. Biochemistry 45:14764–14771
6. Kunzelmann S, Webb MR (2010) A fluorescent, reagentless biosensor for ADP based on tetramethylrhodamine-labeled ParM. ACS Chem Biol 5:415–425
7. Gilardi G, Zhou LQ, Hibbert L, Cass AEG (1994) Engineering the maltose binding protein for reagentless fluorescence sensing. Anal Chem 66:3840–3847
8. Salins LL, Ware RA, Ensor CM, Daunert S (2001) A novel reagentless sensing system for measuring glucose based on the galactose/glucose-binding protein. Anal Biochem 294:19–26
9. Salins LL, Goldsmith ES, Ensor CM, Daunert S (2002) A fluorescence-based sensing system for the environmental monitoring of nickel using the nickel binding protein from Escherichia coli. Anal Bioanal Chem 372:174–180
10. Sharma BV, Shrestha SS, Deo SK, Daunert S (2006) Biosensors based on periplasmic binding proteins. In: Thompson RB (ed) Fluorescence sensors and biosensors. CRC, Boca Raton, FL, pp 45–65
11. Webb MR (2007) Development of fluorescent biosensors for probing the function of motor proteins. Mol Biosyst 3:249–256
12. Webb MR, Reid GP, Munasinghe VRN, Corrie JET (2004) A series of related nucleotide analogues that aids optimization of fluorescence signals to probe the mechanism of P-loop ATPases, such as actomyosin. Biochemistry 43:14463–14471
13. Brune M, Hunter JL, Howell SA, Martin SR, Hazlett TL, Corrie JE et al (1998) Mechanism of inorganic phosphate interaction with phosphate binding protein from Escherichia coli. Biochemistry 37:10370–10380
14. Hirshberg M, Henrick K, Haire LL, Vasisht N, Brune M, Corrie JET et al (1998) The crystal structure of phosphate binding protein labeled with a coumarin fluorophore, a probe for inorganic phosphate. Biochemistry 37:10381–10385
15. Fletcher AN, Bliss DE (1978) Laser dye stability. Part 5. Effect of chemical substituents of bicyclic dyes upon photodegradation parameters. Appl Phys 16:289–295
16. Jameson DM, Eccleston JF (1997) Fluorescent nucleotide analogs: synthesis and applications. Methods Enzymol 278:363–390
17. Forgacs E, Cartwright S, Kovacs M, Sakamoto T, Sellers JR, Corrie JE et al (2006) Kinetic mechanism of myosinV-S1 using a new fluorescent ATP analogue. Biochemistry 45:13035–13045
18. Toseland CP, Martinez-Senac MM, Slatter AF, Webb MR (2009) The ATPase cycle of PcrA helicase and its coupling to translocation on DNA. J Mol Biol 392:1020–1032
19. Raghunathan S, Kozlov AG, Lohman TM, Waksman G (2000) Structure of the DNA binding domain of *E. coli* SSB bound to ssDNA. Nature Struct Biol 7:648–652
20. Lohman TM, Ferrari ME (1994) *Escherichia Coli* single-stranded DNA-binding protein: multiple DNA-binding modes and cooperativities. Annu Rev Biochem 63:527–570
21. Bujalowski W, Lohman TM (1986) *Escherichia coli* single-strand binding protein forms multiple, distinct complexes with single-stranded DNA. Biochemistry 25:7799–7802
22. Kunzelmann S, Morris C, Chavda AP, Eccleston JF, Webb MR (2010) Mechanism of interaction between single-stranded DNA binding protein and DNA. Biochemistry 49:843–852
23. Corrie JET (1994) Thiol-reactive fluorescent probes for protein labelling. J Chem Soc Perkin Trans 1, 2975–2982
24. Raghunathan S, Ricard CS, Lohman TM, Waksman G (1997) Crystal structure of the homo-tetrameric DNA binding domain of *Escherichia coli* single-stranded DNA-binding protein determined by multiwavelength x-ray diffraction on the selenomethionyl protein at 2.9-A resolution. Proc Natl Acad Sci USA 94:6652–6657
25. Slatter AF, Thomas CD, Webb MR (2009) PcrA helicase tightly couples ATP hydrolysis to unwinding double-stranded DNA, modulated by the replication initiator protein, RepD. Biochemistry 48:6326–6334
26. Yeeles JTP, Gwynn EJ, Webb MR, Dillingham MS (2011) The AddAB helicase-nuclease catalyses rapid and processive DNA unwinding using a single Superfamily 1A motor domain. Nucleic Acids Res 39:2271–2285
27. Fili N, Mashanov G, Toseland CP, Batters C, Wallace MI, Yeeles JTP et al (2010) Visualizing DNA unwinding by helicases at the single molecule level. Nucleic Acids Res 38:4448–4457
28. Fili N, Toseland CP, Dillingham MS, Webb MR, Molloy JE (2011) A single-molecule approach to visualize DNA unwinding. Methods Mol Biol 778:193–214
29. Soultanas P, Dillingham MS, Papadopoulos F, Phillips SE, Thomas CD, Wigley DB (1999) Plasmid replication initiator protein RepD increases the processivity of PcrA DNA helicase. Nucleic Acids Res 27:1421–1428

Chapter 18

Fluorescent Single-Stranded DNA-Binding Proteins Enable In Vitro and In Vivo Studies

Piero R. Bianco, Adam J. Stanenas, Juan Liu, and Christopher S. Cohan

Abstract

Fluorescent single-stranded DNA-binding proteins (SSB) that have a defined number of fluorophores per tetramer are invaluable tools to understand biochemical mechanism and biological function. Here, we describe the purification of fluorescent SSB chimeras with a unique number of fluorescent subunits incorporated per tetramer. We describe the use of these tetramers to enable clear visualization of SSB in vivo. Purified chimeras also facilitate single molecule studies (Liu et al., Protein Sci 20:1005–1020, 2011).

Key words: SSB, Fluorescence, DNA replication, Recombination, DNA repair, Immobilized Metal Affinity Chromatography, Fluorescence microscopy

1. Introduction

The eubacterial single-stranded DNA-binding protein (SSB) plays essential roles in all aspects of DNA metabolism (1). It functions to both protect unwound single strands of DNA, and to direct the loading of a variety of proteins onto the DNA (2–5). Loading of these proteins requires a functional C terminus in SSB, and deletion of this region renders cells inviable.

Until recently, fluorescent tagging of SSB has been achieved in one of two ways. In the first, cysteine residues are introduced at unique sites that do not perturb biological function (6, 7). Purified cysteine mutant proteins are chemically modified to attach fluorescent dyes at the cysteine residues. In this approach it is extremely difficult to remove unmodified or dark protein so subsequent biochemical assays will always be contaminated. Further, these modified proteins are only useful for in vitro studies. In the second approach, fluorescent protein fusions to SSB have been made, but these have not been well characterized, nor has their construction and utility been carefully laid out (8).

James L. Keck (ed.), *Single-Stranded DNA Binding Proteins: Methods and Protocols*, Methods in Molecular Biology, vol. 922,
DOI 10.1007/978-1-62703-032-8_18,

To circumvent these issues, we developed a dual plasmid system that enables both in vitro and in vivo studies (9). The first plasmid expresses wild-type SSB protein with an N-terminal hexahistidine tag. The subunit expressed from this plasmid has a wild-type C terminus and is therefore able to interact with cognate proteins. The second plasmid expresses an SSB—fluorescent protein fusion. The fluorescent protein is fused to the C terminus of SSB, which has been lengthened to add a six glycine linker positioned between SSB and the fluorescent protein. This design pushes the fluorescent protein away from the tetramer and thereby minimizes the potential for negative effects on DNA binding and or protein-protein interactions. As the subunit does not contain a functional C terminus, it cannot interact with a cognate protein. Therefore, to create chimeric SSB proteins that are both fluorescent and have the ability to interact with partner proteins, these two plasmids are introduced into the same cell and expression induced. Expression levels of the fusion and histidine tagged subunits are approximately equal when the two plasmids are both of high copy number. When the fusion gene is introduced into a low copy number vector and expressed in the same cell as the high copy number plasmid, levels of the fluorescent protein are reduced thereby achieving a lower level of incorporation of fusion subunit into SSB tetramers.

To facilitate purification of chimeric SSB proteins, lysates from cells expressing both proteins are subjected to nickel column chromatography (Fig. 1). Following extensive column washing, proteins are eluted with an imidazole gradient. Tetramers containing four histidine tagged subunits bind to the column very tightly and are eluted at high concentrations of imidazole. In contrast tetramers containing fluorescent fusion subunits bind less tightly and are eluted at lower concentrations of imidazole. Thus, from a single culture five different SSB species can in theory be obtained. These proteins retain the functionality of wild type. Fluorescent tetramers also enable visualization of the protein on single molecules of DNA (9).

In addition to in vitro work, the SSB chimeras can also be used to track the protein in vivo (Fig. 2). Here, and to facilitate a lower ratio of fluorescent to wild-type protein, a dual plasmid system using one low and one high copy number is used. This strategy maximizes the number of critical C-termini per SSB tetramer while simultaneously enabling visualization of SSB protein using fluorescence microscopy. In addition, wild-type, plasmid-borne SSB is used as the histidine tag is unnecessary. Alternatively, the wild-type plasmid can be omitted as the subunit mixing with the chromosomal protein readily occurs (data not shown).

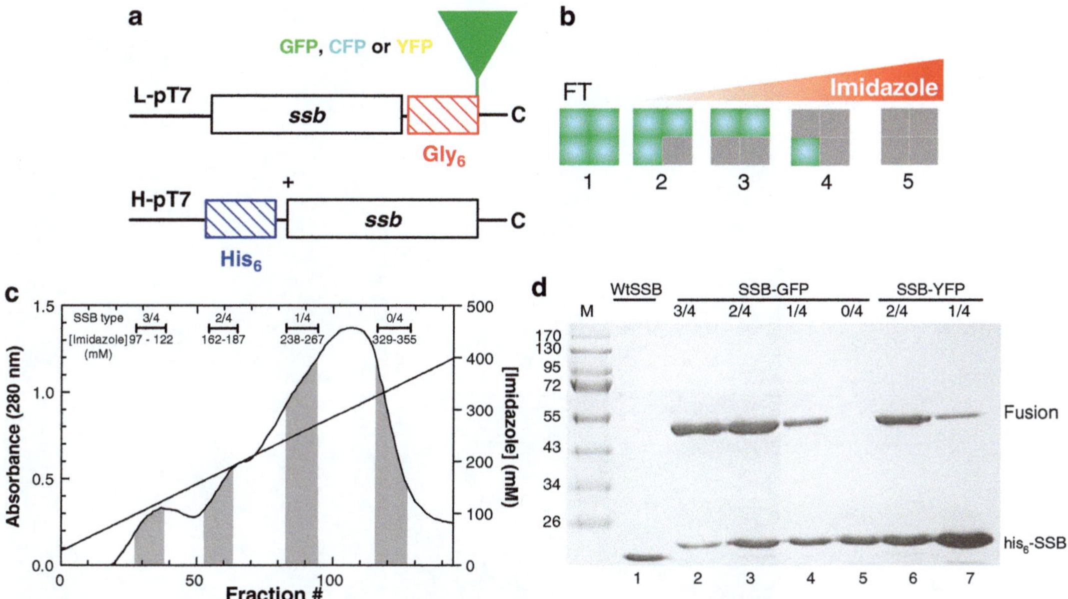

Fig. 1. SDS-PAGE reveals that fluorescent SSB chimeras can be purified to near homogeneity. (**a**) Scheme for production of fluorescent protein-tagged SSB chimeras. Dual plasmid system showing SSB-fluorescent protein fusion and his$_6$-SSB constructs. The designations L-pT7 and H-pT7 refer to low and high copy number plasmids, respectively. (**b**) Purification scheme. The constructs shown in (**a**) are co-expressed in the same cell. Following cell lysis, the heterogeneous population of SSB tetramers is applied to a nickel column and proteins eluted with an imidazole gradient. *Gray squares*, his$_6$-SSB subunits; green, SSB-fluorescent protein fusions. *FT* flow through. Numbers below each tetramer indicate the number of fusion subunits per tetramer. (**c**) SSB chimeras elute at different concentrations of imidazole. A nickel column elution profile from a 1 L cell lysate (his$_6$-SSB/SSB-GFP; H/H) is shown. The elution positions of each chimera as determined by SDS-PAGE are indicated (*gray-shaded regions*). Intervening regions contain mixtures of the two species flanking the region indicated in the elution profile. The amount of contaminating species can be decreased by subjecting each chimera separately to nickel column chromatography. (**d**) SDS-PAGE analysis of the final protein pools of purified and chimeric SSB proteins. Lanes 2–5 are from the column profile shown in panel (**c**); lanes 6 and 7 are from a separate preparation from cells expressing his$_6$-SSB and SSB-YFP fusion protein from H/L plasmid constructs. Components of this figure are used by permission of John Wiley and Sons. They appeared in ref. (9).

2. Materials

2.1. Protein Over-Expression

1. The bacterial strain is a Tuner cell line (Novagen) that contains a mutation in the lac permease enabling expression of IPTG inducible promoter at low concentrations of IPTG. This is important as expression is more uniform, so that a high yield of fluorescent protein is obtained.
2. Plasmids for chimeric protein purification are high copy number pET vectors with different drug markers.
3. The growth medium is LB containing relevant antibiotics.
4. SDS-PAGE electrophoresis equipment and running buffers.

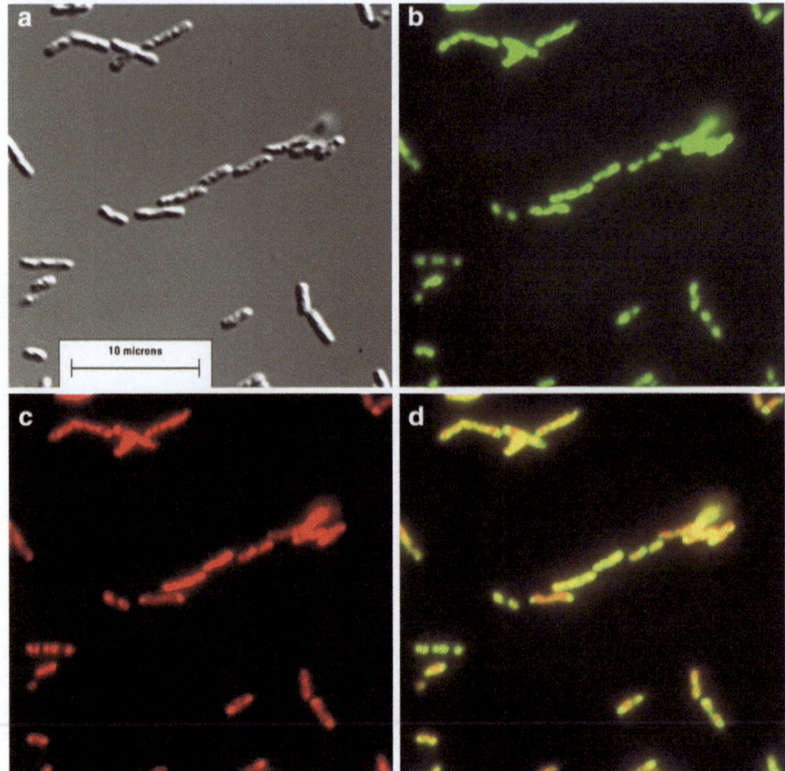

Fig. 2. Fluorescence microscopy of SSB cells reveals that SSB and RecG colocalize. Representative microscopy images showing colocalization of SSB and RecG in exponentially growing cells. (**a**) Differential interference contrast microscopy image; (**b**) the same field of cells viewed in epi-fluorescence using a GFP filter; (**c**) the same cells viewed using a texas red filter and (**d**) merge of panels (**b**) and (**c**). The Pearson's correlation coefficient for this image is 0.93. Expression of both proteins is observed in >90 % of cells.

2.2. Protein Purification

1. The column is a 20 ml nickel sepharose column, sufficient for purification of protein from a 2 L culture.
2. Equilibration and elution buffers containing 30 and 500 mM imidazole, respectively.
3. Wash buffer is equilibration buffer containing 0.2% NP40.
4. SDS-PAGE electrophoresis equipment and running buffers.

2.3. In Vivo Studies Including Fluorescence Microscopy

1. The bacterial strain is a Tuner cell line (Novagen) that contains a mutation in the lac permease enabling expression of IPTG inducible promoter at low concentrations of IPTG.
2. Plasmids for in vivo studies are one low copy number and one high copy number pET vector with different drug markers. The low copy number plasmid expresses the fusion subunit while the high copy number vector expresses the untagged, wild-type subunit.

3. The growth medium is LB containing relevant antibiotics and growth is at 37°C with aeration.
4. Glass microscope slides and glass coverslips.
5. Poly-L-lysine (0.01%, Sigma) to coat the slide for cell adherence.
6. Nail polish to seal slides.
7. Microscope equipped with phase or Differential interference contrast microscopy (DIC) optics and fluorescence imaging capability, including a CCD camera.
8. Image analysis software (Image Pro, Media Cybernetics).
9. Fluorescence spectrophotometer to monitor fluorescence in whole cells.

2.4. Buffers

1. SDS-PAGE running buffer (4×):
   (a) 60 g Tris base.
   (b) 288 g glycine.
   (c) H_2O up to 5 L.
2. SDS-PAGE running buffer (1×):
   (a) 1 L of 4× SDS-PAGE Running buffer.
   (b) 3 L of H_2O.
   (c) 40 ml of 10% SDS.
3. Ni equilibration buffer:
   (a) 20 mM sodium phosphate (pH 7.4).
   (b) 500 mM NaCl.
   (c) 30 mM Imidazole.
   (d) Final pH adjusted to 7.4 with HCl.
4. Ni elution buffer:
   (a) 20 mM sodium phosphate (pH 7.4).
   (b) 500 mM NaCl.
   (c) 500 mM imidazole.
   (d) Final pH adjusted to 7.4 with HCl.

3. Methods

3.1. Protein Over-Expression

1. Wake up the strain from the −80°C freezer onto an LB plate containing ampicillin and kanamycin and 0.2% glucose.
2. Pick several well isolated, single colonies and separately grow these overnight with aeration in LB containing ampicillin, kanamycin, and 0.2% glucose (see Note 1).

3. Dilute the overnights 1/100 into 5 ml of fresh LB containing ampicillin and kanamycin and grow for 2 h at 37°C, with aeration.
4. Add IPTG to 100 μM final, and grow for an additional 3 h.
5. Harvest 1 ml of cells, and make a total cell lysate using 1% SDS.
6. Subject an aliquot of the total cell lysate to electrophoresis in a 15% SDS-PAGE gel to determine the level of expression in the individual cultures.
7. Determine, by visual inspection of the gel, in which culture the highest level of equivalent expression of his-tagged SSB and the SSB fusions is observed.
8. Use the overnight culture from step 3 to inoculate LB containing kanamycin, ampicillin and 0.2% glucose. Grow overnight at 37°C (see Note 2).
9. The following day, using the fresh overnight culture, inoculate a 1 L flask containing LB plus kanamycin and ampicillin.
10. Grow the culture at 37°C with aeration until an OD of 0.4–0.6.
11. Add IPTG to a final concentration of 100 μM, and grow for an additional 4 h.
12. Harvest the cells by centrifugation.
13. Resuspend the cells in Tris–HCl, pH 8.0, 20% sucrose. Add 2 ml of buffer per gram of cell pellet. The cells can be frozen at −80°C or they can be lysed immediately.

3.2. Protein Purification

1. Cell lysis is done at 4°C with constant stirring.
2. Add lysozyme to a final concentration of 1 mg/ml and 250 units of benzonase followed by stirring for 30 min.
3. Add deoxycholate drop wise to a final concentration of 0.5%. Stir for an additional 30 min. At this point, the lysate should be completely liquid as the nucleic acids have been degraded by benzonase.
4. Adjust the sodium chloride to 600 mM and the imidazole concentration to 30 mM. Stir for an additional 10 min.
5. Subject the cells to centrifugation at 35,000 × *g* for 90 min. A tight pellet should form, and the majority of the fluorescence should be observed in the supernatant (see Note 3).
6. Load the cleared cell lysate onto a 20 ml nickel sepharose column (see Note 4).
7. Wash the column with 20 column volumes of Ni equilibration buffer only.
8. Wash the column with 20 column volumes of Ni equilibration buffer containing 0.2% NP 40.

9. Wash the column with 10 column volumes of Ni equilibration buffer only.
10. Elute the proteins with a shallow, linear gradient 20 times the column volume. The Ni elution buffers for the column contain 30 and 500 mM imidazole, respectively.
11. Identify the elution positions of each of the chimeras using 15% SDS-PAGE gels.
12. Pool the fractions corresponding to each of the chimeras carefully as the intervening fractions contain mixtures of the flanking species.
13. Dilute each pool using Ni equilibration buffer, and re-chromatograph each pool separately (see Note 5).
14. Dialyze each pool separately against 20 mM Tris–HCl, pH 8.0, 1 mM EDTA, 500 mM NaCl, followed by dialysis against the same buffer containing 50% glycerol.
15. Typical yields from a 1 L culture of bacteria are 2 mg of 1/4, 5 mg of 2/4, 10 mg of the 3/4, and greater than 20 mg of the 0/4 pool (see Note 6).
16. Determine the levels of nuclease contamination in each pool by incubating 1 μg of protein with 5′-end labeled, single and double-stranded oligonucleotides.
17. Measure the site size of each pool using a fluorescence-based assay that monitors the quenching of the intrinsic fluorescence of the SSB tetramer that occurs on binding to ssDNA (9) (see Note 7).

3.3. In Vivo Studies

3.3.1. Cell Growth

1. Wake up the dual plasmid strain from the −80°C freezer onto an LB plate containing ampicillin and kanamycin and 0.2% glucose. At the same time, wake up the strain containing the wild-type plasmid only as a control for fluorescence measurements.
2. Pick several well-isolated, single colonies from each plate and grow these overnight in LB containing ampicillin, kanamycin, and 0.2% glucose.
3. Dilute the overnights 1/100 into 5 ml of fresh LB containing ampicillin and kanamycin and grow for 2 h at 37°C, with aeration.
4. Add IPTG to 100 μM and grow for an additional 3 h.
5. Harvest 1 ml of cells, and make a total cell lysate using 1% SDS. Use the remaining 4 ml to measure in vivo fluorescence (see step 10, below).
6. Use an aliquot of the total cell lysate for electrophoresis in a 15% SDS-PAGE gel to determine the level of expression in the individual cultures.

7. Determine, by visual inspection of the gel, in which culture the highest level of expression of wild-type SSB and the SSB fusion subunits is observed.
8. Centrifuge the remaining 4 ml of cells at 5,000 × *g* for 10 min. Discard the supernatant.
9. Resuspend the cell pellet in 400 μl of 10 mM $MgSO_4$.
10. Measure the fluorescence present in the culture using the wild-type SSB only cells as a negative control for background fluorescence.
11. Select the culture that exhibits the highest level of fluorescence from step 10 and in which the highest levels of protein expression are observed (step 7).
12. Use the overnight of these cultures from step 2 to inoculate LB containing kanamycin, ampicillin, and 0.2% glucose. Grow overnight at 37°C.
13. The following day, using the fresh overnight, inoculate 5 ml of LB plus kanamycin and ampicillin.
14. Grow the culture at 37°C with aeration until an OD of 0.4–0.6 (typically 90 min).
15. Add IPTG to the desired, final concentration and grow for an additional 4 h (see Note 8).
16. Harvest the cells by centrifugation at 5,000 × *g* for 10 min.
17. Resuspend the cells in 500 μl of 10 mM $MgSO_4$ and place on ice.

3.3.2. Fluorescence Microscopy

1. Mark the center of a coverslip with an "X" using an UltraFine sharpie pen.
2. Deposit 20 μl of poly-L-lysine onto the coverslip immediately adjacent to the "X" and allow binding for 10 min.
3. Remove the remaining liquid by aspiration.
4. Wash the coverslip three times with water and aspirate the remaining water.
5. Deposit 20 μl of cells in 10 mM $MgSO_4$ onto the poly-L-lysine-coated surface and allow to bind for 10 min (see Note 9).
6. Aspirate the remaining liquid.
7. Wash the coverslip with water or 10 mM $MgSO_4$ to remove unbound cells.
8. Aspirate the remaining liquid.
9. Add 20 μl of 10 mM $MgSO_4$ onto a clean microscope slide, and gently place the coverslip on top of the liquid so as not to create bubbles in the viewing field.
10. Seal the coverslip/microscope slide border with nail polish and allow to dry.

11. Image the cells in a microscope using DIC or phase first, followed by fluorescence.
12. If different wavelength fluorophores are used such as GFP and mCherry; or GFP and Hoechst 33258, image cells using the longer wavelength fluorophore first.

4. Notes

1. It is always necessary to check several colonies for expression and fluorescence on a small scale prior to proceeding to large-scale preparations or in vivo analyses.
2. Optimal protein purification and fluorescence microscopy results are obtained when a minimal number of cell passages are performed.
3. Following cell lysis and centrifugation to remove cell debris, the pellet typically exhibits a significant amount of fluorescence color depending on the fusion being purified, but it is typically discarded as the majority of the chimeric proteins are soluble.
4. Once the cleared cell lysate has been loaded onto the nickel column, the color of the column changes significantly with the color determined by the fluorescent protein being purified. There should also be significant fluorescence in the flow through. This is attributed to the homo-tetrameric SSB fusion proteins which do not bind to the resin as they do not contain a histidine tag.
5. A second round of chromatography of each pool done separately followed by careful pooling ensures near homogeneous 1/4, 2/4, 3/4, and 0/4 pools (fusion:WT).
6. The expected mass ratio of fluorescent to his_6-SSB subunits is approximately 7:1 for 3/4 chimera (i.e., 3 fusion and one his-tag subunits); 2:1 for the 2/4, and 1:1.4 for the 1/4 protein. These values are obtained by densitometric analysis of SDS-PAGE gels. These ratios are only obtained using the H/H plasmid system.
7. Protein purification produces the 1/4, 2/4, 3/4, and 0/4 pools of protein (fluorescent/his-subunit). These proteins bind to ssDNA with site sizes similar to wild type. The affinity of the chimeras for ssDNA is affected by the presence of the N-terminal histidine tag, reducing the salt-titration midpoint by a factor of two for the 1/4 protein (9).
8. The standard concentration of IPTG used for microscopy is 100 μM. This can be adjusted depending on the experiment. Due to leaky expression, fluorescence is observed in the absence of IPTG in exponentially growing cultures.
9. If the efficiency of cell binding to the microscope coverslip is high, it may be necessary to perform serial dilutions in 10 mM $MgSO_4$ and adsorb 20 μl of the serial dilutions to the coverslip instead.

References

1. Shereda RD, Kozlov AG, Lohman TM, Cox MM, Keck JL (2008) SSB as an organizer/mobilizer of genome maintenance complexes. Crit Rev Biochem Mol Biol 43:289–318
2. Raghunathan S, Kozlov A, Lohman T, Waksman G (2000) Structure of the DNA binding domain of E. coli SSB bound to ssDNA. Nat Struct Biol 7:648–652
3. Shereda RD, Bernstein DA, Keck JL (2007) A central role for SSB in Escherichia coli RecQ DNA helicase function. J Biol Chem 282:19247–19258
4. Buss JA, Kimura Y, Bianco PR (2008) RecG interacts directly with SSB: implications for stalled replication fork regression. Nucleic Acids Res 36:7029–7042
5. Suski C, Marians KJ (2008) Resolution of converging replication forks by RecQ and topoisomerase III. Mol Cell 30:779–789
6. Dillingham M, Tibbles K, Hunter J, Bell J, Kowalczykowski S, Webb M (2008) Fluorescent single-stranded DNA binding protein as a probe for sensitive, real-time assays of helicase activity. Biophys J 95:3330–3339
7. Roy R, Kozlov A, Lohman T, Ha T (2009) SSB protein diffusion on single-stranded DNA stimulates RecA filament formation. Nature 461:1092–1097
8. Reyes-Lamothe R, Possoz C, Danilova O, Sherratt DJ (2008) Independent positioning and action of Escherichia coli replisomes in live cells. Cell 133:90–102
9. Liu J, Choi M, Stanenas AG, Byrd AK, Raney KD, Cohan C, Bianco PR (2011) Novel, fluorescent, SSB protein chimeras with broad utility. Protein Sci 20:1005–1020

Chapter 19

Use of Fluorescently Tagged SSB Proteins in In Vivo Localization Experiments

Rodrigo Reyes-Lamothe

Abstract

The time and place of DNA replication in cells provides invaluable insight into the cell cycle and DNA metabolism. An effective means of obtaining this information is through fluorescence microscopy. The abundance of *S*ingle-*S*trand *B*inding protein, SSB, at the replication fork makes it a good reporter of DNA replication. In this chapter I describe how to observe replication of the *Escherichia coli* chromosome in a strain that synthesizes a fluorescent derivative of SSB. This methodology provides information about the position and dynamics of DNA replication through epifluorescence.

Key words: DNA replication, SSB, GFP, Fluorescence, *E. coli*

1. Introduction

The last decade has changed our perception and ability to observe events in live bacterial cells mainly due to methodologies using green fluorescent protein (GFP) and fluorescence microscopy. Studies and advances on issues related to DNA replication have flourished due to the utilization of these new tools. Several reports have described the position, dynamics, and stoichiometry of replisome (1–5) and its relation with other proteins involved in DNA metabolism (6, 7) in different bacteria. However, these studies are just the beginning, there are still many unanswered questions on DNA replication within the cellular environment, ensuring that the subject will continue to unfold in the future.

Although fluorescent derivatives of any replisome component can be detected under the microscope, the use of those components with higher stoichiometries offer the advantage of higher signal. One needs to bear in mind that the pattern of each replisome component in the cell reflects its function; and therefore to answer a specific question the use of a particular component might be needed.

James L. Keck (ed.), *Single-Stranded DNA Binding Proteins: Methods and Protocols*, Methods in Molecular Biology, vol. 922,
DOI 10.1007/978-1-62703-032-8_19,

However for most cases, where the aim is detecting the position and timing of DNA replication, a brighter fluorescent spot translates into easier detection and a higher number of observations before bleaching. In this context SSB is a good candidate for enquiring about the state of replication and it has been used successfully for this purpose in the past (4, 8).

In this chapter I describe techniques to prepare and observe *E. coli* cells carrying SSB–YFP fusions. To reduce the complexity in the data obtained, the cells are grown in conditions that result in non-overlapping cell cycles. Two related strategies are described: snapshot and time-lapse analysis. The first involves taking the cells from liquid culture, placing them on a microscope slide and immediately observing them under the microscope. The time-lapse experiments involve spotting the cells on a slide under conditions that allow them to continue growing. While the former strategy provides only spatial information the latter permits one to extract information about the position and dynamics of the replisome.

2. Materials

For simplicity I assume that you have obtained a strain carrying SSB tagged with YFP and that you have transferred this allele into the strain of interest. I nevertheless include some thoughts about strain construction (see Note 1) in case you are considering making a new construct.

All solutions are prepared with ultrapure water.

2.1. Growth Culture and Slide Preparation

1. Luria Broth: 10 g Yeast Extract, 5 g Tryptone, 5 g NaCl in 1 L of water. Adjust pH to 7. Sterilize by autoclaving.
2. 10× M9 Salts: 60 g Na_2HPO_4 (anhydrous), 30 g KH_2PO_4, 5 g NaCl, 10 g NH_4Cl in 1 L of water. Add in salts in this order and dissolve each before adding the next. Sterilize by autoclaving.
3. M9-Gly: 10 mL 10× M9 Salts, 1 mL 20% glycerol, 100 μL 1 M $MgSO_4$, 100 μL 100 mM $CaCl_2$. Adjust to 100 mL with water.
4. PBS: 5 Phosphate buffered saline Dulbecco A tablets (Sigma) in 500 mL of water. Sterilize by autoclaving.
5. 1% agarose in PBS: Add 1 g of agarose to 100 mL of PBS. Heat up in the microwave oven until agarose is completely dissolved.
6. 2× M9-Gly: Same as M9-Glycerol but adjusted to a final volume of 50 mL.
7. 2% agarose: Add 1 g of agarose to 50 mL of water. Heat up in the microwave oven until agarose is completely dissolved.
8. 1% agarose in M9-Gly: Add 500 μL of 2× M9-Gly to an eppendorff tube, melt 2% Agarose and add 500 μL to the same tube. Mix by pipetting and immediately transfer to a slide.

9. Spectrophotometer.
10. Microfuge.
11. Microwave oven.
12. Microscope slides: VWR, catalog number 631-1554.
13. Coverslips: 24 × 50 mm Borosilicate glass. Thickness No. 1.5 (VWR, catalog number 631-0147).
14. DAPI: 4′,6-Diamidino-2-phenylindole (Invitrogen, catalog number D1306). Make stock at 1 mg/mL in water.
15. Gene Frame: 1.7 × 2.8 cm frames (Thermo Scientific, catalog number AB-0578).

2.2. Microscopy

1. Microscope: Nikon CFI Plan Apo 100× objective on a Nikon Eclipse TE2000-U inverted microscope.
2. Fluorescence filter sets: CFP/YFP/mCherry-ET (Chroma, catalog number 89006); DAPI/FITC/TRITC (Chroma, catalog number 61000v2).
3. Neutral density filter: Chroma (catalog number 22000).
4. Light source: Lumen200 (Prior) illumination system using 200-W metal arc lamp.
5. Camera: Photometrics Cool-SNAP HQ CCD.
6. Incubator: Purpose design acrylic chamber (Solent Scientific). Heater unit controlling the temperature inside of the incubator chamber (Solent Scientific).
7. Immersion oil: Lenzol Immersion oil Gurr (VWR, catalog number 361023N).
8. Software: Microscope controlled by Metamorph 6.2 (Molecular Devices). Analysis done using ImageJ (9).

3. Methods

Streak cells from a glycerol stock onto an agarose plate with respective antibiotics and grow at 37°C. Do not use cells from plates more than 2 weeks old.

In all cases described below cells were incubated at 37°C under constant aeration.

3.1. Determination of Position of SSB by Snapshots

3.1.1. Sample Preparation

1. Pick a single colony from a plate carrying the strain with fluorescently tagged SSB. Resuspend in 5 mL of LB and incubate for 5 h. The culture must become cloudy.
2. Transfer 5 μL of the culture to a new tube with 5 mL of M9-Gly (1:1,000 dilution) (see Note 2). Incubate overnight.

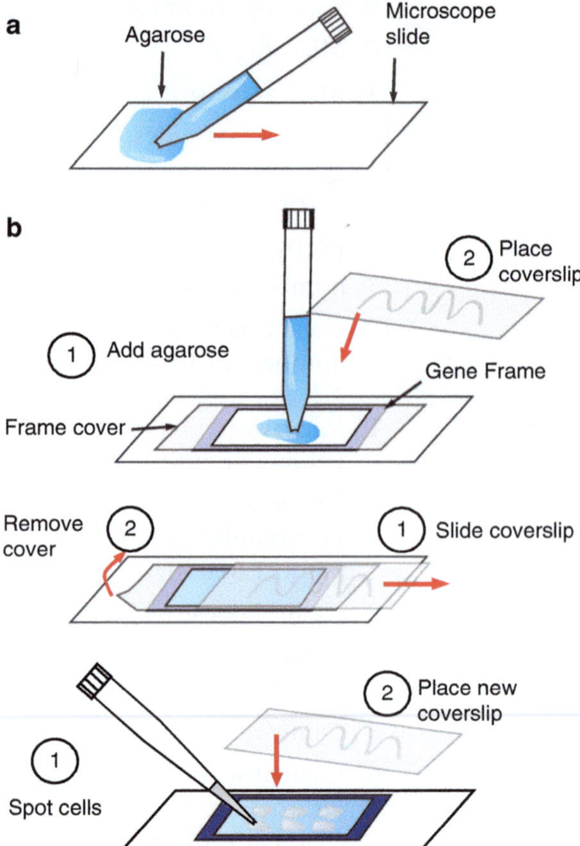

Fig. 1. Slide preparation for snapshot (**a**) and time-lapse (**b**) experiments. In the first case the agarose is spread as it is deposited on the slide using the tip of the pipette. For time-lapse a Gene Frame is placed on a slide, still with its open cover on, and M9-Gly agarose is added inside of it; a coverslip is immediately placed on top of it (**b**, top). After leaving agarose to set, slide coverslip and carefully remove the plastic cover (**b**, middle). Finally, spot the cells on the agarose pad, let the drops dry for a couple of minutes and then place a new coverslip (**b**, bottom).

3. Take 1 mL aliquot of the culture and determine its optical density. Make a dilution in fresh M9-Gly so an initial OD_{600} ~0.05 is obtained.
4. Incubate culture for 2–3 h. The OD_{600} must be between 0.1 and 0.2.
5. Melt 1% agarose in PBS in a microwave oven (see Note 3). Spread 600 μL of the hot agarose solution on a slide with a 1 mL pipette tip (Fig. 1a). Let it set for 5 min.
6. Take 500 μL of culture and add 0.5 μL of DAPI (see Note 4). Mix by tapping the tube.
7. Spin the cells down by centrifugation at 7,800 × *g* for 2 min.
8. Resuspend in 50 μL of M9-Gly and spot 5 μL on the agarose pad without touching it with the tip. Place a coverslip on top.

3.1.2. Microscopy

1. Add a drop of immersion oil on the objective lens. Place the slide on the microscope stage, coverslip facing the objective, and carefully raise the objective lens with the focusing knob until it touches the coverslip. Continue focusing until the outline of the cells is observed.
2. Take a picture of the cells
3. Change excitation and emission filters to acquire YFP signal (see Note 5). Take a picture collecting light for 1 s (see Note 6).
4. Change filters detect DAPI and take a picture collecting light for 100 ms (Fig. 2a).

3.2. Time-Lapse

3.2.1. Sample Preparation

1. Repeat steps 1–4 from Subheading 3.1 (see Note 7).
2. Remove the plastic rectangular cover from the Gene Frame sticker (see Note 8) and carefully paste it on a clean slide (Fig. 1b). Avoid leaving wrinkles.
3. Transfer 500 μL of the 2× M9-Gly into an eppendorf tube.
4. Melt the 2% agarose in water in the microwave oven (see Note 9). Do it in 10 s pulses so it does not boil over.
5. Transfer 500 μL of the hot agarose into the tube containing the 2× M9-Gly. Mix by pipetting and immediately transfer 500 μL of this solution into the center of the rubber frame. Place a coverslip on top and press to remove excess agarose. Let it stand for few minutes.
6. Remove the coverslip by sliding and let the agarose dry for a couple more minutes.
7. Transfer 500 μL of culture to a tube.
8. Spin the cells down by centrifugation at 8krpm for 2 min.
9. Spot 5 μL of the cell suspension distributing them throughout the slide. Let it stand for few minutes more until the drops have dried completely (see Note 10).
10. Remove the top plastic cover from the Gene Frame. Carefully place a coverslip on the exposed sticky face of the frame a coverslip. Avoid the formation of air pockets by making the coverslip contact one of the sides first and spread to the opposite side.

3.2.2. Microscopy

1. Turn on the temperature controller at least a couple hours prior to the start of the experiment. Set it at 37°C.
2. Attenuate the intensity of the excitation light by placing a 4× neutral density filter in the light path (see Note 11).
3. Add a drop of immersion oil to the objective lens and place the slide on top.
4. Focus the cells using phase contrast. Find a field of view on the slide where there are cells evenly distributed through it but

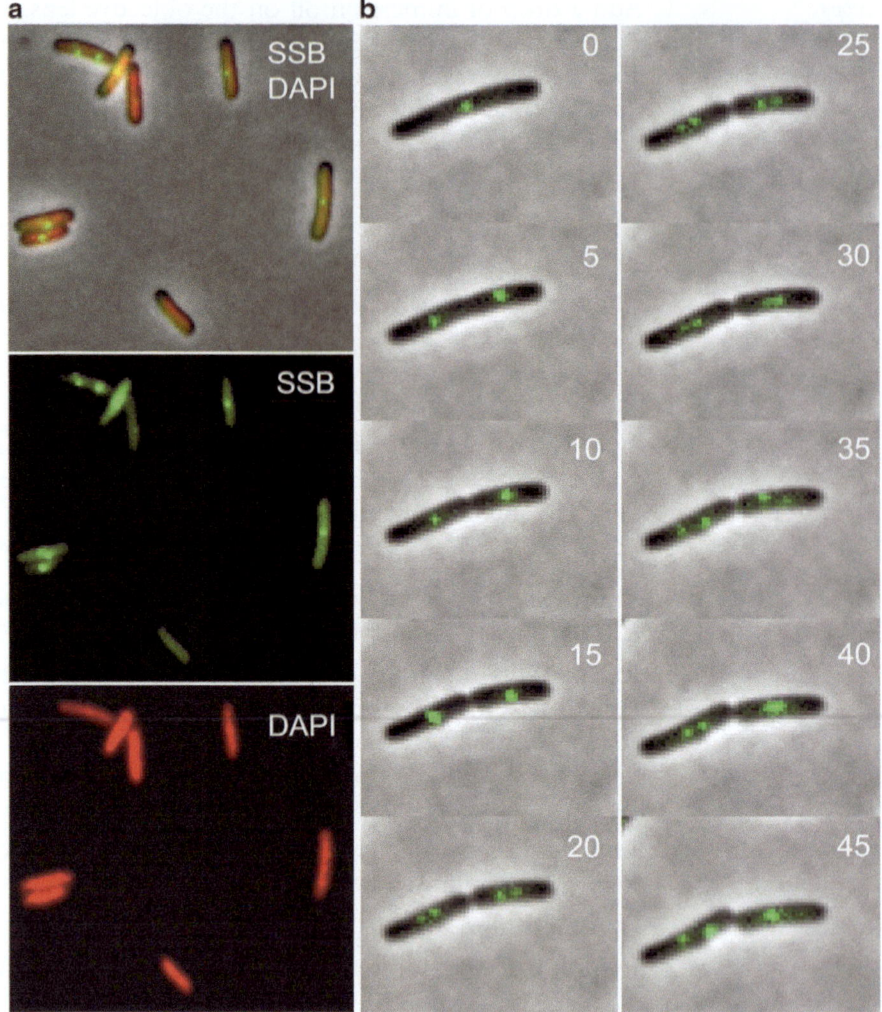

Fig. 2. Examples of snapshot (**a**) and time-lapse (**b**) experiments for SSB. The channels for phase contrast, DAPI, and YFP are shown as a composite image (**a**, top), and separately in the case of YFP (**a**, middle) and DAPI (**a**, bottom). In (**b**) a cell followed through a time-lapse experiment, using 5 min intervals, is shown. The time corresponding to each sampling is shown in the *left top corner* of each picture.

have few microns of distance between each other. Typically this will contain 20 or 30 cells. Take a picture.

5. Incubate the slide for 30 min. Focus and take another picture. When compared to the initial picture, cell elongation should be evident.
6. Start the experiment by taking a picture of the cells using phase contrast. Change the filters and take a picture of YFP capturing light for 1 s.
7. Repeat previous step every 5 min (see Note 12) until the cells stop growing, the fluorescence has bleached or enough data has been collected (see Note 13). If information about cell cycle is

needed it is desirable to follow the cell for more than one generation time (100 min) (see Note 14) (Fig. 2b).

4. Notes

1. Since both N- and C-terminus of SSB are essential all fluorescent fusions will affect its function to some extent. Cells expressing SSB fused to a fluorescent protein in its C-terminus are viable and have similar growth parameters as the parental strain, but only if there is an extra copy of wt SSB. Therefore, I advise the construction of a strain carrying both wt *ssb* (at its original locus) and *ssb-ypet* as an ectopic chromosomal copy driven by the promoter of *ssb*. This should result in the production of equal amounts of wt and fused proteins. Typically a short flexible linker (6–10 residues of Glycine, Alanine, and Serine) is used in between SSB and the fluorescent protein.
2. Cells grown in M9-Gly have a generation time of about 100 min; this gives enough time for replication and division to occur without the need of having overlapping cycles, resulting in the presence of only two replication forks. An alternative is to use M9-Acetate, which increases the generation time to over 150 min and spaces consecutive replication events even further apart.

   It is important that the density of the overnight culture does not reach the late stationary phase (OD_{600} ~1), since this would increase lag phase by several hours. It is therefore recommended to set two dilutions of the culture to have enough cells but not too far into the stationary phase (ideally OD_{600} 0.3–0.5).
3. For most purposes dissolving agarose in PBS is enough to observe the state of the cells, especially when there is only a short delay from the liquid culture to the observation of the cells. An alternative is the use of Agarose in M9-Gly, which may perturb cells less.
4. DAPI is only used when one does not want the cells to continue growing, and is not recommendable for time-lapse experiment. This is because it will bind to DNA, and most likely affect the processes carried out on it, and also because the excitation light is close to UV making it toxic for cells.
5. If the purpose of the experiment is to look for colocalization with another protein then using a fusion of SSB with mCherry (or another red fluorescent protein) is advisable. Fusions with YFP derivatives are usually brighter than other fluorophores. However, since the SSB stoichiometry at the fork is high,

weaker fluorophores still permit to detect it. YFP can then be used to label the second protein to produce brighter foci.

6. The length of exposure may vary according to the microscope setup used.
7. Since time-lapse experiments are more demanding, they should only be done after having characterized the strain through snapshots.
8. Gene Frame is used to avoid desiccation of the agarose pad when the slide gets incubated at 37°C. Other alternatives, such as the use of wax to make a close chamber or the use of a different product, can also be used.
9. Although 2% agarose can be let to solidify in the bottle and be used in later experiments, it is advisable not to use it for more than four experiments since the concentration will change significantly.
10. Excess water in the agarose pad will prevent cells from sticking to its surface; this is why it is important to wait for few minutes before and after spreading the cells on the slide.
11. Two considerations have to be taken into account when deciding how much light to use to excite the sample in a time-lapse experiment: bleaching and phototoxicity. The fluorescence in cells will bleach as the number of pictures taken increases, therefore one should start the experiment by using just enough excitation light as to detect the foci to save fluorescence for later time points. On the other hand, cells themselves are sensitive to light (the more as it gets closer to UV), making it another reason to reduce the intensity and length of light used to excite fluorophores. Neutral density filters can be used to regulate the intensity of light to which the cells are exposed.
12. The interval of time between frames is another factor to consider when planning an experiment. As in the previous note, it is good to economize the fluorescence of the cells. Taking pictures every 5 min is a good way to ensure that the fluorescence will last for over a generation time, while still detecting key events in the cell cycle. Shorter intervals can be used but will reduce the length of the experiment.
13. Unless the conditions used do not permit growth (as when using inhibitors of transcription), only the information obtained from growing cells should be taken in account for analysis.
14. Analysis can be done using different programs (I have used Metamorph, ImageJ, and Matlab), depending on the objective of the experiment. A discussion on microscope analysis falls out of the scope of this chapter. Nevertheless, a basic procedure to qualitatively assess the data is to make an overlay of the

brightfield and fluorescence channels (in ImageJ using the function Merge Channels). Other useful commands in ImageJ when working with overlays are the "Channels tool" and "Brightness/Contrast."

Acknowledgment

I am grateful to A. Badrinarayanan, C. Lesterlin, N. Paramio, M. Tolmasky, and D. Sherratt for valuable comments on the manuscript. This work was by supported by the Welcome Trust and a research fellowship of New College (Oxford).

References

1. Lemon KP, Grossman AD (1998) Localization of bacterial DNA polymerase: evidence for a factory model of replication. Science 282:1516–1519
2. Jensen RB, Wang SC, Shapiro L (2001) A moving DNA replication factory in *Caulobacter crescentus*. EMBO J 20:4952–4963
3. Su'etsugu M, Errington J (2011) The replicase sliding clamp dynamically accumulates behind progressing replication forks in Bacillus subtilis cells. Mol Cell 41:720–732
4. Reyes-Lamothe R, Possoz C, Danilova O, Sherratt DJ (2008) Independent positioning and action of *Escherichia coli* replisomes in live cells. Cell 133:90–102
5. Reyes-Lamothe R, Sherratt DJ, Leake MC (2010) Stoichiometry and architecture of active DNA replication machinery in Escherichia coli. Science 328:498–501
6. Lecointe F, Serena C, Velten M, Costes A, McGovern S, Meile JC, Errington J, Ehrlich SD, Noirot P, Polard P (2007) Anticipating chromosomal replication fork arrest: SSB targets repair DNA helicases to active forks. EMBO J 26:4239–4251
7. Simmons LA, Davies BW, Grossman AD, Walker GC (2008) Beta clamp directs localization of mismatch repair in Bacillus subtilis. Mol Cell 29:291–301
8. Bates D, Kleckner N (2005) Chromosome and replisome dynamics in *E. coli*: loss of sister cohesion triggers global chromosome movement and mediates chromosome segregation. Cell 121:899–911
9. Abramoff MD, Magalhaes PJ, Ram SJ (2004) Image processing with ImageJ. Biophotonics Int 11:36–42

Index

A

Ammonium persulfate (APS) 179
Ammonium sulfate co-precipitation
  materials 152
  methods 152
  single-strand DNA 151
  SSB-Ct interaction 151
Analytical ultracentrifugation (AUC)
  absorbance optics 134, 145
  Beckman Optima XL-A 135
  DNA polymerase III 134
  *Eco*SSB 136
  equipment and optics 135
  gravitational force 135
  SEC 134
  sedimentation coefficient reflect 136
Anisotropy. *See* Fluorescence intensity and anisotropy
Antigens 164
Anti-phosphotyrosine antibody 206
Antisera 165
AUC. *See* Analytical ultracentrifugation (AUC)

B

Bacterial SSB proteins
  bacterial tyrosine kinase 206
  B. subtilis cells 206
  His-tagged SSB proteins 207–208
  immunochromatographic technique 206
  monoclonal antibody 4G10 206
  phosphorylation sites determination
    materials 208
    methods (*see* Phosphorylation sites determination)
  protein synthesis/His-tagged SSB proteins
    materials 207–208
    methods 209–210
  ribbon representation 206, 207
  Ser/Thr phosphorylation 206
  Western blotting 208
    materials 208
    methods 210–211
Bacterial tyrosine kinase 206
Beer's law 50
Biacore Sensor Chip SA 17
Bio-Dot Apparatus 167
Bloom syndrome protein (BLM) 12
Bromophenol blue (BPB) 179

C

Collaborative Computation Project, Number 4 (CCP4) 24

D

Deinococcus radiodurans single-strand DNA-binding protein (*Dr*SSB)
  dialysis 65
  isothermal titration calorimetry 38–39
  MBDF analysis 72
  model-independent analysis 73
  proteins and ssDNA 58
  stoichiometric conditions 63
  titration experiments 62
  titration isotherms 66
Diethylaminocoumarin 220
Differential interference contrast (DIC) microscopy 239
Diffusion detection assays 93
Dimethylformamide (DMF) 124, 130
Dimethyl sulfoxide (DMSO) 187, 189
DNA-binding core (DBC) 102
Double-stranded DNA (dsDNA) 219
Dulbecco's Modified Eagle Medium (DMEM) 195

E

EDTA. *See* Ethylenediamine tetraacetate (EDTA)
Electrophoretic transfer components 163
Equilibrium binding model 59–60
Escherichia coli ssDNA binding protein (*Eco*SSB) *See also* Protein–Protein interactions
    binding modes 38
  data analysis 47
  dialysis 41
  energetics and kinetics 58
  fluorescence anisotropy 58
  materials 40–41
  $(SSB)_{65}$ mode 57
  proteins and ssDNA 58
  ssDNA interactions 78

James L. Keck (ed.), *Single-Stranded DNA Binding Proteins: Methods and Protocols*, Methods in Molecular Biology, vol. 922, DOI 10.1007/978-1-62703-032-8, © Springer Science+Business Media, LLC 2012

Escherichia coli ssDNA binding protein (*EcoSSB*) (*cont.*)
    stoichiometric conditions 63
    thermodynamics 38
Ethylenediamine tetraacetate (EDTA) 179
Eubacterial single-strand DNA-binding protein
    database 25
    structural features
        amino acid sequence 27, 28
        antiparallel β-sheet 26, 27
        clamping of dimer 26
        C-terminal domain 24
        DNA-binding domain 26
        N-terminal domain 25–26
        quaternary structure 26
        SSB tetramere 26
        water bridges 27
    variability, quaternary association
        AD (and BC) interface 30
        dimer, mutual orientation of 29, 30
        monomer-monomer interface 29
        monomer, mutual orientation of 29, 31
        oligomerization buried 29, 30
        tetrameric SSBs 28
Exonuclease I (ExoI) 155–157, 186

F

Far Western blotting
    affinity purification
        materials 164
        methods 165
    antigens 164
    electrophoretic transfer components 163
    immunoblotting component 164
    immunodot blotting 165–166
    prey protein 162
    protein–protein interactions 162
    SDS-PAGE 161, 167
Fluorescence anisotropy (FA)
    F-SSB-Ct 155–156
    heterologous proteins 155
    materials 156
    methods 157–158
    SSB-Ct function 155
Fluorescence intensity and anisotropy
    buffer solution 58
    data analysis 69–70, 75–76
    data fitting 66, 73–75
    equilibrium binding model 59–60, 66
    extrinsic fluorophore 58
    G-factor 77
    ITC method 56
    MBDF analysis 66, 71–73
    $(SSB)_{35}$ mode 57
    monitor binding 60–62
    polarization 77
    protein/ssDNA dialysis 58–59, 65–67
    titration procedure 67–69
Fluorescence microscopy 238–239, 242–243
Fluorescent derivatives, SSB
    DNA replication 245
    growth culture and slide preparation 246–247
    microscopy 247
    non-overlapping cell cycles 245
    SSB snapshots determination
        microscopy 249
        sample preparation 247–248
    time-lapse
        microscopy 249–251
        sample preparation 249
Fluorescent polarization (FP) 128, 186
Fluorescent single-stranded DNA-binding proteins
    buffers 239
    DNA metabolism 235
    dual plasmid system 236
    fluorescence microscopy 236, 238
    fluorescent protein 236
    nickel column chromatography 236–237
    protein over-expression
        materials 237–238
        methods 239–240
    protein purification
        materials 237
        methods 240–241
    SSB chimeras 236, 238
    tetramers 236
    in vivo study
        materials 238–239
        methods 241–243

G

Green fluorescent protein (GFP) 245

H

Helix destabilization assays 92–93
High-throughput inhibitor screen 186–188
Homology-directed repair (HDR) 12

I

Immunoblotting 194, 202–203
Immunochromatographic technique 206
Immunofluorescence 194, 203
Isothermal titration calorimetry (ITC)
    binding affinity/enthalpy 37
    buffer solution 39–40
    DNA titrations 42, 45
    DrSSB/EcoSSB 42, 44, 45
    EcoSSB titration data 44, 48

n-independent and identical sites model ... 44, 46, 47
protein/DNA dialysis ... 41–43
protein-DNA interaction ... 38
proteins/ssDNA ... 40
$(SSB)_{35}/(SSB)_{65}$ binding mode ... 38
SSB/DNA estimate ... 41, 42
SSB titration ... 43–45
stoichiometric affinity ... 38
stoichiometry/binding enthalpy ... 42, 44, 46
titration data ... 45
weaker binding affinity ... 42, 46
wiseman isotherms ... 42, 44, 45

L

Liquid chromatography (LC) ... 208, 212

M

Macromolecular crystallography method ... 24
Macromolecule binding density function (MBDF) analysis ... 66, 71–73
2-mercaptoethane-sulfate (MESNA) ... 225, 228
Methanococcus jannaschii, ... 2
Multistage activation (MSA) ... 216
Mycobacterium tuberculosis single-strand DNA-binding protein(*Mt*SSB)
database ... 25
structural features ... 26
variability, quaternary association ... 30, 32

N

Native gels
DNA metabolism ... 175
electrophoresis ... 181
materials ... 179–180
native-PAGE ... 175, 176
polyacrylamide gel preparation ... 180
protein–protein interactions ... 175
RecR and RecO interactions ... 178
SSB–RecO interaction ... 176–177
Non-overlapping cell cycles ... 245
Nucleotide excision repair (NER) pathway ... 12

O

Oigonucleotide/oligosaccharide-binding (OB) fold ... 102

P

Panvera Beacon 2000 fluorescence anisotropy system ... 156
Phosphate Buffered Saline (PBS) ... 195
Phosphorylation sites determination
in-solution protein digestion ... 211
liquid chromatography ... 212
mass spectrometry ... 212–214
TiO_2 Chromatography ... 211–212
Polyacrylamide gel electrophoresis (PAGE) ... 161, 175
Polyethyleneglycol (PEG)
aminosilanization ... 89
PEGylation ... 89
precleaning/surface activation ... 88–89
Protein data bank (PDB) ... 24
Protein–protein interactions
analytical ultracentrifugation
absorbance optics ... 134
Beckman Optima XL-A ... 135
DNA polymerase III ... 134
*Eco*SSB ... 136
fluorescence detection system ... 134–135
free ligand ... 136
Rayleigh interference optics ... 134
reaction boundary ... 136
SEC ... 134
sedimenting species ... 135
s-value ... 136–137
AUC experiments
materials ... 141
methods ... 142–143
binding parameters determination
c(s) method ... 137, 138
C-terminus ... 139
*Eco*SSB ... 137, 138
isotherms binding ... 139, 140
SEDFIT ... 137
SSB ... 140
wild-type interaction ... 140
data analysis
materials ... 141
methods ... 143–144
*Eco*SSB ... 133–134
immunodot blotting ... 162
sample preparation
materials ... 141
methods ... 142
static fluorimeter ... 148
Pyrococcus furiosus, ... 4

R

Reagentless biosensor
characterization ... 225
Cy3B-SSB and TIRFM assay ... 223–224
DCC-SSB considerations ... 221–223
DNA/RNA manipulation ... 219

Reagentless biosensor (*cont.*)
G26C SSB preparation
ammonium sulfate precipitation 227
dialysis 227–228
E. coli expression 225–226
heparin column chromatography 227, 231–232
polyethyleneimine precipitation 227
protein extraction 226, 231
SSB protein purification 226
helicase assay 225, 230–232
IDCC and labeling
characterization 229–230
labeling 228
purification 228–229
labeling 224–225
SSB protein preparation 224
ssDNA 221
RecO protein
buffer solution 124
cloning 125
crystallization screens/plates
materials 124
methods 126
Dimethylformamide (DMF) 124, 130
DNA 124
peptide binding
materials 124
methods 127–128
purification
materials 124
methods 125
structure determination 126–127
Replication protein A (RPA)
cell culture
materials 195
methods 197–198
electrotransfer immunoblotting 201
gradient gel purification 199–200
immunoblotting components
materials 196–197
methods 202–203
immunofluorescence staining components
materials 195
methods 198–199
N-terminal and C-terminal of 193
optimal buffer solubility screening
materials 103–104
methods 106–110
posttranslational detection 194
sample preparation and electrophoresis 200–201
SAXS/SANS purification
materials 105–106
methods 114–116
SDS polyacrylamide gel 196
SEC-MALS
concentration series 105, 112–113
preparion/equilibration/ calibration 105, 111–112
time series 105, 113
single-strand DNA-binding protein
DNA interfaces 10–11
DNA repair interfaces 11–13
Response units (RUs) 169

S

Saccharomyces cerevisiae, ... 11
Sedimentation velocity experiments
c(s) method 137, 138
C-terminus 139
*Eco*SSB 137, 138
isotherms binding 139, 140
SEDFIT 137
SSB 140
wild-type interaction 140
Sednterp software 49
Selenomethionine (SeMeth) 125
Single molecule Förster (or fluorescence) resonance energy transfer (smFRET) 85
force spectroscopy
anti-digoxigenin-coated beads preparation 94
data acquisition protocol 94–95
materials 88
mechanical regulation of SSB diffusion 97
probing SSB dissociation 95–97
TIR microscopy
helix destabilization assays 92–93
materials 87
sample preparation 90
SSB binding assay 90–91
SSB diffusion 91–92
Single-strand DNA-binding protein (SSBs)
Bacillus subtilis, ... 15
Bacteriophage T7 gp2.5 7
DBD-B 3, 6
D. radiodurans, ... 15
ecSSB protein 3, 5
eubacterial SSB (*see* Eubacterial single-strand DNA-binding protein) gp2.5 and gp32 binding mode 7
hydrophobic and ionic interaction 5–6
N-terminal domain 7
posttranslational modifications 13–14
protein arrangement 7
protein interactions
archaeal SSB 13
protein interfaces 10

recombination interface 9
repair interfaces 10
replication interface 8–9
RPA 10–13
$(SSB)_{35}$ /$(SSB)_{65}$ binding mode 5–6
structural diversity
bacterial/eukaryotic distribution 2
bacterial function 2, 3
DNA-binding domain 4
eukaryotic RPAs 3, 4
gp32 core protein structure 3, 5
gp2.5 dimer structure 3, 5
heterologus protein 2
M. jannaschii, ... 4
N-terminal region 4
OB fold 2
oligomeric protein 2, 3
phosphate backbone 3, 4
RPA 2
three-helix bundle 3, 4
in vitro condition/protein 2
zinc finger motif 4
structural studies (*see* RecO protein)
Single-stranded DNA (ssDNA)
Bacteriophage T7 gp2.5 7
DBD-B 3, 6
DNA oligonucleotides 86–87
ecSSB protein 3, 5
flow chamber assembly 89
gp2.5 and gp32 binding mode 7
hydrophobic and ionic interaction 5–6
N-terminal domain 7
optical tweezers 85
PEG surface preparation
aminosilanization 89
PEGylation 89
precleaning/surface activation 88–89
protein arrangement 7
reagent and buffer 87–88
RPA (*see* Replication protein A (RPA))
smFRET
force spectroscopy 88, 94–97
TIR microscopy 87, 90–93
$(SSB)_{35}$ /$(SSB)_{65}$ binding mode 5–6
Size-exclusion chromatography-multi-angle light scattering (SEC-MALS)
concentration series
materials 105
methods 112–113
preparion/equilibration/calibration
materials 105
methods 111–112
time series
materials 105
methods 113
Small-angle neutron scattering (SANS)
materials 105–106
methods 115–116
Small-angle X-ray scattering (SAXS)
materials 105
methods 114–115
Sodium dodecyl sulfate (SDS) 161, 175
SPR. *See* Surface plasmon resonance (SPR)
SSB–protein interactions.*See also* Surface plasmon resonance (SPR)
bacterial SSBs function 183
buffers 185
Escherichia coli, ... 184
exonuclease I protein purification 186
high-throughput inhibitor screen 186–188
in house 184–185
protein–protein interactions 183
retesting hits 188
small molecule hits identification 188
small molecule screening 185
SSB-Ct sequence 184
Step-size determination assays 93–94
Sulfolobus solfataricus, ... 2, 4
Surface plasmon resonance (SPR)
biotin–streptavidin interaction 170
equilibrium and kinetic binding 169
materials 170
methods
data analysis 172
sensor chip 171
SIP injection 171–172
solution preparation 170–171
SSB removal 172
protein–protein interactions 169, 170
response units 169, 170
SSB–ssDNA complex 170

T

Thermus thermophilus (Tth), ... 180
Titanium Oxide (TiO_2) Chromatography 208, 211
Tobacco etch virus (TEV) protease 124
Total internal reflectance fluorescence microscopy (TIRFM) 223
Translesion synthesis (TLS) 12
Tris buffered saline (TBS) 197
Tris(2-carboxyethyl)phosphine (TCEP) 48

U

Uracil DNA glycosylase (UDG) 10

W

Werner syndrome protein (WRN) 12
Wiseman isotherms 45

MIX
Papier aus verantwortungsvollen Quellen
Paper from responsible sources
FSC® C105338

If you have any concerns about our products, you can contact us on
ProductSafety@springernature.com

In case Publisher is established outside the EU, the EU authorized representative is:
Springer Nature Customer Service Center GmbH
Europaplatz 3, 69115 Heidelberg, Germany

Printed by Libri Plureos GmbH
in Hamburg, Germany